COOPERATIVE GAMES ON COMBINATORIAL STRUCTURES

THEORY AND DECISION LIBRARY

General Editors: W. Leinfellner (*Vienna*) and G. Eberlein (*Munich*)

Series A: Philosophy and Methodology of the Social Sciences

Series B: Mathematical and Statistical Methods

Series C: Game Theory, Mathematical Programming and Operations Research

Series D: System Theory, Knowledge Engineering and Problem Solving

SERIES C: GAME THEORY, MATHEMATICAL PROGRAMMING AND OPERATIONS RESEARCH

VOLUME 26

Scope: Particular attention is paid in this series to game theory and operations research, their formal aspects and their applications to economic, political and social sciences as well as to socio-biology. It will encourage high standards in the application of game-theoretical methods to individual and social decision making.

COOPERATIVE GAMES ON COMBINATORIAL STRUCTURES

by

JESÚS MARIO BILBAO

University of Seville

Spain

KLUWER ACADEMIC PUBLISHERS

BOSTON / DORDRECHT / LONDON

Distributors for North, Central and South America:
Kluwer Academic Publishers
101 Philip Drive
Assinippi Park
Norwell, Massachusetts 02061 USA
Telephone (781) 871-6600
Fax (781) 871-6528
E-Mail <kluwer@wkap.com>

Distributors for all other countries:
Kluwer Academic Publishers Group
Distribution Centre
Post Office Box 322
3300 AH Dordrecht, THE NETHERLANDS
Telephone 31 78 6392 392
Fax 31 78 6546 474
E-Mail <orderdept@wkap.nl>

Electronic Services <http://www.wkap.nl>

Library of Congress Cataloging-in-Publication Data
Bilbao, Jesús Mario.
Cooperative games on combinatorial structures / by Jesús Mario Bilbao.
p.cm. -- (Theory and decision library. Series C, Game theory, mathematical programming, and operations research; v.26)
Includes bibliographical references and index.
ISBN 0-7923-7782-6 (acid free paper)
1. Game theory. 2. Combinatorial analysis. I. Title. II. Series.

QA269.B45 2000
519.e--dc21

00-020450

Printed on acid-free paper.

Printed in the United States of America

Contents

Preface ix

1 Structures **1**
- 1.1 Cooperative games . . . 1
- 1.2 Graph theory . . . 7
- 1.3 Examples of cooperative games . . . 10
- 1.4 Closure spaces and convex geometries . . . 15
- 1.5 Matroids, greedoids and antimatroids . . . 18
- 1.6 Project: Bicooperative games . . . 23

2 Linear optimization methods **27**
- 2.1 Polyhedra . . . 27
- 2.2 Linear programming . . . 32
- 2.3 Integer and combinatorial optimization . . . 38
- 2.4 Balanced games . . . 46
- 2.5 The greedy algorithm . . . 52
- 2.6 Polymatroids . . . 57

3 Discrete convex analysis **65**
- 3.1 Fenchel conjugation . . . 65
- 3.2 Subgradients and subdifferentials . . . 70
- 3.3 Indirect functions of cooperative games . . . 75
- 3.4 The least increment function . . . 80
- 3.5 The Lovász extension of games . . . 82

4 Computational complexity **89**
- 4.1 Models of computation . . . 90
- 4.2 The complexity classes P and NP . . . 91
- 4.3 NP-completeness and #P-completeness . . . 98
- 4.4 Complexity in cooperative game theory . . . 101
- 4.5 Project: Locally concave (convex) games . . . 105

5 Restricted games by partition systems **111**
5.1 Partition systems 112
5.2 $\mathcal{F}$-restricted games 113
5.3 The core of restricted games 114
5.4 Hart and Mas-Colell potential 119
5.5 The Shapley and Banzhaf values 122

6 Restricted games by union stable systems **129**
6.1 Union stable systems 130
6.2 Components of sets 133
6.3 The Myerson value 135
6.4 The position value 142
6.5 Convexity of the conference game 150
6.6 Project: Solution continuation games 154

7 Values for games on convex geometries **157**
7.1 Games on convex geometries 158
7.2 Probabilistic values for games 160
7.3 Efficiency and compatible-order values 165
7.4 Axioms for the Shapley value 170
7.5 Axioms for the Banzhaf index 173
7.6 Project: Marginalist and efficient values 180

8 Values for games on matroids **181**
8.1 The rank game 181
8.2 The Shapley value for games on matroids 185
8.3 Quasi-probabilistic values 193
8.4 The probabilistic Shapley value 198

9 The core, the selectope and the Weber set **205**
9.1 The core 205
9.2 The selectope 208
9.3 The Weber set 217
9.4 Convex and quasi-convex games 225
9.5 Marginal operators 233

10 Simple games on closure spaces **237**
10.1 Minimal winning coalitions 237
10.2 The core 240
10.3 The Weber set 242
10.4 Stable sets 244
10.5 Bargaining sets 246

11 Voting power **253**
11.1 Weighted voting games . . . 253
11.2 Generating functions . . . 255
11.3 Computational complexity . . . 258
11.4 Algorithms with Mathematica . . . 260
11.5 Power in 2-weighted voting games . . . 265

12 Computing values with Mathematica **271**
12.1 Computing values of restricted games . . . 271
12.2 Computing convex Shapley values . . . 277
12.3 Computing convex Banzhaf indices . . . 283
12.4 The Banzhaf power in the European Union . . . 288
12.5 Computing the number of subtrees . . . 290
12.6 The Myerson value in the European Union . . . 296

Bibliography **303**

Index **321**

Preface

The aim of this book is to analyze conflict situations in which two or more players can make coalitions and obtain prizes and penalties. This approach to situations of competition and cooperation was given in the seminal treatise by John von Neumann and Oskar Morgenstern *Theory of Games and Economic Behavior* [242, 1944]. Cooperative game theory has succeeded in providing a lot of applications of game theory. I quote from the Robert Aumann's interview realized by Eric van Damme [52]:

> "Cooperative theory is actually doing quite well. I've already said in this interview that many of the most interesting applications of game theory come from the cooperative side ... Mike Maschler discussed over 30 significant contributions to the cooperative theory that have been produced over the last few years ... On the cooperative side, there are three or four central solution concepts —value, core, nucleolus, stable sets— but much less flexibility in constructing the model. The model is much better defined."

The classical hypothesis in cooperative game theory is that all coalitions of players are possible. However, this assumption is not realistic because cultural, ideological and structural issues will prevent the formation of some coalitions. Models of cooperative games in which cooperation among players depends on their communication were proposed by Myerson [182], Owen [195], and Borm, Owen and Tijs [26]. These models provide a worth for the feasible coalitions $S \in \mathcal{F}$ and define a *restricted game* on all coalitions $S \in 2^N$ by additive extension with the maximal components of S. The paper *Cooperation and communication restrictions: a survey* [27] by Borm, van den Nouweland and Tijs, is a guide for the above model in the context of imperfect or incomplete information and communication in economic organizations. Myerson [183] introduced hypergraphs in communication situations and van den Nouweland, Borm and Tijs [188] defined the *interaction sets* and obtained the axiomatic characterizations of the *position* and *Myerson values.*

A more general model, the *partitioning games*, was developed by Kaneko and Wooders [138]. They present the superadditive extension of a game on feasible coalitions, by using maximal components. There is another extension by Faigle and Kern [78], the cooperative games under precedence constraints. In this model the games are defined on distributive lattices of subsets of players. The *core* of a game $v : \mathcal{F} \to \mathbb{R}$ was investigated by Faigle [77], who obtained necessary and sufficient conditions for the existence of the core from a general sandwich theorem for set functions.

Shapley [212, 1953] established an allocation rule satisfying a few simple axioms (see a survey in Roth [208]) and introduced [217] the class of *convex games*. For these games, the optimal allocation in the core can be obtained by the greedy algorithm, when the criterion is linear (see Fujishige [96]). With respect to the optimization methods, Grötschel, Lovász and Schrijver [116] observed that:

> "Historically, there is a close connection between geometry and optimization. This is illustrated by methods like the gradient method and the simplex method, which are associated with clear geometric pictures. In combinatorial optimization, however, many of the strongest and most frequently used algorithms are based on the discrete structure of the problems: the greedy algorithm, shortest path and alternating path methods, branch-and-bound, etc. In the last several years geometric methods, in particular polyhedral combinatorics, have played a more and more profound role in combinatorial optimization as well."

In this volume, we will analyze games defined on combinatorial structures, i.e., a set system over a set of players. In many situations we will work in a *closure space*. Let $N = \{1, \ldots, n\}$ be a set of players. By a closure operator $- : 2^N \to 2^N$ we mean an operator satisfying the standard closure axioms: $A \subseteq \overline{A}$, $\overline{\overline{A}} = \overline{A}$ and $A \subseteq B$ implies $\overline{A} \subseteq \overline{B}$. A set $S \in 2^N$ is said to be closed if $\overline{S} = S$. The pair $(N, -)$ is a closure space [113]. Examples of closure operators are the spanning operator of linear algebra and all convex hull operators. *Matroids* were introduced in 1935 by Whitney who conceived a matroid as a generalization of a matrix. The closure operator of a matroid and the spanning operator of linear algebra satisfies the *Steinitz-MacLane exchange property*:

$$x, y \notin \overline{A} \text{ and } y \in \overline{A \cup x} \text{ imply } x \in \overline{A \cup y}, \quad \text{for every } A \subseteq N.$$

Convex geometries are closure spaces with the *anti-exchange property*:

$$x, y \notin \overline{A} \text{ and } y \in \overline{A \cup x} \text{ imply } x \notin \overline{A \cup y}, \quad \text{for every } A \subseteq N.$$

The last property is a combinatorial abstraction of the convex closure in Euclidean spaces. We call the closed sets in a convex geometry *convex sets*. A common framework for matroids and convex geometries is found by the *greedoids*, structures introduced by Korte, Lovász and Schrader [144].

In a *cost allocation problem* there is a finite set of users who cooperate in a joint project. The problem is how to allocate the cost among the players in terms of a rule with properties such as efficiency, anonymity and stability (see Curiel [51]). Vectors in the core of the cost game are the best cost allocations when the core is nonempty. The *min-cost spanning tree game* is a balanced game and a core vector for this game is constructed from a minimal spanning tree by Granot and Huberman [109]. However, this does not mean that we shall be able to characterize all the extreme points of the core. Kuipers [148] showed that in the subclass of *information graph-games,* the extreme points are the marginal allocations of the associated submodular game. Nagamochi, Zeng, Kabutoya and Ibaraki [185] proposed a new minimum forest game on an edge-weighted graph and extended this game to a weighted matroid.

Combinatorial cooperative games are games such that the value $v(S)$ of the coalition $S \in 2^N$ is the optimal value of a combinatorial optimization problem defined by S (see Faigle and Kern [79]). Matching games, simple flow games, traveling salesman games, sequencing games and delivery games (associated with the Chinese postman problem) have been considered.

Chapters 1 through 4 constitute a review of mathematical concepts from the *Cooperative Game Theory, Graph Theory, Linear and Integer Programming, Combinatorial Optimization, Discrete Convex Analysis* and *Computational Complexity.* The table of contents is a short guide to the topics and methods treated in this book. In Chapters 11 and 12, several *notebooks* are presented with the system Mathematica by Wolfram [248] in the contexts of the packages *DiscreteMath* (Skiena [221]) and *Cooperative* (Carter [34]). There will also be found in the book several *research projects.* These are intended to offer new ideas that the reader should consider with caution.

We hope that this book will be of interest to graduate students with some experience in game theory or mathematical programming and professional researchers in game theory, operational research and its applications in Economic Theory, and the Political and Social Sciences. We would also like this volume to be especially useful for professionals who are interested in models for understanding and managing conflicts: management and operational research scientists, political and military scientists as well as professional negotiators.

I wish to thank Peter Borm, Theo Driessen, Paul Edelman, Ulrich Faigle, Hans Peters, Joss Potters and Stef Tijs for their suggestions. I am grateful to several colleagues who read early drafts of the book, in particular, to my good friends Encarnación Algaba Durán, Julio Fernández García, Francisco Ramón Fernández García, Nieves Jiménez Jiménez, Andrés Jiménez Losada, Esperanza Lebrón Rueda, Jorge López Vázquez, Juana Isabel Marín Arrese, Juan Enrique Martínez Legaz, Pedro Pérez Carreras and Justo Puerto Albandoz.

I also offer my warmest thanks to my wife Encarnación Rodríguez Lorenzo who gave me her constant encouragement. Finally, it has been a pleasure to work with Allard Winterink and her/his colleagues from Kluwer Academic Publishers.

Jesús Mario Bilbao Arrese — Seville, Spain,
http://www.esi.us.es/~mbilbao/home.htm — E-mail: mbilbao@matinc.us.es

Chapter 1

Structures

We begin our study with notions from cooperative games, graph theory, and combinatorial structures that will be used throughout this book. As references for several of the topics included in this chapter we mention Curiel [51], Driessen [57], Myerson [184], Swamy and Thulasiraman [228], Grötschel, Lovász and Schrijver [116], and Korte, Lovász and Schrader [144].

1.1 Cooperative games

Game theory began with the works from Borel [25, 1921], von Neumann [241, 1928], and the book *Theory of Games and Economic Behavior* [242, 1944] by von Neumann and Morgenstern. The appeal of game theory in recent years derives from the development of its applications in economic theory. Today, all economists start with a model about utility functions, production functions and endowments of the players. Moreover, abstract game models provide a framework for conflict analysis in theoretical politics. As Ordeshook [192] stated:

> "Indeed, the thesis of this volume is that game theory provides the integrating force of the new political theory ... This perspective ... has its roots in decision theory, and game theory is simply the branch of the decision sciences that seeks to explore how people make decisions if their actions and fates are correlated."

Game theory is a tool to analyze situations in which two or more decision makers are in dispute over some objectives and are conscious that their actions affect each other. The following definition draws the difference between the two theories of games.

Definition 1.1.1 *A cooperative game is a game in which the players can make binding commitments, as opposed to a noncooperative game, in which they cannot.*

However, Rasmusen [204, page 18] has pointed out that

> "the real difference lies in the modelling approach. Both theories start off with the rules of the game, but they differ in the kinds of solution concepts employed. Cooperative game theory is axiomatic, frequently appealing to Pareto optimality, fairness and equity. Noncooperative game theory is economic in flavor, with solution concepts based on players maximizing their own utility functions subject to stated constraints."

Fang, Hipel and Kilgour [76] proposed a classification of decision making methods by using the number of players and objectives. Table 1.1 displays how these methods can be categorized according to the above criteria.

		Objectives	
		One	Two or more
Players	One	*Operational Research*	*Multiobjective Decision*
	Two or more	*Cooperative Games*	*Noncooperative Games*

Table 1.1

In the cooperative game-theoretical approach, solutions are characterized by axioms and the conflict situations can be modeled by a bargaining game (see Peters [199]). Cooperative game theory models a conflict by a set function which is interpreted as the worth that a coalition may obtain by itself.

In a *transferable utility game* only one number is attached to each coalition and the main problem is to fix allocation rules such that the players may redistribute the utilities among themselves.

Definition 1.1.2 *A transferable utility game is a pair* (N, v), *where* N *is a finite set and* $v : 2^N \to \mathbb{R}$, *is a function with* $v(\emptyset) = 0$.

The elements of $N = \{1, 2, \ldots, n\}$ are called *players*, the subsets $S \in 2^N$ *coalitions* and $v(S)$ is the *worth* of S. The worth of a coalition is interpreted as the maximal profit or minimal cost for the players in their own coalition. Function v is called the *characteristic function.* We will consider *cost games* if the characteristic function measures the cost $v(S)$ incurred by S or *profit games* if $v(S)$ measures the profit of the coalition S.

Given a game (N, v) and a coalition $S \subseteq N$, the *subgame* (S, v) is obtained by restricting v to 2^S. By Γ^N we denote the set of all games (N, v). We will use a shorthand notation and write i for the set $\{i\}$, $S \cup i$ for $S \cup \{i\}$, and $S \setminus i$ for $S \setminus \{i\}$.

In a game (N, v), a vector $x \in \mathbb{R}^n$ is called *efficient* if it distributes the worth $v(N)$ among the players, i.e., $\sum_{i \in N} x_i = v(N)$. The set of all efficient vectors is called the *pre-imputation set* and is denoted by $I^*(v)$. The *imputation set* of a profit game (N, v) is defined by

$$I(v) := \{x \in I^*(v) : \; x_i \geq v(i) \;\; \text{for all } i \in N\}.$$

Note that $I(v) \neq \emptyset$ if and only if $v(N) \geq \sum_{i \in N} v(i)$. If (N, c) is a cost game, then its imputation set is

$$I(c) := \{x \in I^*(c) : \; x_i \leq c(i) \;\; \text{for all } i \in N\}.$$

Therefore, $I(c) \neq \emptyset$ if and only if $c(N) \leq \sum_{i \in N} c(i)$. For a cost game (N, c), we consider the following polyhedron

$$P(c) = \{x \in \mathbb{R}^n : x(S) \leq c(S) \;\; \text{for all } S \subseteq N\},$$

where $x(S) = \sum_{i \in S} x_i$ and $x(\emptyset) = 0$. For a profit game (N, v), we define

$$P(v) = \{x \in \mathbb{R}^n : x(S) \geq v(S) \;\; \text{for all } S \subseteq N\}.$$

Note that $P(c) \neq \emptyset$ if and only if $c(\emptyset) \geq 0$, and $P(v) \neq \emptyset$ if and only if $v(\emptyset) \leq 0$. Thus, the polyhedra associated to profit or cost games are nonempty.

Assuming that the coalition N of all players will be formed, a solution concept will prescribe a distribution of the cost $c(N)$ or the profit $v(N)$ among the players. The *core* of a game is the most attractive solution concept in cooperative game theory. The definition is due to Gillies [106], who use it to analyze the stable sets.

Definition 1.1.3 *The core of a cost game c and the core of a profit game v are respectively defined by*

$$\begin{aligned} C(c) &:= \{x \in P(c) : x(N) = c(N)\}, \\ C(v) &:= \{x \in P(v) : x(N) = v(N)\}. \end{aligned}$$

Games with a nonempty core are called *balanced games.* A game is *totally balanced* if each subgame is balanced.

The *dual game* of (N, v) is defined by the dual function $v^* : 2^N \to \mathbb{R}$, where $v^*(S) = v(N) - v(N \setminus S)$, for all $S \subseteq N$. Note that $(v^*)^* = v$. We remark that if (N, v) is a profit (cost) game and (N, v^*) is a cost (profit) game, then $C(v) = C(v^*)$.

Proposition 1.1.1 *Let (N, v) be a profit (cost) game and let (N, v^*) be a cost (profit) game. Then $P(v) \cap P(v^*) = C(v) = C(v^*)$.*

Proof. Let $x \in P(v) \cap P(v^*)$ be a common vector. Then $x(N) \geq v(N)$ and $x(N) \leq v^*(N) = v(N) - v(\emptyset) = v(N)$ hence $x(N) = v(N)$. By using the definitions $x \in C(v) = C(v^*)$. Furthermore, $x \in C(v) = C(v^*)$ implies that $x \in P(v) \cap P(v^*)$. □

Notice that no coalition $S \subseteq N$ should be able to improve upon $x \in C(v)$. Given a profit game (N, v), a vector $x \in \mathbb{R}^n$ and a coalition S, we define $e(S, x) = v(S) - x(S)$. This number is called the *excess* of S at x and can be interpreted as a measure of discontent of S with the payoff vector x. The core of the profit game (N, v) satisfies

$$C(v) = \{x \in I^*(v) : e(S, x) \leq 0 \text{ for all } S \subseteq N\}.$$

Let ε be a real number. The *strong ε-core* is given by

$$C_\varepsilon(v) = \{x \in I^*(v) : e(S, x) \leq \varepsilon \text{ for all } S \neq N, \emptyset\}.$$

We have that $C_\varepsilon(v) \supset C_{\varepsilon'}(v)$ if $\varepsilon > \varepsilon'$. The *least core*, denoted $LC(v)$, is the intersection of all nonempty strong ε-cores (see Maschler, Peleg and Shapley [168]). If $\varepsilon_\star(v)$ is the minimum ε such that $C_\varepsilon(v) \neq \emptyset$, i.e.,

$$\varepsilon_\star(v) = \min_{x \in I^*(v)} \max_{\emptyset \neq S \neq N} e(S, x),$$

then the least core satisfies $LC(v) = C_{\varepsilon_\star(v)}(v)$ and it is always nonempty.

The *multiplicative ε-core* was introduced by Faigle and Kern [79]. It is the set of all pre-imputations such that

$$C^\varepsilon(v) = \{x \in I^*(v) : e(S, x) \leq \varepsilon v(S) \text{ for all } S \neq N, \emptyset\}.$$

Let (N, v) be a profit game such that $v(N) \geq \sum_{i \in N} v(i)$. For each imputation $x \in I(v)$, let $\theta(x)$ be the 2^n-vector whose components are the excesses $e(S, x)$ for all $S \subseteq N$, arranged in nonincreasing order. The *lexicographic order* of such vectors is given by the relation $\theta(x) \precsim \theta(y)$, which holds if and only if $\theta(x) = \theta(y)$ or $\theta(x) \neq \theta(y)$ and for the minimum index i such that $\theta_i(x) \neq \theta_i(y)$ we have $\theta_i(x) < \theta_i(y)$.

Definition 1.1.4 *Let the set $Y \subseteq I^*(v)$. The nucleolus for Y of the profit game (N, v) is the set of pre-imputations in Y that minimize θ in the lexicographic ordering, that is,*

$$\mathcal{N}_Y(v) := \{x \in Y : \theta(x) \precsim \theta(y) \text{ for all } y \in Y\}.$$

The nucleolus for the set of imputations $I(v)$ is called the *nucleolus* of (N, v) and is denoted $\eta(v)$, while the nucleolus for $I^*(v)$ is called the pre-nucleolus of (N, v) and is denoted $\eta^*(v)$. Schmeidler [209] proved that the nucleolus of a game with a nonempty imputation set contains exactly one element. The same is true for the pre-nucleolus. The *lexicographic center*, a geometrical approach by Maschler, Peleg and Shapley [168] is equivalent to Schmeidler's definition.

Note that the core may be empty or quite large. Then, we would like to obtain an allocation recommended by an arbitrator whom the players may wish to consult. Shapley [212] approached this situation with a system of axioms that determine a unique payoff vector for each game. The *Shapley value* for the player $i \in N$ is defined by

$$\Phi_i(N, v) := \sum_{\{S \subseteq N : i \in S\}} \frac{(s-1)!(n-s)!}{n!} (v(S) - v(S \setminus i)), \tag{1.1}$$

where $n = |N|$, and $s = |S|$. This value is an average of *marginal contributions* $v(S) - v(S \setminus i)$ of the player i to all possible coalitions $S \in 2^N \setminus \{\emptyset\}$. In this value, the sets S of different size get different weight.

Dubey and Shapley [59] suggested the so-called *Banzhaf value*,

$$\beta'_i(N, v) := \sum_{\{S \subseteq N : i \in S\}} \frac{1}{2^{n-1}} (v(S) - v(S \setminus i)), \quad i \in N. \tag{1.2}$$

The Banzhaf value enjoys the symmetry, dummy, and linearity properties that are traditionally used to axiomatize the Shapley value. Only the *efficiency axiom* fails, since we have in general $\sum_{i \in N} \beta'_i(N, v) \neq v(N)$.

Every cooperative game (N, v) is uniquely determined by the collection of its worths $\{v(S) : S \subseteq N,\ S \neq \emptyset\}$. Then, the vector space of all cooperative games, denoted by Γ^N, will be identified with $\mathbb{R}^{2^n - 1}$. For any $T \subseteq N$, $T \neq \emptyset$, we define the *unanimity* game,

$$u_T(S) = \begin{cases} 1, & \text{if } T \subseteq S \\ 0, & \text{otherwise.} \end{cases}$$

Every game is a unique linear combination of unanimity games,

$$v = \sum_{T \subseteq N} d_T(v) u_T, \quad \text{where} \quad d_T(v) = \sum_{S \subseteq T} (-1)^{|T|-|S|} v(S).$$

Following Harsanyi [122], we shall call $d_T(v)$ the *dividend* of T in the game v. The Shapley and Banzhaf values are linear mappings w.r.t. the characteristic function and the images of the unanimity games are respectively

$$\Phi_i(N, u_S) = \begin{cases} 1/|S|, & \text{if } i \in S \\ 0, & \text{otherwise,} \end{cases}$$

$$\beta_i'(N, u_S) = \begin{cases} 1/2^{|S \setminus i|}, & \text{if } i \in S \\ 0, & \text{otherwise.} \end{cases}$$

In terms of dividends, we have

$$\Phi_i(N, v) = \sum_{\{S \subseteq N \,:\, i \in S\}} \frac{d_S(v)}{|S|}, \qquad \beta_i'(N, v) = \sum_{\{S \subseteq N \,:\, i \in S\}} \frac{d_S(v)}{2^{|S \setminus i|}}. \tag{1.3}$$

The *Tijs value* of a game is a compromise between the upper and the lower value for the game introduced by Tijs [231]. The *upper value* of the game (N, v) is the n-vector $M(v)$, where

$$M_i(v) := v(N) - v(N \setminus i), \quad \text{for all} \quad i \in N.$$

The component $M_i(v)$ is called utopia payoff for player i in the grand coalition N. The remainder of $i \in S$ if the coalition S forms and all other players in S obtain their utopia payoff is

$$R^v(S, i) := v(S) - \sum_{j \in S \setminus i} M_j(v).$$

The *lower value* is the vector $m(v) \in \mathbb{R}^n$, defined by

$$m_i(v) := \max_{\{S \subseteq N \,:\, i \in S\}} R^v(S, i).$$

Definition 1.1.5 *A game (N, v) is called quasi-balanced if*

(QB1) $m(v) \leq M(v)$,

(QB2) $\sum_{i=1}^n m_i(v) \leq v(N) \leq \sum_{i=1}^n M_i(v)$.

For a quasi-balanced game v the *Tijs value*, denoted by $\tau(v)$, is the unique pre-imputation (efficient vector) on the closed interval $[m(v), M(v)]$ in $\mathbb{R}^n$. Then we have

$$\tau(v) := m(v) + \lambda \left(M(v) - m(v) \right), \tag{1.4}$$

where $\lambda \in \mathbb{R}$ satisfies $\sum_{i \in N} \tau_i(v) = v(N)$.

1.2 Graph theory

A *graph* $G = (V, E)$ consists of two sets: a finite set V of elements called *vertices* and a finite set E of elements called *edges*. Each edge is identified with an unordered pair of vertices, called its *end vertices*. Two edges are called *parallel* if they have the same pair of end vertices. Furthermore, if the end vertices of an edge are equals, then this edge is called a *loop*.

A graph is called a *simple graph* if it does not have parallel edges or loops. The number $|V|$ of vertices is the *order* of G, while the number $|E|$ of edges is the *size*. An edge is said to be *incident on* its end vertices. Two vertices are *adjacent* if they are the end vertices of some edge. The set of edges incident on a vertex $v \in V$ is denoted by $\delta(v)$ and the *degree* of the vertex v is the number $d(v) = |\delta(v)|$. A vertex of degree 1 is called a *pendant vertex*. A vertex of degree 0 is called an *isolated vertex*.

Consider a graph $G = (V, E)$. If $W \subseteq V$ then $G - W$ denotes the graph obtained after removing from G the vertices in W and all the edges incident on a vertex in W. The subgraph of G induced by the vertex set $W \subseteq V$, denoted as $G[W]$, is the graph $G - (V \setminus W)$. For $F \subseteq E$, the graph $G - F$ is the graph that results after removing from G the edges in F. Note that the end vertices of the edges in F are not removed from G and $G - F = (V, E \setminus F)$. The graph $\overline{G}$ is called *complement* of a simple graph G if it has the same vertex set and two vertices are adjacent in $\overline{G}$ if and only if they are not adjacent in G.

The *complete graph* of order n, denoted K_n, has all possible edges, that is, its size is $\binom{n}{2}$. If every vertex of G has exactly the same degree r, then G is called *r-regular*. A cubic graph is regular of degree 3.

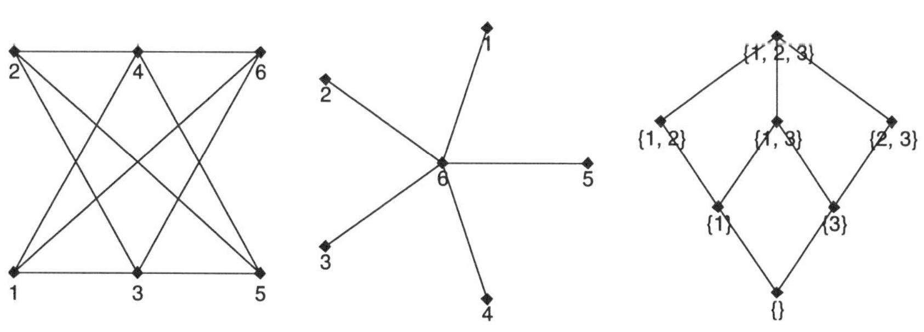

Figure 1.1 Graphs

A graph $G = (V, E)$ is *bipartite* if V can be partitioned into two nonempty disjoint sets V_1 and V_2 with $V_1 \cup V_2 = V$ such that each edge of E has one end vertex in V_1 and another in V_2. If the graph G is simple and bipartite, with $|V_1| = m$, $|V_2| = n$, and all mn possible edges are present, then G is called *complete bipartite* and is denoted by $K_{m,n}$.

A *matching* in a graph G is a set of edges M such that no two of them have a common vertex. The edges of a matching are called *independents.* A vertex is said to be *saturated* in a matching M if it is an end vertex of an edge in M. A matching is called *perfect* if every vertex is saturated.

A *walk* in a graph G is a finite sequence of vertices $v_0, v_1, \dots, v_k$, such that v_{i-1} is adjacent to v_i for $i = 1$ to k. This walk is called a $v_0 - v_k$ walk and the vertices v_0, v_k are called its end vertices. The number k of edges used is the *length* of the walk. Note that in a walk, edges and vertices may be repeated. A *trail* is a walk in which there are no repeated edges, and a *path* has no repeated vertices (and so it is a trail).

A walk is called closed if it has nonzero length and its end vertices are identical. An *Eulerian tour* in a graph G is a closed walk which contains each edge of G at most once. An Eulerian graph is a graph containing an Eulerian tour. A *cycle* is a walk with at least three different vertices that has no repeated vertices except the first and last one. A cycle is *even* if it is of even length, otherwise it is *odd.* An *acyclic* graph has no cycles. A cycle of length n in a graph of order n is called a *Hamiltonian cycle.* A graph that contains a Hamiltonian cycle is called Hamiltonian.

The *distance* between two vertices u and v in G is the length of a shortest $u - v$ path. The *diameter* of a graph is the maximum of the distances between all pairs of vertices.

A graph G is *connected* if any two vertices can be joined by a path. A maximal connected subgraph of G is a *component* of G. A *cutvertex* is a vertex whose removal increases the number of components, and a *bridge* is an edge with the same property. A graph is *2-connected* if it is connected, has at least 3 vertices and contains no cutvertex. A subgraph B of a graph G is a *block* of G if either B is a bridge or else it is a maximal 2-connected subgraph of G. A *clique* in a graph is a vertex set such that it induces a complete subgraph.

A *chord* of a cycle is an edge joining nonconsecutive vertices of the cycle. A *chordal graph* is a graph in which every cycle of length strictly greater than 3 has a chord. A vertex is called *simplicial* if its neighbor vertices form a clique.

A *tree* is a connected, acyclic graph. A tree of a graph G is a connected, acyclic subgraph of G. A *spanning tree* of a graph G is a tree of G containing all the vertices of G. An acyclic graph is called a *forest.*

Let $G = (V, E)$ be a graph. If $W \subseteq V$ then $\delta(W)$ denotes the set of edges with one end vertex in W and the other end vertex in $V \setminus W$. Any edge set of the form $\delta(W)$ for some $W \subseteq V$, $\emptyset \neq W \neq V$ is called a *cut* of G. We use this concept to prove the following characterization.

Proposition 1.2.1 *A graph $G = (V, E)$ is connected if and only if there is no vertex set $W \subseteq V$, $\emptyset \neq W \neq V$, with $\delta(W) = \emptyset$.*

Proof. Suppose that there exists $W \subseteq V$, $\emptyset \neq W \neq V$, such that $\delta(W) = \emptyset$. If $w \in W$ and $v \in V \setminus W$, then there is no $w - v$ path and G is not connected. To prove the converse, suppose that G is not connected and let W be a component of G. Then $\emptyset \neq W \neq V$, and $\delta(W) = \emptyset$. □

The graph representation of a traffic or telecommunication network requires directed graphs. A *digraph* $D = (V, A)$ consists of two sets: a finite set V of *nodes* and a set A of *arcs*. Each arc a is identified with an ordered pair (u, v) of nodes, called its *end nodes*: u is the initial or *tail* and v is the terminal or *head* of a. We say that a leaves u and enters v and that u is a *predecessor* of v and that v is a *successor* of u. The undirected graph that results after deleting the orientations of a digraph D is called the *underlying graph* of D. Induced subgraphs of a digraph is defined as in the undirected case.

If $v \in V$ then the set of arcs having v as tail is denoted by $\Gamma^+(v)$ and the set of arcs having v as head is denoted by $\Gamma^-(v)$, that is,

$$\Gamma^+(v) = \{w : (v, w) \in A\}, \quad \Gamma^-(v) = \{w : (w, v) \in A\}.$$

We define $\Gamma(v) := \Gamma^+(v) \cup \Gamma^-(v)$. The following numbers

$$d^+(v) := \left|\Gamma^+(v)\right|, \; d^-(v) := \left|\Gamma^-(v)\right|, \; d(v) := |\Gamma(v)|,$$

are called *out-degree*, *in-degree* and *degree* of v.

An *orientation* of an undirected graph G is an assignment of exactly one direction to each of the edges of G.

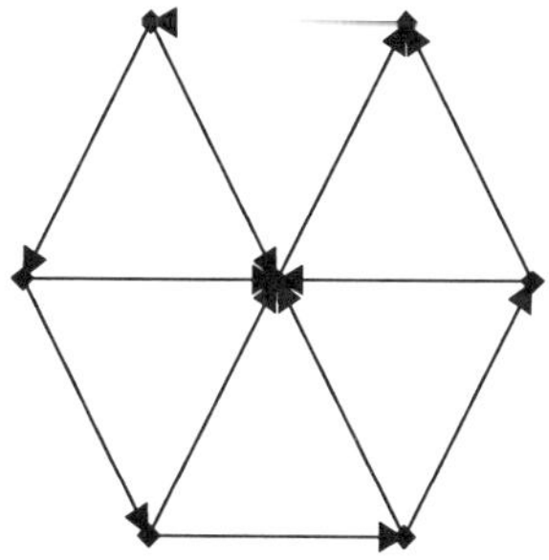

Figure 1.2 An orientation of a wheel

A node v in a digraph D is a *root* of D if there are directed paths from v to all the others nodes of D. A digraph is a tree if the underlying graph is a tree. A digraph is an *arborescence* if it is a tree and has a root.

1.3 Examples of cooperative games

Some instructive examples that arise from combinatorial optimization problems (see Curiel [51]) are the following.

Example: Assignment and matching games (Shapley and Shubik [218]).
We consider a partition of the players N into sellers N_1 and buyers N_2. A seller i and a buyer j can obtain a benefit $c(\{i,j\}) \geq 0$ and we suppose that each seller can only make one agreement and the same holds for buyers. Taking a bipartite graph $G = (N_1 \cup N_2, E)$ and a weight function $c : E \to \mathbb{R}_+$, the *assignment game* is

$$v(S) = \max\left\{\sum_{e\in M} c(e) : M \text{ is a perfect matching in the subgraph } G[S]\right\}.$$

Shapley and Shubik showed that the core of this game is nonempty. The *matching game* is defined by the above formula for matchings on a edge-weighted graph and its core is nonempty if the graph is bipartite.

Example: Min-cost spanning tree games (Granot and Huberman [109]).
Situations in which there is one common supplier 0 (water, electricity, cable, telecommunication) and a set $N = \{1,\ldots,n\}$ of users (houses, companies, cities) can be modeled by a complete graph $G = K_{n+1} = (N \cup \{0\}, E)$ and a nonnegative edge-weight function $w : E \to \mathbb{R}_+$. The associated *min-cost spanning tree game* evaluates the cost $c(S)$ of any coalition $S \subseteq N$ as the costs of a subgraph that connects all users of S to the supplier at minimum cost. Then, the game is defined by

$$c(S) = \min\left\{\sum_{e\in T_S} w(e) : T_S \text{ is a spanning tree of } G\,[S \cup \{0\}]\right\}.$$

Claus and Kleitman [40] and Bird [18] showed that the MCST game has a nonempty core. Furthermore, Granot and Huberman [109] proved that the following allocation rule $x \in \mathbb{R}^n$ is a point of the core: Find a minimum cost spanning tree T_N of the complete graph K_{n+1} and x_i is the weight of the first edge that i encounters on the unique path in T_N from i to the supplier 0.
The *survivable network* design problem was introduced by Gabow, Goemans and Williamson [100] and Grötschel, Monma and Stoer [115] and it is a good exercise to define the corresponding *survivable games.*

Example: Linear production games (Owen [194]).
In these games each player owns certain amounts of m resources given by a vector $b^i = \left(b_1^i, b_2^i, \ldots, b_m^i\right) \in \mathbb{R}^m,\ \ i = 1, 2, \ldots, n$. These resources can be used

to produce p goods, and a unit of the j-th good requires a_{kj} units of the k-th resource, and can be sold at a price c_j. The players can pool their resources and every coalition $S \subseteq N$ will have $b_k(S) = \sum_{i \in S} b_k^i$ units of the k-th resource. The linear production game (N, v) is defined by

$$v(S) = \max \left\{ \sum_{j=1}^{p} c_j x_j : \sum_{j=1}^{p} a_{kj} x_j \leq b_k(S), \;\; k = 1, \ldots, m, \;\; x \geq 0 \right\}.$$

Owen proved that a linear production game is totally balanced, in the sense that each subgame is balanced and proposed a method to obtain points in the core. He considers the dual of the above linear program

$$v(S) = \min \left\{ \sum_{k=1}^{m} b_k(S) y_k : \sum_{k=1}^{m} a_{kj} y_k \geq c_j, \;\; j = 1, \ldots, p, \;\; y \geq 0 \right\}.$$

Let $y^* = (y_1^*, \ldots, y_m^*)$ be an optimal solution for $S = N$. Then, we have $\sum_{k=1}^{m} b_k(N) y_k^* = v(N)$, and $\sum_{k=1}^{m} b_k(S) y_k^* \geq v(S)$, for every $S \subset N$. Consider now the vector $u = (u_1, \ldots, u_n)$ defined by $u_i = \sum_{k=1}^{m} b_k^i y_k^*, \quad i = 1, \ldots, n$. Hence for each $S \subset N$, we have

$$\begin{aligned} u(S) &= \sum_{k=1}^{m} b_k(S) y_k^* \geq v(S), \\ u(N) &= v(N), \end{aligned}$$

and therefore $u \in C(v)$. The components of the optimal solution y^* can be interpreted as equilibrium prices for the m resources. Each of the players is then paid for his/her resources according to the equilibrium price vector y^* and the resulting payments u_i will give a vector in the core. However, not all core elements can be obtained from the solution of the dual problem.

Feltkamp, van den Nouweland, Borm, Tijs and Koster [88] studied linear production games with transfer possibilities. In these games, each coalition tries to produce a bundle of maximal worth with the resources it controls, possibly transporting resources, products and technologies.

Example: Tax games (Tijs and Driessen [233]).

Let (N, v) be a profit game and let S be a proper coalition, $\emptyset \neq S \neq N$. Tijs and Driessen introduced a taxed game so that the worth is reduced to $v(S) - \varepsilon \left(v(S) - \sum_{i \in S} v(i) \right)$. Therefore the ε-tax core is

$$\begin{aligned} C\left(v^{\varepsilon}\right) &= \left\{ x \in I^*(v) : x(S) \geq v(S) - \varepsilon \left(v(S) - \sum_{i \in S} v(i) \right) \text{ for all } S \neq N, \emptyset \right\} \\ &= \left\{ x \in I^*(v) : e(S, x) \leq \varepsilon \left(v(S) - \sum_{i \in S} v(i) \right) \text{ for all } S \neq N, \emptyset \right\}. \end{aligned}$$

The ε-tax core coincides with the multiplicative ε-core for games v with $v(i) = 0$ for all $i \in N$ (0-normalized games).

Example: Bin packing games (Faigle and Kern [79]).

Suppose that there are n items of sizes $a_1, \ldots, a_n \in \mathbb{R}_+$ and m bins of sizes $b_1, \ldots, b_m \in \mathbb{R}_+$. We assume that each item i fits into any bin j, that is $a_i \leq b_j$ for all i, j. The set N of players of the *bin packing game* consists of all items and all bins. A coalition S is formed in the following way: $S = A \cup B$, where $A \subseteq \{1, \ldots, n\}$ is a set of items and $B \subseteq \{1, \ldots, m\}$ is a set of bins. The characteristic function $v : 2^N \to \mathbb{R}_+$ is defined as the maximum total size of items in A that can be packed into bins in B,

$$v(S) = \max \sum_{j=1}^{|B|} \sum_{i \in A_j} a_i,$$

where the maximum is taken over all pairwise disjoint subsets $A_1, \ldots, A_{|B|}$ of A such that

$$\sum_{i \in A_j} a_i \leq b_j.$$

Example: Simple flow games (Kalai and Zemel [136], Reijnierse, Maschler, Potters and Tijs [206]).

A *flow network* is defined by a digraph $D = (V, A)$ with two distinguished nodes (source and sink) and a capacity function $c : A \to \mathbb{R}_+$. It is allowed that several arcs have the same end nodes. Let N be a set of players and let us define the *ownership function* $o : 2^N \to 2^A$ which assigns to each player the arcs she/he owns and to each coalition the arcs of its members. Arcs not owned by a player are called *public arcs.* If all the arcs have capacity 1 and each player owns exactly one arc, we say that the flow network is *simple.* A game v is said to be a simple flow game if there exists a simple flow network such that for each $S \in 2^N$ its worth $v(S)$ is the maximum flow from source to sink that uses only arcs owned by S and public arcs.

Example: Permutation games (Tijs, Parthasarathy, Potters and Rajendra Prasad [232]).

Consider the situation in which there are n players $N = \{1, \ldots, n\}$ and each player i has a job J_i and a machine M_i. We assume that any machine M_j can process at most a job J_i, with cost k_{ij}. The *permutation game*, corresponding to the cost matrix $K = [k_{i\,j}]_{1 \leq i,j \leq n}$ is defined by

$$c(S) = \min \sum_{i \in S} k_{i\,\pi(i)},$$

where the minimum is taken over all permutations $\pi : S \to S$. Note that a permutation π corresponds to a scheme in which job J_i is processed on the machine $M_{\pi(i)}$.

Example: Traveling salesman games (Potters, Curiel and Tijs [200]).
A professor is required to visit the universities $N = \{1, 2, \ldots, n\}$. He has to start at his university 0, visit each university exactly once, and then return to his university. Suppose we consider a complete graph K_{n+1} on the vertex set $N \cup \{0\}$. The travel cost between two universities may be represented as a non-negative weight associated with the corresponding edge. The *traveling salesman game* (N, c) is defined by

$$c(S) = \text{cost of a minimal Hamiltonian cycle in } K_{|S|+1}, \quad \text{for all } S \subseteq N.$$

Example: Delivery games (Hamers [117]).
Consider the problem faced by a postman who must deliver mail along each edge (street) of a graph and return to his starting point (post office). A postman tour of G is a closed walk which contains each edge of G at least once. The length c_e of an edge is a non-negative number and the length of a tour is the sum of the lengths of its edges. The Chinese postman problem is to find the minimum length postman tour of a connected graph G. Note that if there is an Euler tour in the graph, then it solves the Chinese postman problem.

Let $G = (N, E)$ be a connected graph. In a delivery situation, the players are the edges $e \in E$, and we assign delivery costs $d : E \to \mathbb{R}_+$ and travel costs $t : E \to \mathbb{R}_+$. For every coalition $S \subseteq E$, a postman S-tour is a closed walk which contains each edge of S at least once and starts in the post office. The *delivery game* (N, c) is defined by

$$c(S) = \text{cost of a minimal postman } S\text{-tour}, \quad \text{for all } S \subseteq N.$$

Example: Sequencing games (Curiel, Pederzoli and Tijs [49]).
A set of customers $N = \{1, \ldots, n\}$ is standing in a queue. The original position of each customer is given by a permutation $\pi_0 : N \to N$. It is assumed that each customer i has a service time $s_i > 0$ and a cost function $c_i : \mathbb{R}_+ \to \mathbb{R}$ defined by $c_i(t) = \alpha_i t + \beta_i$, with $\alpha_i > 0$. A *sequencing situation* as described above is denoted by (N, π_0, s, α) where π_0 is a permutation, $s = (s_i)_{i \in N} \in \mathbb{R}^n_+$ and $\alpha = (\alpha_i)_{i \in N} \in \mathbb{R}^n_+$. Note that the vector $\beta = (\beta_i)_{i \in N}$ is omitted since it represents fixed costs that are independent of the order of the customers.

Given a permutation $\pi : N \to N$, the total time for costumer i is given by $S(\pi, i) = \sum_{j \in P(\pi, i)} s_j + s_i$, where $P(\pi, i) = \{j : \pi(j) < \pi(i)\}$ is the set of predecessors of i according to π. Then, the total cost of N is

$$C(\pi) = \sum_{i=1}^{n} c_i \left(S\left(\pi, i\right)\right) = \sum_{i=1}^{n} \left(\alpha_i S\left(\pi, i\right) + \beta_i\right).$$

We will define *sequencing games* (N, v), where the characteristic function $v : 2^N \to \mathbb{R}$ is given by

$$v(S) = \max_{\pi \in \Sigma_S} \sum_{i \in S} \left(S\left(\pi_0, i\right) - S\left(\pi, i\right)\right),$$

where $\Sigma_S = \{\pi : P(\pi, i) = P(\pi_0, i) \text{ for all } i \in N \setminus S\}$ is the set of admissible orderings for the coalition S.

Example: Restricted games (Myerson [182] and Owen [195]).

In cooperative game theory it is generally assumed that the whole group of players decides to cooperate. However, the classical model seems to be inappropriate in modelling certain situations. So, the hypothesis of the Shapley value (the probability of a coalition depends on its size, with the total probability of each size being the same) or the Banzhaf value (all coalitions are equally possible) may be unrealistic. Because of this, Myerson [184, page 444] proposed:

> "the term *cooperation structure* to refer to any mathematical structure that describes which coalitions (within the set of all $2^n - 1$ possible coalitions) can negotiate or coordinate effectively in a coalition game."

A *communication structure* is a graph $G = (N, E)$ whose vertex set is the set of players. The edge set E is defined by the following relation: $\{i, j\} \in E$ if and only if there exists communication between the players i and j. Then, the collection of feasible coalitions is

$$\mathcal{F} := \left\{S \in 2^N \ : \ \text{the subgraph } G\left[S\right] \text{ is connected}\right\}.$$

Myerson [182] introduced the *graph-restricted* games and the *Myerson value*, i.e., the Shapley value of the graph-restricted game. These games and this value were also investigated in Owen [195], who obtained a formula for computing the dividends in restricted games, where the graph is a tree. Borm, Owen and Tijs [26] defined the *position value* for communication structures and provided a new axiomatic characterization of the Myerson value.

The goal of this book is to develop a theoretical framework in which cooperative games are analyzed where only certain coalitions are allowed to form. Thus, we will study the following combinatorial structures: *convex geometries*, *matroids*, *greedoids* and *antimatroids*.

1.4 Closure spaces and convex geometries

Let $N = \{1, \dots, n\}$ be a finite set. A *set system* over N is a pair $(N, \mathcal{F})$, where $\mathcal{F} \subseteq 2^N$ is a collection of subsets of N.

Definition 1.4.1 *A closure space [113] is a set system $(N, \mathcal{C})$ which satisfies the following properties:*

1. *$\emptyset \in \mathcal{C}$ and $N \in \mathcal{C}$,*

2. *If $A \in \mathcal{C}$ and $B \in \mathcal{C}$, then $A \cap B \in \mathcal{C}$.*

Let $(N, \mathcal{C})$ be a closure space. Then, for any subset S of N we define the *closure* of S, denoted by $\overline{S}$, to be $\overline{S} := \bigcap\{C : C \in \mathcal{C},\ C \supseteq S\}$.

The map $- : 2^N \to 2^N$ is a closure operator which satisfies $A \subseteq \overline{A}$, $\overline{\overline{A}} = \overline{A}$ and $A \subseteq B$ implies $\overline{A} \subseteq \overline{B}$, for all $A, B \subseteq N$, with the additional condition that $\overline{\emptyset} = \emptyset$. The subsets in $\mathcal{C}$ or, equivalently, those subsets of N such that $\overline{S} = S$, are called *closed sets.* A closure space ordered by inclusion is a complete lattice, with the meet $\wedge$ and join $\vee$ operations defined as $A \wedge B := A \cap B$, and $A \vee B := \overline{A \cup B}$.

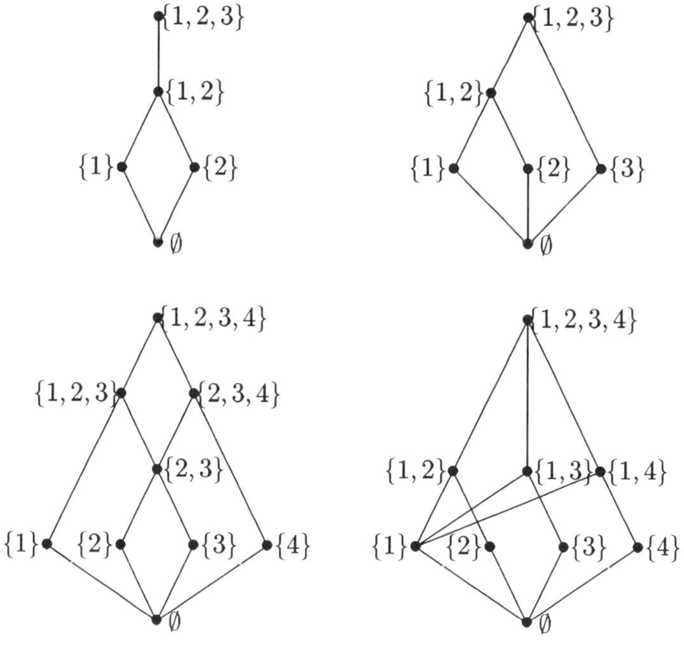

Figure 1.3 Four closure spaces

Convex geometries are a combinatorial abstraction of convex sets introduced by Edelman and Jamison [65].

Definition 1.4.2 *The set system $\mathcal{L} \subseteq 2^N$ is a convex geometry over N if it satisfies the properties:*

(C1) $\emptyset \in \mathcal{L}$,

(C2) *$\mathcal{L}$ is closed under intersection,*

(C3) *If $C \in \mathcal{L}$ and $C \neq N$, then there exists $j \in N \setminus C$ such that $C \cup j \in \mathcal{L}$.*

The elements of $\mathcal{L}$ are called *convex sets.* Note that property (C3) implies $N \in \mathcal{L}$. A maximal chain of $\mathcal{L} \subseteq 2^N$ is an ordered collection of convex sets that is not contained in any larger chain. From property (C3) and by induction, Edelman and Jamison [65] showed that every maximal chain contains $n+1$ convex sets $\emptyset = S_0 \subset S_1 \subset \cdots \subset S_{n-1} \subset S_n = N$, and the cardinal $|S_k| = k$, for all $k = 0, 1, \ldots, n$.

An element i of a convex set $C \in \mathcal{L}$ is an *extreme point* of C if $C \setminus i \in \mathcal{L}$. The set of extreme points of C is denoted by $\text{ex}(C)$. The convex geometries are the closure spaces satisfying the finite Minkowski-Krein-Milman property: *Every convex set is the closure of its extreme points* [144, Theorem 1.1].

Furthermore, every convex geometry satisfies the following anti-exchange property (see Edelman and Jamison [65]),

$$\text{For all } S \subseteq N, \;\; i, j \notin \overline{S}, \;\; j \in \overline{S \cup i} \;\; \text{imply} \;\; i \notin \overline{S \cup j}.$$

This property is a combinatorial abstraction of the convex closure defined in Euclidean spaces. That is, in Figure 1.4, the points i and j are not in the *convex hull* of the set S, where the convex hull of $X \subseteq \mathbb{R}^n$ is the smallest convex set containing X. Thus, if j is in $\text{conv}(S \cup i)$ then i is outside $\text{conv}(S \cup j)$.

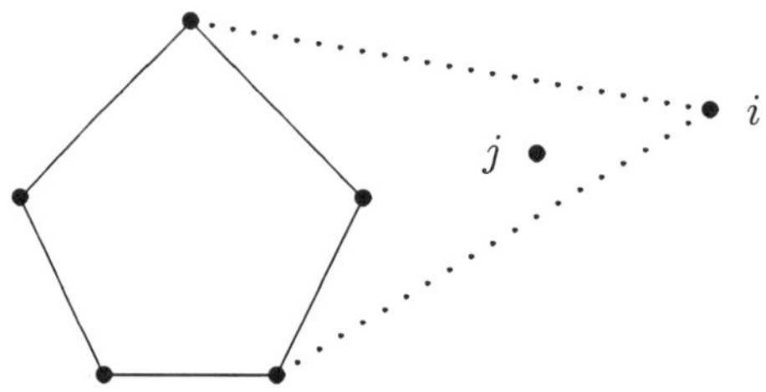

Figure 1.4 The anti-exchange property

Example: A graph G is a *block graph* if every block is a complete graph. The block graphs are called *cycle-complete* graphs in van den Nouweland and Borm

[187]. If G is a forest or a tree, then G is a block graph. Jamison [65] has shown that: *G=(N,E) is a connected block graph if and only if the collection of subsets of N which induces connected subgraphs is a convex geometry.*

Example: The concept of convexity is essential for pattern recognition and image processing in the bit-map-display models (see Hertel [126]). Let (X, d) be a finite metric space and let us define:

1. The point $z \in X$ is *between* x and y if $d(x, z) + d(z, y) = d(x, y)$.
2. The *segment* xy is the set of all points z between x and y.
3. The subset A of X is called *d-convex* if for each $x, y \in A$ the segment xy belongs to A.

Then the family of all d-convex sets in X is a closure system on X.

Example: A subset S of a partially ordered set $(P, \leq)$ is *convex* whenever $a \in S$, $b \in S$ and $a \leq b$ imply that the interval $[a, b] = \{x \in P : a \leq x \leq b\} \subseteq S$. The convex subsets of any poset P form a convex geometry $Co(P)$ (Birkhoff and Bennett [20, Theorem 3]).

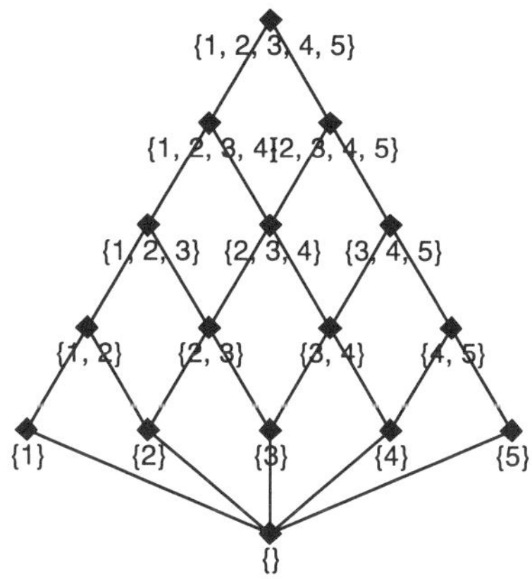

Figure 1.5 The convex geometry $Co(\{1 < 2 < 3 < 4 < 5\})$

Example: Let $(P, \leq)$ be a poset. For any $X \subseteq P$,

$$X \longmapsto \overline{X} := \{y \in P : y \leq x \text{ for some } x \in X\},$$

defines a closure operator on $(P, \leq)$. Its closed sets are the *order ideals* of $(P, \leq)$, and we use $J(P)$ to denote this collection. Since the union and intersection of order ideals is again an order ideal, it follows that $J(P)$ is a convex geometry closed under set-union and ex(S) is the set of all maximal points $Max(S)$ of the subposet $S \in J(P)$.

1.5 Matroids, greedoids and antimatroids

Matroids were introduced by Whitney [246, 1935] who axiomatized the properties of linear dependence of vectors in Euclidean spaces and cycles in a graph.

Definition 1.5.1 *Let N be a finite set. The set system $(N, \mathcal{M})$ is a matroid if the following properties are satisfied:*

(M1) $\emptyset \in \mathcal{M}$.

(M2) *If $S \in \mathcal{M}$ and $T \subseteq S$, then $T \in \mathcal{M}$.*

(M3) *If $T, S \in \mathcal{M}$ and $|S| = |T| + 1$, then there exists $i \in S \setminus T$ such that $T \cup i \in \mathcal{M}$.*

The sets belonging to $\mathcal{M}$ are called *independent sets*. A subset of N not belonging to $\mathcal{M}$ is called *dependent*. A set system $(N, \mathcal{I})$ satisfying (M1) and (M2) is called an *independence system*. Now, we present some examples.

Example: Let F be a finite subset of a vector space E and let $\mathcal{M}$ be the set of linearly independent subsets of vectors of F. Then $(F, \mathcal{M})$ is a matroid.

Example: Let E be the set of edges of a graph G and let $\mathcal{M} = \mathcal{M}(G)$ be a family that consists of those subsets of E that do not contain a cycle of G. The matroid $(E, \mathcal{M})$ is called a *graphic matroid* or *forest matroid* of G.

Example: For $k \in \mathbb{N}$, $0 < k \leq n$, the family $\{S \subseteq N : |S| \leq k\}$ of all subsets of N of size at most k is the uniform matroid denoted by U_n^k. In particular, 2^N is called a *free matroid*.

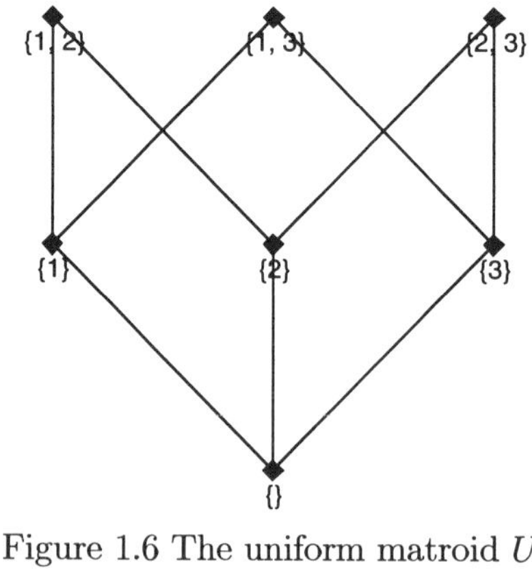

Figure 1.6 The uniform matroid U_3^2

Example: A pair of distinct elements i, j of N are *parallel* in the matroid $(N, \mathcal{M})$ if $\{i, j\}$ is a minimal dependent set. If $i, j \in N$, and $i \neq j$, the family

$$\mathcal{M}_n(i \| j) = \{S \subseteq N : \{i, j\} \not\subseteq S\}$$

is a matroid.

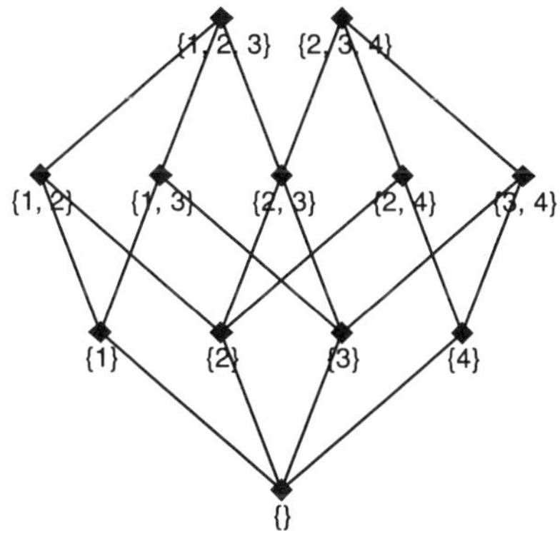

Figure 1.7 The matroid $\mathcal{M}_4(1\|4)$

We also need to introduce the following definitions and results (note the analogy with vector spaces).

Definition 1.5.2 *For a given set system $(N, \mathcal{M})$ and $S \subseteq N$, a subset $B \subseteq S$, $B \in \mathcal{M}$ is called a basis of S if $B \cup i \notin \mathcal{M}$ for all $i \in S \setminus B$.*

That is, B is a basis of S when B is a maximal feasible subset of S. The above definition can be used to characterize a matroid (see Welsh [245]).

Theorem 1.5.1 *A set system $(N, \mathcal{M})$ is a matroid if and only if it satisfies (M1), (M2), and the following property (M4): For every $S \subseteq N$, all bases of S have the same cardinality.*

Proof. Suppose that property (M4) fails to hold. Then there are bases B_1, B_2 of a set $S \subseteq N$ with $|B_1| < |B_2|$. We take $T \subseteq B_2$ with $|T| = |B_1| + 1$. Property (M2) implies that $T \in \mathcal{M}$, and therefore, by (M3) the set $B_1 \cup i \in \mathcal{M}$ for some $i \in T \setminus B_1 \subseteq S \setminus B_1$, contradicting the assumption that B_1 is a basis of S.

To see the opposite implication, suppose that (M3) is not true. Then there are $T, S \in \mathcal{M}$ with $|S| = |T| + 1$ such that $T \cup i \notin \mathcal{M}$ for all $i \in S \setminus T$. This implies that T is a basis of $S \cup T$. Also, there exists a basis B of $S \cup T$ with $S \subseteq B$. Then $|T| \neq |B|$, in contradiction to property (M4). $\square$

We call the bases of N the *bases* of the matroid $(N, \mathcal{M})$. The *rank* of the matroid $(N, \mathcal{M})$ is the cardinality of every basis. Then the family of the bases is $\mathcal{B}(\mathcal{M}) = \{B \in \mathcal{M} : |B| = r\}$, where r is the rank of $(N, \mathcal{M})$.

Remark 1.5.1 *Note that the length of all the maximal chains of a matroid is its rank.*

We can consider a matroid $\mathcal{M} \neq 2^N$ split into influence zones, where the bases of N are the ones at the summits of the different influence zones. Note that these influence zones are not disjointed.

Definition 1.5.3 *The influence set of $S \in \mathcal{M}$ is the set*

$$\mathcal{B}_S(\mathcal{M}) = \{B \in \mathcal{B}(\mathcal{M}) : S \subseteq B\},$$

and its influence worth is the ratio b_S/b, where $b_S = |\mathcal{B}_S(\mathcal{M})|$ and $b = |\mathcal{B}(\mathcal{M})|$.

We denote b_i when $S = \{i\}$ is considered. Note that $\mathcal{M} = 2^N$ if and only if $b = 1$.

Example: Minimum base game on matroids (Nagamochi, Zeng, Kabutoya and Ibaraki [185]).

We consider a matroid $(E, \mathcal{M})$ with a weight function $w : E \to \mathbb{R}$, and the characteristic function $c : 2^E \to \mathbb{R}$ of the *minimum base game* is for any $S \subseteq E$,

$$c(S) = \min \left\{ \sum_{e \in B(S)} w(e) : B(S) \text{ is a basis of } S \right\}.$$

If the matroid is the graphic matroid of a graph $G = (N, E)$ then we obtain the *minimum forest game,* defined by

$$c(S) = \min \left\{ \sum_{e \in F_S} w(e) : F_S \text{ is a maximal forest on } S \right\},$$

for all $S \subseteq E$.

Remark 1.5.2 *Note that in this game the players are the edges in G and it is different from the min-cost spanning tree game defined in page 10.*

Definition 1.5.4 *A minimal dependent set C, that is, a dependent set C such that $C \setminus e \in \mathcal{M}$, is called a circuit. We say that a circuit C is an all-negative circuit if $w(e) < 0$ for all $e \in C$.*

Nagamochi *et al.* [185] showed the following characterizations.

Theorem 1.5.2 *Let (E, c) be the minimum base game on a weighted matroid $((E, \mathcal{M}), w)$. Then the following statements are equivalent:*

1. *The characteristic function $c : 2^E \to \mathbb{R}$ is subadditive.*
2. *The game (E, c) is balanced.*
3. *The game (E, c) is quasi-balanced.*
4. *The weighted matroid $((E, \mathcal{M}), w)$ has no all-negative circuits.*
5. *The set $E^- = \{e \in E : w(e) < 0\}$ is independent.*

We consider now an extension of the matroids, namely:

Definition 1.5.5 *A greedoid is a set system $(N, \mathcal{G})$ such that:*

(G1) $\emptyset \in \mathcal{G}$,

(G2) *For every nonempty $S \in \mathcal{G}$ there is an element $i \in S$ such that $S \setminus i \in \mathcal{G}$.*

(G3) *For $S, T \in \mathcal{G}$ with $|T| = |S| + 1$, there is an element $j \in T \setminus S$ such that $S \cup j \in \mathcal{G}$.*

A set $S \in \mathcal{G}$ is called *feasible.* The family of independent sets in a matroid satisfies these properties, hence every matroid is a greedoid. Note that the difference between matroids and greedoids is that every subset of an independent set is independent in a matroid (M2), but a feasible set in a greedoid will have non feasible subsets. Properties (G1) and (G2) of greedoids imply that for any feasible set $S \in \mathcal{G}$ there is a chain of feasible sets

$$\emptyset = S_0 \subset S_1 \subset \cdots \subset S_{|S|-1} \subset S_{|S|} = S,$$

such that each S_k is obtained by adding one element to S_{k-1}, $1 \leq k \leq |S|$.

Example: Let $D = (V, A)$ be a digraph and let $r \in V$ be a distinguished node called the root. If $\mathcal{G}$ is the set of all arborescences in D rooted at r then $(A, \mathcal{G})$ is a greedoid.

Greedoids include matroids and the set systems called *antimatroids.*

Definition 1.5.6 *An antimatroid $(N, \mathcal{A})$ is a greedoid which satisfies $S, T \in \mathcal{A}$ implies $S \cup T \in \mathcal{A}$.*

A *loop* in a greedoid is an element of N which does not appear in any feasible set. We only consider antimatroids with no loops, i.e., $N \in \mathcal{A}$. Antimatroids are dual to convex geometries. For an antimatroid $(N, \mathcal{A})$ we consider the system of complements $\mathcal{L} = \{C \subseteq N : N \setminus C \in \mathcal{A}\}$.

Theorem 1.5.3 *$(N, \mathcal{A})$ is an antimatroid if and only if $(N, \mathcal{L})$ is a convex geometry.*

Proof. Since $\emptyset, N \in \mathcal{A}$, we have that $\emptyset, N \in \mathcal{L}$ and conversely. $\mathcal{A}$ is closed under union if and only if $\mathcal{L}$ is closed under intersection and property (G2) for $\mathcal{A}$ is equivalent to property (C3) for $\mathcal{L}$. We only have to prove that if $\mathcal{L}$ is a convex geometry then the system of complements $\mathcal{A}$ satisfies (G3).

Let $S, T \in \mathcal{A}$ with $|S| < |T|$. Note that $S \cup j \in \mathcal{A}$ for some $j \in T \setminus S$, if and only if $T \cap \text{ex}(N \setminus S) \neq \emptyset$. If we assume that this intersection is empty, then $\text{ex}(N \setminus S) \subseteq N \setminus T$. Since both $N \setminus S$ and $N \setminus T$ are convex, the Minkowski-Krein-Milman property implies

$$N \setminus S = \overline{\text{ex}(N \setminus S)} \subseteq \overline{N \setminus T} = N \setminus T,$$

contradicting that $|S| < |T|$. □

Example: Let $(P, \leq)$ be a partially ordered set and we define the collection

$$J(P) = \{I \subseteq P : I \text{ is an order ideal}\}.$$

Since the union and intersection of order ideals is again an order ideal, it follows that $J(P)$ is an antimatroid and a convex geometry.

Example: Let $G = (V, E)$ be a forest and let

$$\mathcal{A} = \{S \subseteq V : G - S \text{ has the same number of connected components as } G\}.$$

Then $(V, \mathcal{A})$ is an antimatroid.

Example: Let $E \subseteq \mathbb{R}^n$ be a finite set of points and let

$$\mathcal{A} = \{\{x_1, \dots x_k\} \subseteq E : x_i \notin \text{conv}\,(E \setminus \{x_1, \dots x_i\}) \text{ for all } i \leq k\},$$

where the convex hull of a set $X \subseteq \mathbb{R}^n$ is

$$\text{conv}(X) = \left\{ \sum_{i=1}^{t} \lambda_i x_i : t \geq 0,\ \sum_{i=1}^{t} \lambda_i = 1,\ \lambda_i \geq 0,\ x_i \in X \text{ for all } i = 1, \dots, t \right\}.$$

Then $(E, \mathcal{A})$ is an antimatroid.

Example: Let $G = (V, E)$ be a chordal graph and so G has at least two simplicial vertices (see Jamison [131]). We can get an antimatroid $\mathcal{A}$ on the vertex set V by repeatedly deleting simplicial vertices. Then, $S \in \mathcal{A}$ if there is some ordering $\{v_1, \dots, v_s\}$ of the elements of S so that v_i is simplicial in the subgraph $G - \{v_1, \dots, v_{i-1}\}$ for all $i = 1, \dots, s$.

1.6 Project: Bicooperative games

Let $N = \{1, 2, \ldots, n\}$ be a set of players. Aubin [6, Section 13.2] introduced the concept of *generalized coalition* as a function $c : N \to [-1, 1]$ which associates each player i with his/her level of participation $c(i) \in [-1, 1]$. A positive level is interpreted as attraction of the player i for the coalition, and a negative level as repulsion. In particular, the set 2^N of coalitions is the set of functions $c : N \to \{0, 1\}$.

We will introduce a natural generalization of the concept of cooperative game. First, we consider the set of all the ordered pairs of disjoint coalitions, that is the set of all *signed coalitions* given by

$$3^N = \{(S, T) : S, T \subseteq N,\ S \cap T = \emptyset\}.$$

Each signed coalition (S, T) of 3^N can be identified with the $\{0, \pm 1\}$-vector $\mathbf{1}_{(S,T)}$ defined by

$$\mathbf{1}_{(S,T)}(i) = \begin{cases} 1, & \text{if } i \in S \\ -1, & \text{if } i \in T \\ 0, & \text{otherwise,} \end{cases}$$

for all $i \in N$. The support of a signed coalition (S, T) of N is $S \cup T$. A partial order $\sqsubseteq$ on 3^N is defined by

$$(S_1, T_1) \sqsubseteq (S_2, T_2) \Longleftrightarrow S_1 \subseteq T_1 \text{ and } S_2 \subseteq T_2$$

for all $(S_1, T_1), (S_2, T_2) \in 3^N$.

Example: Felsenthal and Machover [86] introduced *ternary voting games* on a finite set N. This concept is a generalization of voting games which recognizes *abstention* as an option alongside *yes* and *no* votes. These games are given by mappings $u : 3^N \to \{-1, 1\}$ satisfying the following three conditions:

1. $u(N, \emptyset) = 1$,
2. $u(\emptyset, N) = -1$,
3. If $\mathbf{1}_{(S,T)}(i) \leq \mathbf{1}_{(S',T')}(i)$ for all $i \in N$, then $u(S, T) \leq u(S', T')$.

A negative outcome, -1, is interpreted as defeat and a positive outcome, 1, as passage of a bill.

Definition 1.6.1 *A bicooperative game is a pair (N, b), where N is a finite set and $b : 3^N \to \mathbb{R}$ is a function with $b(\emptyset, \emptyset) = 0$.*

We have two binary operations, reduced union $\sqcup$ and intersection $\sqcap$, on 3^N defined as

$$\begin{aligned}(S_1,T_1)\sqcup(S_2,T_2) &= ((S_1\cup S_2)\setminus(T_1\cup T_2),(T_1\cup T_2)\setminus(S_1\cup S_2)),\\ (S_1,T_1)\sqcap(S_2,T_2) &= (S_1\cap S_2,T_1\cap T_2).\end{aligned}$$

Definition 1.6.2 *A bicooperative game $c:3^N\to\mathbb{R}$ is called bisubmodular if it satisfies*

$$c((S_1,T_1)\sqcup(S_2,T_2))+c((S_1,T_1)\sqcap(S_2,T_2))\leq c(S_1,T_1)+c(S_2,T_2)$$

for all $(S_1,T_1),(S_2,T_2)\in 3^N$. A bicooperative game $b:3^N\to\mathbb{R}$ is called bisupermodular if $-b$ is bisubmodular and bimodular if the above inequality holds with equality.

The bisubmodular inequality has been introduced by Chandrasekaran and Kabadi [35]. Fujishige [96, Section 3.5c], Ando and Fujishige [5] studied bisubmodular systems $(\mathcal{F},f)$ where $\mathcal{F}\subseteq 3^N$ is a family of signed subsets closed with respect to the reduced union $\sqcup$ and intersection $\sqcap$, and $f:\mathcal{F}\to\mathbb{R}$ is bisubmodular on $\mathcal{F}$. *Delta-matroids* are essentially equivalent structures considered by Bouchet [28].

The bisubmodular/bisupermodular polyhedra associated with these games are respectively

$$\begin{aligned}P_*(c) &= \left\{x\in\mathbb{R}^n : x(S)-x(T)\leq c(S,T) \text{ for all } (S,T)\in 3^N\right\},\\ P_*(b) &= \left\{x\in\mathbb{R}^n : x(S)-x(T)\geq b(S,T) \text{ for all } (S,T)\in 3^N\right\}.\end{aligned}$$

These polyhedra were introduced by Dunstan and Welsh [62] and Bouchet and Cunningham [29] showed that the convex hull of the set of points of a *jump system* is an integral bisubmodular polyhedron.

If $P_*(c)$ is nonempty then $-c(\emptyset,S)\leq x(S)\leq c(S,\emptyset)$ for all $S\subseteq N$, where $x\in P_*(c)$. Therefore, we obtain $c(\emptyset,S)+c(S,\emptyset)\geq 0$ for all $S\subseteq N$. Fujishige [96, Theorem 3.63] showed that this necessary condition is also sufficient.

We remark that if $c:3^N\to\mathbb{R}$ is a bisubmodular game then

$$c(\emptyset,S)+c(S,\emptyset)\geq 2c(\emptyset,\emptyset)=0,$$

and this implies that every bisubmodular polyhedron is nonempty. The same holds for the class of bisupermodular polyhedra.

A pair $(S,T)\in 3^N$ such that its support $S\cup T=N$ is called an *orthant*. The nonempty face of $P_*(c)$ defined by

$$B_{(S,T)}(c)=\{x\in P_*(c) : x(S)-x(T)=c(S,T)\}$$

is called the base polyhedron of c in the orthant (S,T). Moreover, Fujishige [96, Section 3.5b] showed that the set of extreme points (vertices) of $P_*(c)$ satisfies

$$\mathrm{ex}(P_*(c)) = \bigcup \{\mathrm{ex}(B_{(S,T)}(c)) : (S,T) \text{ is an orthant}\}.$$

Example: If $c : 3^N \to \mathbb{R}$ is a bisubmodular game and (S,T) an orthant, we define the associated cooperative game $c_{ST} : 2^N \to \mathbb{R}$ by

$$c_{ST}(X) = c(S \cap X, T \cap X)$$

for all $X \subseteq N$. The bisubmodular inequality implies that for all $X, Y \subseteq N$,

$$\begin{aligned} c_{ST}(X \cup Y) + c_{ST}(X \cap Y) &= c(S_1 \cup S_2, T_1 \cup T_2) + c(S_1 \cap S_2, T_1 \cap T_2) \\ &= c((S_1,T_1) \sqcup (S_2,T_2)) + c((S_1,T_1) \sqcap (S_2,T_2)) \\ &\leq c(S_1,T_1) + c(S_2,T_2) \\ &= c_{ST}(X) + c_{ST}(Y), \end{aligned}$$

where $S_1 = S \cap X$, $S_2 = S \cap Y$, $T_1 = T \cap X$, $T_2 = T \cap Y$. Hence the cooperative game c_{ST} is concave. Note that if b is a bisupermodular game and (S,T) an orthant, then the cooperative game b_{ST} is convex.

Example: Let $v : 2^N \to \mathbb{R}$ be a convex profit game, i.e., for all $S, T \subseteq N$, we have $v(S \cup T) + v(S \cap T) \geq v(S) + v(T)$. We consider the associated bicooperative game $b : 3^N \to \mathbb{R}$ defined by

$$b(S,T) := v(S) - v^*(T) = v(S) + v(N \setminus T) - v(N).$$

We compute for all $(S_1,T_1), (S_2,T_2) \in 3^N$,

$$\begin{aligned} b((S_1,T_1) \sqcup (S_2,T_2)) &= v(S \setminus T) + v(N \setminus (T \setminus S)) - v(N) \\ &= v(S \cap (N \setminus T)) + v(S \cup (N \setminus T)) - v(N) \\ &\geq v(S) + v(N \setminus T) - v(N), \end{aligned}$$

where $S = S_1 \cup S_2$ and $T = T_1 \cup T_2$. Thus

$$\begin{aligned} b((S_1,T_1) \sqcup (S_2,T_2)) &\geq v(S_1 \cup S_2) + v((N \setminus T_1) \cap (N \setminus T_2)) - v(N), \\ b((S_1,T_1) \sqcap (S_2,T_2)) &= v(S_1 \cap S_2) + v((N \setminus T_1) \cup (N \setminus T_2)) - v(N). \end{aligned}$$

Since v is supermodular, the sum satisfies

$$b((S_1,T_1) \sqcup (S_2,T_2)) + b((S_1,T_1) \sqcap (S_2,T_2)) \geq b(S_1,T_1) + b(S_2,T_2)$$

and the bicooperative game b is bisupermodular. Furthermore, the core of the profit game (N,v) is the bisupermodular polyhedron associated with b, that is,

$$P_*(b) = \{x \in \mathbb{R}^n : x(S) \geq v(S) \text{ for all } S \subseteq N,\ x(N) = v(N)\} = C(v).$$

Chapter 2

Linear optimization methods

In order to present the most important facts about linear optimization methods, we need to become familiar with the theory of convex polyhedra and the methods of linear and integer programming. We recommend the books by Cook, Cunningham, Pulleyblank and Schrijver [45], Martin [163], Nemhauser and Wosley [186], Schrijver [210] and Ziegler [250].

2.1 Polyhedra

A *(convex) polyhedron* is the intersection of a finite number of closed halfspaces in some $\mathbb{R}^n$. Then, a set $P \subseteq \mathbb{R}^n$ is a polyhedron if there exists a matrix A and a column vector b such that

$$P = \{x \in \mathbb{R}^n : Ax \leq b\}.$$

A finite set of vectors $\{x_i : i \in I\} \subseteq \mathbb{R}^n$ is affinely independent if the system of equations

$$\sum_{i \in I} \lambda_i x_i = 0, \ \sum_{i \in I} \lambda_i = 0 \ \text{ imply } \ \lambda_i = 0 \ \text{ for all } \ i \in I.$$

The *dimension* of a set $P \subseteq \mathbb{R}^n$ is one less than the maximum cardinality of an affinely independent set $\{x_i : i \in I\} \subseteq P$. A polyhedron $P \subseteq \mathbb{R}^n$ of dimension n is said to be of full dimension and denoted by n-polyhedron. The points of an n-polyhedron $P \subseteq \mathbb{R}^n$ affinely span $\mathbb{R}^n$, that is,

$$\text{aff}\,(P) = \left\{ \sum_{i \in I} \lambda_i x_i : \sum_{i \in I} \lambda_i = 1, \ \{x_i\}_{i \in I} \subset P \right\} = \mathbb{R}^n.$$

The minimal representation of points in a convex set of $\mathbb{R}^n$ can be linked to its dimension.

Theorem 2.1.1 *(Carathéodory) Let $X \subseteq \mathbb{R}^n$ be a set with dimension d. Then for each point $x \in \operatorname{conv}(X)$ there exist $d+1$ points $\{x_1,, \ldots, x_{d+1}\} \in X$ such that $x \in \operatorname{conv}(\{x_1, \ldots, x_{d+1}\})$.*

Proof. See Stoer and Witzgall [227, Section 2.2]. □

An inequality $c^T x \leq d$ is *valid* for a polyhedron P if $P \subseteq \{x : c^T x \leq d\}$. A hyperplane $H = \{x : c^T x = d\}$ is called a *supporting hyperplane* of P if $c^T x \leq d$ is valid for P and $P \cap H \neq \emptyset$. This is equivalent to $\max\{c^T x : x \in P\} = d$.

A proper *face* F of an n-polyhedron P is the intersection of P with a supporting hyperplane. F itself is a polyhedron of some lower dimension. If the dimension of F is k we call F a k-face of P.

The empty set and the polyhedron P are trivial faces, 0-faces of P are called *vertices*, 1-faces are called *edges* and $(n-1)$-faces are called *facets* .

Given an n-polyhedron P, we denote by $f_k(P)$ the number of k-faces of P. The vector $(f_0(P), f_1(P), \ldots, f_n(P))$ is called the f-vector of P. The *Euler formula* gives the following relation between the numbers of vertices, edges and 2-faces of every bounded polyhedron $P \subseteq \mathbb{R}^3$,

$$f_0(P) - f_1(P) + f_2(P) = 2.$$

The set of vertices and bounded edges of P are a graph called the graph of P and denoted by $G(P)$.

Example: The full dimensional simplex (Kalai [137]).

The simplex S_n of dimension n is the convex hull of $n+1$ affinely independent vectors in $\mathbb{R}^n$. The faces of S_n are themselves simplices. Further, the convex hull of every subset of vertices of a simplex is a face so that $f_k(S_n) = \binom{n+1}{k+1}$. Thus, the graph $G(S_n)$ is the complete graph K_{n+1}.

Example: The imputation set (Tijs [234]).

Let (N, v) be a game with $|N| = n$ and $v(N) > \sum_{i \in N} v(i)$. We will show that the imputation set $I(v)$ is the convex hull of the set $\{a^1, \ldots, a^n\} \subseteq \mathbb{R}^n$, where

$$a_j^i = \begin{cases} v(j), & \text{if } j \neq i \\ v(N) - \sum_{j \in N \setminus i} v(j), & \text{if } j = i. \end{cases}$$

Proposition 2.1.2 *If $v(N) > \sum_{i \in N} v(i)$, then the set $\{a^1, \ldots, a^n\} \subseteq \mathbb{R}^n$ is affinely independent and $I(v)$ is the simplex of dimension $n-1$ generated by the vectors $\{a^1, \ldots, a^n\}$.*

Proof. Let $\lambda_1, \ldots, \lambda_n$ be numbers, not all 0, such that $\sum_{i=1}^n \lambda_i = 0$. If $\sum_{i=1}^n \lambda_i a^i = 0$, then $\sum_{i=1}^n \lambda_i a_j^i = 0$ for all $j = 1, \ldots, n$. Hence

$$\begin{aligned} 0 &= \sum_{i=1}^n \lambda_i a_j^i = \left(\sum_{i \neq j} \lambda_i \right) v(j) + \lambda_j \left(v(N) - \sum_{i \in N \setminus j} v(i) \right) \\ &= \lambda_j \left(v(N) - \sum_{i \in N} v(i) \right) \end{aligned}$$

implies $\lambda_j = 0$ for all $j = 1, \ldots, n$.

Since the imputation set is convex and $\left\{a^1, \ldots, a^n\right\} \subseteq I(v)$, we conclude that conv $\left\{a^1, \ldots, a^n\right\} \subseteq I(v)$. To obtain the reverse inclusion, let $x \in I(v)$. For each $j = 1, \ldots, n$ we take

$$\lambda_j = \frac{x_j - v(j)}{v(N) - \sum_{i \in N} v(i)}.$$

Since $x(N) = v(N)$ and $x_j \geq v(j)$ for each j, so that $\sum_{j=1}^n \lambda_j = 1$ and $\lambda_j \geq 0$ for each j. Moreover,

$$\sum_{i=1}^n \lambda_i a_j^i = (1 - \lambda_j) v(j) + \lambda_j \left(v(N) - \sum_{i \in N \setminus j} v(i) \right) = v(j) + x_j - v(j) = x_j$$

for all $j = 1, \ldots, n$. $\square$

Let (N, v) be a game with $v(N) = 1$ and $v(i) = 0$ for all $i \in N$. The imputation set is the standard $(n-1)$-simplex $\triangle_{n-1}$ generated by the canonical vectors,

$$\triangle_{n-1} = \left\{ x \in \mathbb{R}^n : \sum_{i=1}^n x_i = 1, \;\; x \geq 0 \right\} = \text{conv} \left\{ e_1, \ldots, e_n \right\}.$$

Example: The full dimensional cube (Kalai [137]).

The cube C_n of dimension n is the set of all points $(x_1, \ldots, x_n) \in \mathbb{R}^n$ that satisfy $0 \leq x_i \leq 1$, for all $i = 1, \ldots, n$. The vertices of C_n are the $\{0, 1\}$-vectors of $\mathbb{R}^n$ and the f-vector is given by $f_k(C_n) = 2^{n-k}\binom{n}{k}$. Two vertices are adjacent in the graph $G(C_n)$ if they agree in all but one coordinates.

Definition 2.1.1 *A polyhedron P is a polytope if there exist vectors $l, u \in \mathbb{R}^n$ such that $l \leq x \leq u$ for all $x \in P$.*

Proposition 2.1.3 *Let $S = \{s_1, s_2, \ldots, s_t\} \subseteq \mathbb{R}^n$ be a finite set and let $c \in \mathbb{R}^n$. Then* $\max\left\{c^T x : x \in S\right\} = \max\left\{c^T x : x \in \text{conv}(S)\right\}$, *where*

$$\text{conv}(S) = \left\{\sum_{i=1}^{t} \lambda_i s_i : \sum_{i=1}^{t} \lambda_i = 1,\ \lambda_i \geq 0 \ \textit{for all} \ \ i = 1, \ldots, t\right\}.$$

Proof. Let $u \in S$ be a vector such that $c^T u = \max\left\{c^T x : x \in S\right\}$ and let v be such that $c^T v = \max\left\{c^T x : x \in \text{conv}(S)\right\}$. First, $S \subseteq \text{conv}(S)$ implies that $c^T u \leq c^T v$. Next, if $v \in \text{conv}(S)$, then $v = \sum_{i=1}^t \lambda_i s_i$, for some $s_1, \ldots, s_t \in S$, and nonnegative coefficients such that $\sum_{i=1}^t \lambda_i = 1$. We therefore deduce that $c^T v = \sum_{i=1}^t \lambda_i c^T s_i \leq \left(\sum_{i=1}^t \lambda_i\right) c^T u = c^T u$. □

If the polyhedron P has at least one vertex, then P is called *pointed*. The polyhedron $\{(x_1, x_2) : x_1 \leq 0\}$ is not pointed. For the following results see Cook *et al.* [45, Chapter 6] and Schrijver [210, Chapter 8].

1. Let $P = \{x \in \mathbb{R}^n : Ax \leq b\}$ and let $v \in P$. Then v is a vertex of P if and only if $v \notin \text{conv}\,(P \setminus v)$.
2. A polytope is always pointed and is the convex hull of its vertices.
3. A set $P \subseteq \mathbb{R}^n$ is a polytope if and only if there exist $x_1, \ldots, x_t \in \mathbb{R}^n$ such that $P = \text{conv}\,\{x_1, \ldots, x_t\}$.
4. The polyhedron $P = \{x \in \mathbb{R}^n : Ax \leq b\}$ is bounded if and only if
$$\{x \in \mathbb{R}^n : Ax \leq 0\} = \{0\}.$$
5. The polyhedron $P = \{x \in \mathbb{R}^n : Ax \leq b\}$ is pointed if and only if
$$\{x \in \mathbb{R}^n : Ax = 0\} = \{0\}.$$

Proposition 2.1.4 *The core of a balanced game is a polytope.*

Proof. Let (N, v) be a profit game with $C(v) \neq \emptyset$. This polyhedron is a polytope if there exist vectors $l, u \in \mathbb{R}^n$ such that $l \leq x \leq u$ for all $x \in C(v)$. If $x \in C(v)$, then for each $i \in N$, we have $x_i \geq v(i)$ and also

$$x_i = x(N) - x(N \setminus i) = v(N) - x(N \setminus i) \leq v(N) - v(N \setminus i).$$

If (N, c) is a cost game and $x \in C(c)$, then $x_i \leq c(i)$ for all $i \in N$. Further,

$$x_i = c(N) - x(N \setminus i) \geq c(N) - c(N \setminus i)$$

for all $i \in N$. □

Remark 2.1.1 *The core of a balanced game is a bounded set of* $\mathbb{R}^n$ *which is also closed. Hence the core is a compact set of* $\mathbb{R}^n$.

Let $P = \{x : Ax \leq b\}$ be a nonempty polyhedron, where A has order $m \times n$. It can be shown that for any face F of P there exists a subsystem $A'x \leq b'$ of $Ax \leq b$ such that $F = \{x \in P : A'x = b'\}$. The faces of P are ordered by inclusion and the following theorem by Hoffman and Kruskal [127] characterizes the special structure of the minimal faces.

Theorem 2.1.5 *A nonempty set* F *is a minimal face of* $P = \{x : Ax \leq b\}$ *if and only if* $F = \{x : A'x = b'\}$ *for some subsystem* $A'x \leq b'$ *of* $Ax \leq b$. *Moreover,* $\text{rank}\,(A') = \text{rank}\,(A)$.

All minimal faces have the same dimension $n - \text{rank}(A)$. Only if $\text{rank}(A) = n$, does P have vertices which are its minimal faces. Then, $v \in P$ is a vertex of P if and only if $A'v = b'$ for some subsystem $A'x \leq b'$ of $Ax \leq b$, with $\text{rank}(A) = n$. This matrix A' is called a *basis* for the vertex v.

We can further show that two vertices v_1 and v_2 are adjacent in the graph $G(P)$ if and only if v_1 and v_2 have bases A^1 and A^2 such that $A^1x \leq b^1$ and $A^2x \leq b^2$, so that they have precisely $n-1$ inequalities in common.

An inequality $c^Tx \leq d$ induces a facet F of P if $F = \left\{x \in P : c^Tx = d\right\}$. A system of inequalities and equations is *minimal* (or irredundant) if no inequality can be made into an equation without reducing the solution set and no inequality or equation can be removed without enlarging the solution set.

If P is an n-polyhedron then P has a unique (up to positive scalar multiples) minimal defining system $P = \{x : Ax \leq b\}$ such that each of the inequalities $a_i^Tx \leq b_i$ defines a facet of P and such that each facet of P is defined by exactly one of these inequalities.

Moreover, if $P = \{x : Ax \leq b\}$ is an n-polyhedron, then $Ax \leq b$ is minimal if and only if for each pair $a_i^Tx \leq b_i$ and $a_j^Tx \leq b_j$ there is a vector x' satisfying $a_i^Tx' = b_i$ and $a_j^Tx' < b_j$.

Definition 2.1.2 *An* n*-polyhedron* P *is called simple if every vertex of* P *belongs to exactly* n *edges.*

Proposition 2.1.6 *Let* P *be a simple* n*-polytope and let* v *be a vertex of* P. *Every set of* k *edges adjacent to* v *determines a* k*-dimensional face of* P *which contains the vertex* v. *In particular, there are* $\binom{n}{k}$ *faces of dimension* k *containing* v.

Proof. Let H be a hyperplane which intersects the interior of P near a vertex v. Since P is simple, the intersection $P \cap H$ is an $(n-1)$-dimensional simplex S. The vertices of S are the intersections of edges of P which contain v with H and the $(k-1)$-dimensional faces of S are the intersections of k-faces of P with H. □

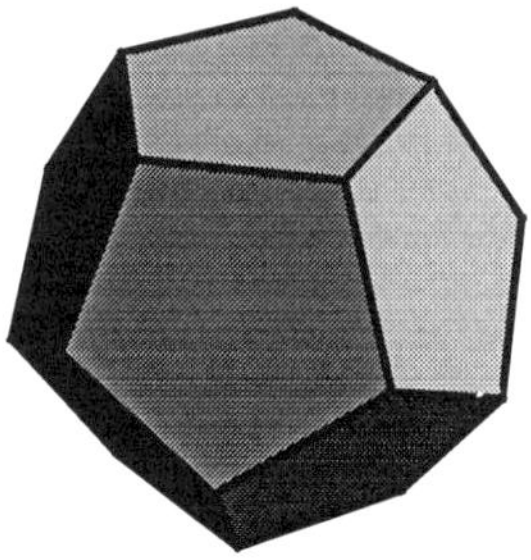

Figure 2.1 A simple 3-polytope

2.2 Linear programming

Linear programming concerns the problem of maximizing or minimizing a linear function over a polyhedron. Examples are

$$\max\left\{c^T x : Ax \leq b\right\}, \quad \min\left\{c^T x : Ax \geq b\right\}.$$

We first state the duality theorem of linear programming, proved by von Neumann in 1947, as a consequence of the *Farkas' lemma* for inequalities. Kuhn [147] showed that this variant of the Farkas' lemma can be derived from an elimination procedure for systems of linear inequalities known as Fourier-Motzkin elimination [177].

Let $Ax \leq b$ be a system of m linear inequalities and n variables, then we can use positive scalar multiples to obtain the equivalent system

$$\begin{aligned} x_1 + (a_i')^T x' &\leq b_i, \quad i = 1, \ldots, m_1 \\ -x_1 + (a_i')^T x' &\leq b_i, \quad i = m_1 + 1, \ldots, m_2 \\ (a_i')^T x' &\leq b_i, \quad i = m_2 + 1, \ldots, m \end{aligned}$$

where the vector $x' = (x_2, \ldots, x_n)^T$ and $(a_i')^T$ is the i-th row of A with the first entry a_{i1} removed. Since the first m_2 inequalities are equivalent to

$$\max_{m_1+1 \leq j \leq m_2} \left((a_j')^T x' - b_j \right) \leq x_1 \leq \min_{1 \leq i \leq m_1} \left(b_i - (a_i')^T x' \right),$$

the unknown x_1 can be eliminated. This gives that the original system is equivalent to the system

$$\begin{aligned} (a_j')^T x' - b_j &\leq b_i - (a_i')^T x', \quad i = 1, \ldots, m_1 \text{ and } j = m_1 + 1, \ldots, m_2 \\ (a_i')^T x' &\leq b_i, \quad i = m_2 + 1, \ldots, m. \end{aligned}$$

Writing this reduced system in the usual form we have that $Ax \leq b$ has a solution if and only if the system

$$\begin{aligned}\left(a_i' + a_j'\right)^T x' &\leq b_i + b_j, \;\; i = 1, \ldots, m_1 \text{ and } j = m_1 + 1, \ldots, m_2 \\ \left(a_i'\right)^T x' &\leq b_i, \;\; i = m_2 + 1, \ldots, m\end{aligned}$$

has a solution.

Hence we have eliminate variable x_1 with a system with many more inequalities. Every elimination step may transform m inequalities into up to $\lfloor m^2/4 \rfloor$ (see Ziegler [250, Chapter 1] for computational aspects).

Theorem 2.2.1 (Farkas' lemma 1) *Let A be an $m \times n$ matrix and let $b \in \mathbb{R}^m$. Then the system $Ax \leq b$ has a solution $x \in \mathbb{R}^n$ if and only if $y^T b \geq 0$ holds for each vector $y \geq 0$ with $y^T A = 0$.*

Proof. For the necessity of the condition, we suppose $Ax \leq b$ has a solution x. If there exists a vector $y \geq 0$ with $y^T A = 0$ and $y^T b < 0$, then

$$0 = \left(y^T A\right) x = y^T (Ax) \leq y^T b$$

which is a contradiction.

To prove sufficiency, first define $P = \{x \in \mathbb{R}^n : Ax \leq b\}$, and

$$Q = \left\{(x_0, x) \in \mathbb{R}^{n+1} : -bx_0 + Ax \leq 0\right\}.$$

Note that $P \neq \emptyset$ if and only if Q contains a point with $x_0 > 0$. We eliminate the variables $x_1, \ldots, x_n$ from Q by using the Fourier-Motzkin elimination. Then every inequality in the eliminated system is the sum of positive multiples of previous inequalities, and iterating the elimination we get the system

$$\left\{(x_0, x) \in \mathbb{R}^{n+1} : -Cbx_0 + CAx \leq 0\right\},$$

where C is a nonnegative matrix with $CA = \mathbf{0}$, since all the variables other than x_0 have been eliminated. Next suppose that $Ax \leq b$ has no solution, i.e., the set $P = \emptyset$, so that also $Q \subseteq \left\{(x_0, x) \in \mathbb{R}^{n+1} : x_0 \leq 0\right\}$. We apply the Fourier-Motzkin elimination to obtain

$$\left\{(x_0, x) \in \mathbb{R}^{n+1} : -Cbx_0 \leq 0\right\} \subseteq \left\{(x_0, x) \in \mathbb{R}^{n+1} : x_0 \leq 0\right\},$$

and hence the eliminated system contains an inequality $\alpha_{i0} x_0 \leq 0$ with $\alpha_{i0} > 0$. Let $y^T \geq 0$ be the row of C that yields this, then we have $y^T b = -\alpha_{i0} < 0$ and $y^T A = 0$. □

Corollary 2.2.2 (Farkas' lemma 2) *Let A be an $m \times n$ matrix and let $b \in \mathbb{R}^m$. Then the system $Ax = b$ has a solution $x \geq 0$ if and only if $y^T b \geq 0$ holds for each vector $y \in \mathbb{R}^m$ with $y^T A \geq 0$.*

Proof. The system $Ax = b$ has a solution $x \geq 0$ if and only if

$$Ax \leq b, \ \ (-A)\,x \leq -b, \ \ -x \leq 0,$$

has a solution $x \in \mathbb{R}^n$. The equivalent system is $A'x' \leq b'$, where

$$A' = \begin{pmatrix} A \\ -A \\ -I \end{pmatrix}, \quad b' = \begin{pmatrix} b \\ -b \\ 0 \end{pmatrix}$$

and I denotes the identity matrix. Application of Farkas' lemma 1 implies that $A'x' \leq b'$ has a solution x' if and only if $(y_1 - y_2)^T\, b \geq 0$ holds for all vectors $y_1 \geq 0$, $y_2 \geq 0$, $z \geq 0$ satisfying $(y_1 - y_2)^T\, A - z = 0$. If we define $y = y_1 - y_2 \in \mathbb{R}^m$ then we obtain the corollary. □

Corollary 2.2.3 *Suppose that $Ax \leq b$ has at least one solution. Then every solution x of $Ax \leq b$ satisfies $c^T x \leq \delta$, if and only if there exists $y \geq 0$ such that $y^T A = c^T$ and $y^T b \leq \delta$.*

Proof. Clearly, if x satisfies $Ax \leq b$ and there exists a vector $y \geq 0$ such that $y^T A = c^T$ and $y^T b \leq \delta$, then

$$c^T x = y^T Ax \leq y^T b \leq \delta.$$

Conversely, suppose such a vector y does not exist. Then the following system

$$\left(y^T, \alpha\right) \begin{pmatrix} A & b \\ 0 & 1 \end{pmatrix} = \left(c^T, \delta\right)$$

has no solution $\left(y^T, \alpha\right) \geq (0,0)$. By Farkas' lemma 2, this implies that there exists a vector $\begin{pmatrix} z \\ \gamma \end{pmatrix}$ satisfying

$$\begin{pmatrix} A & b \\ 0 & 1 \end{pmatrix} \begin{pmatrix} z \\ \gamma \end{pmatrix} \geq 0 \text{ and } \left(c^T, \delta\right) \begin{pmatrix} z \\ \gamma \end{pmatrix} < 0.$$

If $\gamma = 0$ then $Az \geq 0$ and $c^T z < 0$. By assumption, $Ax \leq b$ has at least one solution x_0 and we take $\alpha \in \mathbb{R}_+$ so that $A\,(x_0 - \alpha z) \leq b$ and $c^T\,(x_0 - \alpha z) > \delta$. This gives a contradiction, since the solution $x = x_0 - \alpha z$ satisfies $c^T x > \delta$.

If $\gamma > 0$ then we can take $x = -\gamma^{-1} z$ and again we obtain a contradiction, since $Ax \leq b$ and $c^T x > \delta$. □

Theorem 2.2.4 (Duality Theorem of LP) *Let A be an $m \times n$ matrix and let $b \in \mathbb{R}^m$ and $c \in \mathbb{R}^n$. Then*

$$\max\left\{c^T x : Ax \le b\right\} = \min\left\{y^T b : y \ge 0,\ y^T A = c^T\right\},$$

provided that both sets are nonempty.

Proof. By hypothesis, there are feasible solutions x and $y \ge 0$ such that $Ax \le b$ and $y^T A = c^T$. Then $c^T x = y^T Ax \le y^T b$, for all feasible x, y and hence

$$\delta = \sup\left\{c^T x : Ax \le b\right\} \le \inf\left\{y^T b : y \ge 0,\ y^T A = c^T\right\}.$$

Since every solution of $Ax \le b$ satisfies $c^T x \le \delta$, it follows from Corollary 2.2.3 that there exists $y \ge 0$ such that $y^T A = c^T$ and $y^T b \le \delta$. This implies

$$\inf\left\{y^T b : y \ge 0,\ y^T A = c^T\right\} = \min\left\{y^T b : y \ge 0,\ y^T A = c^T\right\} = \delta.$$

To complete the proof we have to show that $\max\left\{c^T x : Ax \le b\right\} = \delta$, i.e., that there exists x such that $Ax \le b$ and $c^T x \ge \delta$, which is equivalent to

$$\begin{pmatrix} A \\ -c^T \end{pmatrix} x \le \begin{pmatrix} b \\ -\delta \end{pmatrix}.$$

Suppose on the contrary that the above system has no solution. By Farkas' lemma 1, there exists a vector $z \ge 0$ and a number $\gamma \ge 0$ such that

$$(z^T, \gamma)\begin{pmatrix} A \\ -c^T \end{pmatrix} = 0, \quad (z^T, \gamma)\begin{pmatrix} b \\ -\delta \end{pmatrix} < 0.$$

Then $z^T A = \gamma c^T$ and $z^T b < \gamma\delta$. By assumption, there exists x such that $Ax \le b$. If $\gamma = 0$ then $0 = z^T Ax \le z^T b < 0$ yields a contradiction and hence $\gamma > 0$. Now define $y = \gamma^{-1} z$. Then $y \ge 0$, $y^T A = c^T$ and $y^T b < \delta$. But this is impossible because δ is the minimum of these vectors. □

Corollary 2.2.5 *Let A be an $m \times n$ matrix and let $b \in \mathbb{R}^m$ and $c \in \mathbb{R}^n$. Then*

1. $\max\left\{c^T x : x \ge 0,\ Ax \le b\right\} = \min\left\{y^T b : y \ge 0,\ y^T A \ge c^T\right\}$,
2. $\max\left\{c^T x : x \ge 0,\ Ax = b\right\} = \min\left\{y^T b : y^T A \ge c^T\right\}$,
3. $\min\left\{c^T x : x \ge 0,\ Ax \ge b\right\} = \max\left\{y^T b : y \ge 0,\ y^T A \le c^T\right\}$,
4. $\min\left\{c^T x : Ax \ge b\right\} = \max\left\{y^T b : y \ge 0,\ y^T A = c^T\right\}$,

provided that both sets are nonempty.

Proof. See Schrijver [210, page 91]. □

The following principle of *complementary slackness* is a useful optimality condition for linear programming problems.

Theorem 2.2.6 *Let A be an $m \times n$ matrix and let $b \in \mathbb{R}^m$ and $c \in \mathbb{R}^n$. Let x^* be a feasible solution to the primal problem*

$$\max\left\{c^T x : Ax \leq b\right\}$$

and let y^* be a feasible solution for the dual problem

$$\min\left\{y^T b : y \geq 0,\ y^T A = c^T\right\}.$$

Then x^* and y^* are optimal for the primal and dual problems, respectively, if and only if $y_i^* > 0$ implies $a_i^T x^* = b_i$ for each $i = 1, \ldots, m$ (equivalently $a_i^T x^* < b_i$ implies $y_i^* = 0$).

Proof. Suppose that x^* and y^* are optimal for the primal and dual problems. The Duality Theorem 2.2.4 implies that

$$\begin{aligned} 0 &= (y^*)^T b - c^T x^* \\ &= (y^*)^T b - (y^*)^T Ax^* \\ &= (y^*)^T (b - Ax^*) \\ &= \sum_{i=1}^m y_i^* \left(b_i - a_i^T x^*\right), \end{aligned}$$

where $y_i^* \geq 0$ and $b_i - a_i^T x^* \geq 0$ for each $i = 1, \ldots, m$. We deduce that $y_i^* \left(b_i - a_i^T x^*\right) = 0$ so that either $y_i^* = 0$ or $a_i^T x^* = b_i$ for each $i = 1, \ldots, m$.

Conversely, the complementary slackness conditions imply that

$$(y^*)^T b - c^T x^* = \sum_{i=1}^m y_i^* \left(b_i - a_i^T x^*\right) = 0.$$

Since x^* and y^* are feasible solutions, it follows that

$$\begin{aligned} c^T x^* &\leq \max\left\{c^T x : Ax \leq b\right\} \\ &= \min\left\{y^T b : y \geq 0,\ y^T A = c^T\right\} \\ &\leq (y^*)^T b, \end{aligned}$$

and therefore the equality $(y^*)^T b = c^T x^*$ implies that x^* and y^* are optimal for the primal and dual problems. □

The *simplex method*, designed by Dantzig in the late 1940s, is the method used most often for linear programming. We wish to solve $\max\left\{c^T x : Ax \leq b\right\}$, where the polyhedron $P = \{x : Ax \leq b\}$ is pointed, i.e., $\text{rank}(A) = n$.

In geometrical terms, the simplex method is to make a trip, going from one vertex x_k of P to an adjacent vertex x_{k+1} so that in each step the value of the objective function is increased. The specific way to choose x_{k+1} given x_k is called the *pivot rule.*

First suppose that we know a vertex x_0 of the feasible set P. Thus the process can be described by the subsystems

$$A_0 x \leq b_0,\ A_1 x \leq b_1,\ A_2 x \leq b_2, \ldots$$

where each $x_k = A_k^{-1} b_k$ is a vertex of P, where $A_k x \leq b_k$ differs in exactly one constraint from $A_{k+1} x \leq b_{k+1}$ (x_k and x_{k+1} are adjacent vertices) and where

$$c^T x_0 \leq c^T x_1 \leq c^T x_2 \leq \cdots$$

Suppose that we find A_k, x_k. If $c^T A_k^{-1} \geq 0$ then x_k is a optimal solution, since for each x such that $Ax \leq b$ we have $A_k x \leq b_k$ and hence

$$c^T x = \left(c^T A_k^{-1}\right) A_k x \leq \left(c^T A_k^{-1}\right) b_k = c^T x_k.$$

If $c^T A_k^{-1} \not\geq 0$ then choose the smallest index i for which the component $\left(c^T A_k^{-1}\right)_i < 0$, and let $z = -A_k^{-1} e_i$ (the i-th column of $-A_k^{-1}$). Note that for $\lambda \geq 0$, $x_k + \lambda z$ traverses a face of dimension 1 of P, or it is outside of P for all $\lambda > 0$. Furthermore, $c^T z = -c^T A_k^{-1} e_i = -\left(c^T A_k^{-1}\right)_i > 0$.

We now consider two cases:

If $Az \leq 0$ then $A\left(x_k + \lambda z\right) = b_k + \lambda Az \leq b_k$ for all $\lambda \geq 0$, so that $x_k + \lambda z \in P$ for all $\lambda \geq 0$. Hence, we obtain that $\max\left\{c^T x : Ax \leq b\right\} = \infty$.

If $Az \not\leq 0$, let λ_0 the largest λ such that $x_k + \lambda z \in P$, that is

$$\lambda_0 = \min\left\{\frac{b_j - a_j^T x_k}{a_j^T z} : j = 1, \ldots, m,\ \ a_j^T z > 0\right\}.$$

Let j be the smallest index attaining this minimum. Replacing the i-th inequality in the subsystem $A_k x \leq b_k$ by inequality $a_j^T x \leq b_j$, we obtain the required subsystem $A_{k+1} x \leq b_{k+1}$, where $x_{k+1} = x_k + \lambda_0 z$.

If P is a simple n-polyhedron then each vertex has exactly one basis (non-degenerate case) and $x_{k+1} \neq x_k$ for all $k \geq 0$. This implies that $c^T x_{k+1} > c^T x_k$ and the process stops in a vertex for which the maximum of $c^T x$ is attained.

However, it can happen that $x_{k+1} = x_k$ for some k. Note that cycling can appear only if the problem is degenerate.

2.3 Integer and combinatorial optimization

A vector or matrix is called rational (integral) if its entries are all rationals (integers). A system of linear equations $Ax = b$ or linear inequalities $Ax \leq b$ is rational (integral) if A and b are rational (integral). A *rational polyhedron* is a polyhedron that can be defined by rational linear inequalities. The integer hull P_I of a polyhedron P is defined by $P_I = \text{conv}\{x \in P : x \text{ integral}\}$. Meyer [176] has shown that if P is rational then P_I is a polyhedron again. Note that this is obvious if P is a polytope.

Definition 2.3.1 *A rational polyhedron P is called an integral polyhedron if $P = P_I$.*

For a rational polyhedron P, the following statements are equivalent:

1. P is an integral polyhedron.
2. Every nonempty face of P contains an integral vector.
3. Every minimal face of P contains an integral vector.
4. The optimal value of $\max\left\{c^T x : x \in P\right\}$ is attained by an integral vector, for each c, for which the maximum is finite.
5. The optimal value of $\max\left\{c^T x : x \in P\right\}$ is an integer, for each integral vector c, for which the maximum is finite.

The last characterization was shown by Edmonds and Giles [73]. For this proof see Schrijver [210, Section 22.1].

Integer linear programming studies linear programming problems in which the variables are restricted to integers. The general *ILP* problem is: Given a rational matrix A, and rational vectors b and c, determine

$$\max\left\{c^T x : Ax \leq b,\ x \text{ integral}\right\}.$$

The *incidence vector* or *characteristic vector* of a set $S \subseteq N$ is the $\{0,1\}$-vector in $\mathbb{R}^n$, denoted by $\mathbf{1}_S$, satisfying

$$\mathbf{1}_S(i) = \begin{cases} 1, & \text{if } i \in S \\ 0, & \text{if } i \in N \setminus S. \end{cases}$$

Most combinatorial optimization problems are given by a finite ground set N, a collection $\mathcal{F}$ of subsets in N and a linear objective function $c : N \to \mathbb{R}$. The task is to solve the problem

$$\max\left\{\sum_{i \in S} c(i) : S \in \mathcal{F}\right\} = \max\left\{c^T \mathbf{1}_S : S \in \mathcal{F}\right\}.$$

Proposition 2.1.3 implies that this combinatorial problem is equivalent to $\max\left\{c^T x : x \in P\right\}$, where $P = \operatorname{conv}\{\mathbf{1}_S : S \in \mathcal{F}\}$ is an integral polytope. We know that $P = \{x : Ax \leq b\}$, but we must determinate them in order to apply LP algorithms like the simplex method.

Given an integer linear programming $\max\left\{c^T x : Ax \leq b,\ x \text{ integral}\right\}$, the linear program $\max\left\{c^T x : Ax \leq b\right\}$ is called its *LP-relaxation*. We may also consider the dual of this LP-relaxation and the associated linear program. In general, no satisfactory analogues of Farkas' lemma has been found for ILP problems. If the optima involved exist, then we have the following inequalities:

$$\begin{array}{llllll} \max c^T x & \leq & \max c^T x & = & \min y^T b & \leq \ \min y^T b \\ Ax \leq b & & Ax \leq b & & y^T A = c^T & \quad\ y^T A = c^T \\ x \text{ integral} & & & & y \geq 0 & \quad\ y \geq 0 \\ & & & & & \quad\ y \text{ integral} \end{array}$$

In general we will have strict inequalities in the above duality relations.

Definition 2.3.2 *The rational system $Ax \leq b$ is totally primal integral (TPI) if the first inequality holds with equality for each vector c, for which the maximum is finite.*

Note that $Ax \leq b$ is a TPI-system if and only if $P = \{x : Ax \leq b\}$ is an integral polyhedron.

Definition 2.3.3 *The rational system $Ax \leq b$ is totally dual integral (TDI) if the second inequality holds with equality for each integral vector c, for which the minimum is finite.*

Edmonds and Giles [73] showed that the total dual integrality of $Ax \leq b$ implies total primal integrality, if b is integral.

Proposition 2.3.1 *If the vector b is integral and the system $Ax \leq b$ is totally dual integral, then it is also totally primal integral.*

Proof. If b is integral and $Ax \leq b$ is a TDI-system, then the duality equation

$$\max\left\{c^T x : Ax \leq b\right\} = \min\left\{y^T b : y \geq 0,\ y^T A = c^T\right\}$$

implies that the maximum is an integer for all integral vectors c. Therefore, the polyhedron $P = \{x : Ax \leq b\}$ is integral. □

Definition 2.3.4 *A matrix A is called totally unimodular if every square submatrix of it has determinant $0, 1$, or -1.*

In particular, each entry in a totally unimodular matrix is $0, 1$, or -1. A connection between total unimodularity and integer linear programming is given by the following results.

Theorem 2.3.2 *Let A be a totally unimodular matrix and let b be an integral vector. Then the polyhedron $P = \{x : Ax \leq b\}$ is integral.*

Proof. Let F be a minimal face of P. By Theorem 2.1.5, $F = \{x : A'x = b'\}$ for some subsystem $A'x \leq b'$ of $Ax \leq b$, with A' having full row rank. Then $A' = (A_1, A_2)$, with A_1 nonsingular (possibly after permuting columns). The total unimodularity of A implies that A_1 is integral and $\det(A_1) = \pm 1$. By Cramer's rule we obtain that A_1^{-1} is integral, which implies that the vector

$$x = \begin{pmatrix} A_1^{-1}b' \\ 0 \end{pmatrix}$$

is an integral vector in F. □

Corollary 2.3.3 *Let A be a totally unimodular matrix, and let b and c be integral vectors. Then both problems in the duality equation*

$$\max\left\{c^T x : Ax \leq b\right\} = \min\left\{y^T b : y \geq 0,\ y^T A = c^T\right\}$$

have integral optimal solutions.

Proof. We deduce this directly from the previous theorem, since also the matrix

$$\begin{pmatrix} -I \\ A^T \\ -A^T \end{pmatrix}$$

is also totally unimodular. □

Hoffman and Kruskal [127] proved the following characterization of totally unimodular matrices.

Theorem 2.3.4 *An integral matrix A is totally unimodular if and only if for each integral vector b the polyhedron $\{x : x \geq 0,\ Ax \leq b\}$ is integral.*

Proof. See Schrijver [210, Section 19.1]. □

Corollary 2.3.5 *An integral matrix A is totally unimodular if and only if for all integral vectors b and c, both problems in the duality equation*

$$\max\left\{c^T x : x \geq 0,\ Ax \leq b\right\} = \min\left\{y^T b : y \geq 0,\ y^T A \geq c^T\right\}$$

have integral optimal solutions (provided the optimum is finite).

Corollary 2.3.6 *If A is totally unimodular, then $Ax \leq b$ is a TDI-system for each rational vector b.*

Given a $\{0,1\}$-matrix A we consider the polytope associated with the LP-relaxation of the *set packing problem*: $P = \{x \in \mathbb{R}^n : x \geq 0,\ Ax \leq \mathbf{1}_N\}$.

Definition 2.3.5 *The $\{0,1\}$-matrix A is called perfect if the packing polytope*

$$P = \{x \in \mathbb{R}^n : x \geq 0,\ Ax \leq \mathbf{1}_N\}$$

is integral, i.e., $P = P_I$. The $\{0,1\}$-matrix A is balanced if, for every submatrix A' obtained from A by deleting a set of its rows, the associated packing polytope is integral.

Padberg [197] has given a characterization of perfect matrices in terms of forbidden submatrices. The above definition implies that every balanced matrix is perfect. Berge [9] characterized balanced matrices as matrices not containing a square submatrix of odd order with in each row and in each column exactly two ones. That is, if it has no submatrix of form (possibly after permutations of rows or columns)

$$\begin{pmatrix} 1 & 0 & \cdots & 0 & 1 \\ 1 & 1 & \ddots & 0 & 0 \\ \vdots & \ddots & \ddots & \ddots & \vdots \\ 0 & 0 & \ddots & 1 & 0 \\ 0 & 0 & \cdots & 1 & 1 \end{pmatrix}$$

of odd order.

Let C_k be the collection of $\{0,1\}$-square submatrices with order $k \geq 3$, with in each row and in each column exactly two ones, that do no contain the submatrix

$$\begin{pmatrix} 1 & 1 \\ 1 & 1 \end{pmatrix}.$$

Definition 2.3.6 *A $\{0,1\}$-matrix A is totally balanced if A does not contain a submatrix in C_k for any $k \geq 3$.*

Since a $\{0,1\}$-matrix is balanced if it does not contain a submatrix in C_k for any odd $k \geq 3$, we obtain that every totally balanced matrix is balanced. Some simple properties of these matrices are listed in the next proposition.

Proposition 2.3.7 *Let A be a $\{0,1\}$-matrix which is (totally) balanced. Then the following matrices are (totally) balanced:*

1. *The transpose of A.*
2. *Any submatrix of A.*
3. *Any matrix obtained from A by permuting rows or columns.*

Proposition 2.3.8 *Every $\{0,1\}$-totally unimodular matrix is balanced.*

Proof. We suppose that A is not balanced. Then A contains a square submatrix of odd order with in each row and in each column exactly two ones, and hence it has determinant 2 or -2. Since every square submatrix of a totally unimodular matrix has determinant $0, 1$, or -1, we obtain a contradiction. □

The synoptic diagram in Figure 2.2 shows the relationship among these classes of $\{0,1\}$-matrices.

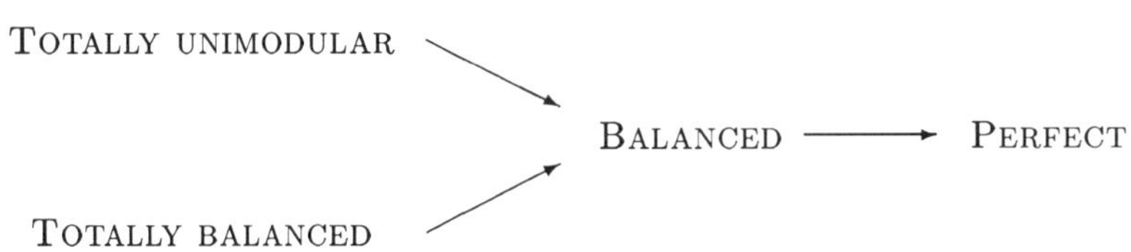

Figure 2.2

Fulkerson, Hoffman and Oppenheim [98] characterized balanced matrices by using LP-duality and proved the following result.

Proposition 2.3.9 *Let A be a balanced matrix. Then the polyhedron*

$$P = \{x \in \mathbb{R}^n : x \geq 0,\ Ax = \mathbf{1}_N\}$$

is integral.

Proof. See Schrijver [210, Theorem 21.7]. □

It follows from Carathéodory's theorem 2.1.1 that if the optima in the LP-duality equation

$$\max\left\{c^T x : Ax \leq b\right\} = \min\left\{y^T b : y \geq 0,\ y^T A = c^T\right\}$$

are finite, there is an optimal dual solution y with at most $r = \text{rank}\,(A)$ positive components.

Definition 2.3.7 *A Hilbert basis is a set of vectors* $\{a_1, \ldots, a_t\} \subseteq \mathbb{R}^n$, *with the property that if an integral vector* x *is a nonnegative linear combination of the vectors* $a_1, \ldots, a_t$ *then it is an integer nonnegative linear combination of them.*

Cook *et al.* [44] showed that if $\{a_1, \ldots, a_t\}$ is a Hilbert basis of integral vectors in $\mathbb{R}^n$, then any integral vector x that is a nonnegative linear combination of the vectors $a_1, \ldots, a_t$ is an integer nonnegative linear combination of at most $2n-1$ of these vectors.

Schrijver [211] derived from this result that if $Ax \leq b$ is a TDI-system in n variables and A is integral, then for any integral vector c the dual problem

$$\min\left\{y^T b : y \geq 0,\ y^T A = c^T\right\}$$

has an *integral optimal* solution y with at most $2n-1$ nonzero components, for each integral vector c for which the minimum is finite.

For a graph $G = (V, E)$, the $E \times V$ incidence matrix $M(G)$ has $|E|$ rows and $|V|$ columns, and an entry $M(G)_{e,v}$ is equal to 1 if v is an end vertex of edge e, and 0 otherwise, i.e.,

$$M(G)_{e,v} = \begin{cases} 1, & \text{if } e \in \delta(v) \\ 0, & \text{otherwise.} \end{cases}$$

Note that $M(G)$ is the incidence matrix of the collection

$$\{\delta(v) : v \in V\} \subseteq \{0,1\}^E.$$

That is, their columns are the incidence vectors $\mathbf{1}_{\delta(v)}$, for all $v \in V$.

If G contains an odd cycle, then the incidence matrix of the cycle is a square submatrix of $M(G)$ with determinant 2 or -2. Further, it is possible to obtain the following characterization by induction:

$M(G)$ *is totally unimodular* $\Longleftrightarrow$ G *contains no odd cycle* $\Longleftrightarrow$ G *is bipartite.*

The total unimodularity of the incidence matrix of a bipartite graph has the following consequences.

Example: The (perfect) matching polytope of a graph (Edmonds [68]).

Let $G = (V, E)$ be a graph, and let $c : E \to \mathbb{R}_+$ be a weight function on the edges of G. Recall that a *matching* in a graph G is a set $M \subseteq E$ of pairwise disjoint edges. A matching M is called *perfect* if every vertex is an end vertex of an edge in M.

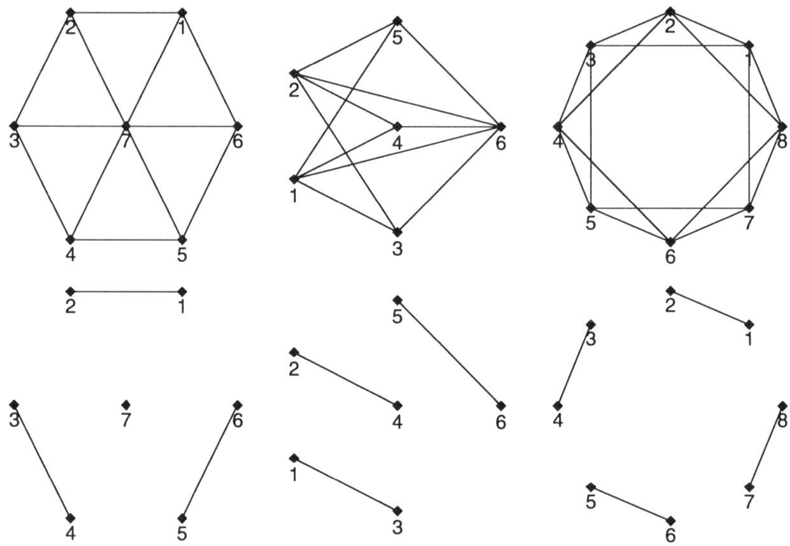

Figure 2.3 Maximal matching and perfect matchings

For a graph $G = (V, E)$ and $W \subseteq V$, we denote

$$\begin{aligned}\delta(W) &= \{e \in E : e \text{ has one end vertex in } W \text{ and the other end in } V \setminus W\}\\ \gamma(W) &= \{e \in E : e \text{ has both end vertices in } W\}.\end{aligned}$$

Suppose we want to solve the problem

$$\begin{aligned}\max\left\{\sum_{e\in M} c(e) : M \text{ matching in } G\right\} &= \max\left\{c^T \mathbf{1}_M : M \text{ matching in } G\right\}\\ &= \max\left\{c^T x : x \in P\right\},\end{aligned}$$

where the set $P = \operatorname{conv}\{\mathbf{1}_M : M \text{ matching in } G\}$ is an integral polytope in $\mathbb{R}^E_+$, called the *matching polytope* of G.

The matching polytope is contained in the set of all vectors $x \in \mathbb{R}^E$ satisfying the following linear inequalities:

$$\begin{aligned} x(e) &\geq 0, \quad e \in E,\\ x(\delta(v)) &\leq 1, \quad v \in V. \end{aligned}$$

If the graph G is bipartite, then the incidence matrix of G is totally unimodular and the converse inclusion follows from Theorem 2.3.2 (the above system is integral).

For non-bipartite graphs, the matching polytope is not completely determined by the system of linear inequalities. Let C be an odd cycle in G. Then the vector $x \in \mathbb{R}^E$ defined by $x(e) = 1/2$ if $e \in C$ and $x(e) = 0$ if $e \notin C$, satisfies the inequalities but x is not in the matching polytope.

Edmonds [68] showed that the matching polytope of a graph $G = (V, E)$ is equal to the set of vectors $x \in \mathbb{R}^E$ such that

$$\begin{aligned} x(e) &\geq 0, \quad e \in E, \\ x(\delta(v)) &\leq 1, \quad v \in V, \\ x(\gamma(S)) &\leq \frac{|S| - 1}{2}, \quad S \subseteq V, \ |S| \geq 3 \text{ and } |S| \text{ odd.} \end{aligned}$$

The polytope conv $\{\mathbf{1}_M : M \text{ perfect matching in } G\}$ is the *perfect matching polytope* of $G = (V, E)$. It is a face of the matching polytope of G. If G is a bipartite graph, then the perfect matching polytope is determined by

$$\begin{aligned} x(e) &\geq 0, \quad e \in E, \\ x(\delta(v)) &= 1, \quad v \in V. \end{aligned}$$

This property is equivalent to the following theorem of Birkhoff [19]: *Each doubly stochastic matrix is a convex combination of permutation matrices.*

Theorem 2.3.10 *Let $G = (V, E)$ be a graph and let*

$$P = \left\{x \in \mathbb{R}_+^E : x(\delta(v)) = 1 \text{ for all } v \in V\right\}.$$

Then x is a vertex of P if and only if $x(e) \subset \left\{0, \frac{1}{2}, 1\right\}$ for all $e \in E$ and the set $\left\{e \in E : x(e) = \frac{1}{2}\right\}$ is a union of disjoint odd cycles.

Proof. See Cook *et al.* [45, Theorem 6.13]. □

The perfect matching polytope of $G = (V, E)$ is equal to the set of vectors $x \in \mathbb{R}^E$ such that

$$\begin{aligned} x(e) &\geq 0, \quad e \in E, \\ x(\delta(v)) &= 1, \quad v \in V, \\ x(\delta(S)) &\geq 1, \quad S \subseteq V, \ |S| \geq 3 \text{ and } |S| \text{ odd.} \end{aligned}$$

2.4 Balanced games

In this section we show some properties of the balanced games. Bondareva [24] and Shapley [215] proved a characterization of *balanced games* by using the Duality Theorem 2.2.4. Faigle [77] obtained a generalization of the theorem of Bondareva and Shapley for games $v : \mathcal{F} \to \mathbb{R}$, where $\mathcal{F} \subseteq 2^N$. We collect the mentioned characterizations in the following theorem.

Theorem 2.4.1 *Let (N, c) be a cost game with dual profit game c^*. Then the following statements are equivalent:*

1. *The core $C(c)$ is nonempty.*
2. *The distance between the polyhedra $P(c)$ and $P(c^*)$ is 0.*
3. *The preimputation set $I^*(c) = \{x \in \mathbb{R}^n : \mathbf{1}_N^T x = c(N)\}$ is a supporting hyperplane of $P(c)$.*
4. *For all $S \in 2^N \setminus \{\emptyset\}$ and $y_S \geq 0$,*
$$\sum_{S \in 2^N \setminus \{\emptyset\}} y_S \mathbf{1}_S = \mathbf{1}_N \text{ implies } c(N) \leq \sum_{S \in 2^N \setminus \{\emptyset\}} y_S \, c(S).$$
5. *For all $S_1, \ldots, S_p \in 2^N \setminus \{\emptyset\}$ and $m \in \mathbb{N}$,*
$$\frac{1}{m} \sum_{i=1}^{p} \mathbf{1}_{S_i} = \mathbf{1}_N \text{ implies } c(N) \leq \frac{1}{m} \sum_{i=1}^{p} c(S_i).$$

Proof. To prove that $1 \Rightarrow 2$, we observe that Theorem 1.1.1 implies that $P(c) \cap P(c^*) = C(c)$ and hence

$$\operatorname{dist}(P(c), P(c^*)) := \min \{\|x - y\| : x \in P(c),\ y \in P(c^*)\} = 0.$$

The inequality $\mathbf{1}_N^T x \leq c(N)$ is valid for $P(c)$. Suppose that 2 holds, and let $x \in P(c) \cap P(c^*)$. Then $x \in P(c) \cap I^*(c)$, and we obtain 3.

Next, $3 \Rightarrow 4$. To prove this, note that 3 implies

$$\max \{\mathbf{1}_N^T x : x \in P(c)\} = c(N),$$

and hence to prove 4 we have only to apply the Duality Theorem 2.2.4:

$$c(N) = \min \left\{ \sum_{S \in 2^N \setminus \{\emptyset\}} y_S \, c(S) : y \geq 0, \sum_{S \in 2^N \setminus \{\emptyset\}} y_S \mathbf{1}_S = \mathbf{1}_N \right\}.$$

Clearly, 4 implies 5. Finally, Faigle [77] obtained the equivalence $5 \Leftrightarrow 1$ from a general sandwich theorem for set functions. □

Remark 2.4.1 *If (N, v) is a profit game then we obtain an equivalent formulation by obvious modifications of the statements.*

Definition 2.4.1 *A game (N, c) is called subadditive if*

$$c(S \cup T) \leq c(S) + c(T)$$

for all S, T in 2^N with $S \cap T = \emptyset$. A game (N, v) is called superadditive if the reverse inequality holds.

Le Breton, Owen and Weber [155] and Kuipers [149] investigated necessary and sufficient conditions for a game to have a nonempty core by using the following concept.

Definition 2.4.2 *In a cost game (N, c) a coalition S is said to be essential (or effective) if $c(S) < \sum_{T \in \mathcal{P}(S)} c(T)$ for each partition $\mathcal{P}(S)$ of S such that it contains at least two nonempty sets.*

For a cost game (N, c) we denote by $\mathcal{E}$ the collection of its nonempty essential coalitions.

Remark 2.4.2 *Note that the collection of essential coalitions contains the empty set $\emptyset$ and coalitions $S = \{i\}$ for each $i \in N$.*

If R is a coalition which is not essential, then there exists a partition of R such that $c(R) \geq \sum_{T \in \mathcal{P}(R)} c(T)$. Let $x \in \mathbb{R}^n$ be a vector which satisfies the inequalities $x(T) \leq c(T)$ for every $T \in \mathcal{P}(R)$. Then we obtain

$$x(R) = \sum_{T \in \mathcal{P}(R)} x(T) \leq \sum_{T \in \mathcal{P}(R)} c(T) \leq c(R).$$

Therefore, the collection $\mathcal{E}$ of the nonempty essential coalitions gives a complete description of the core, i.e.,

$$C(c) = \{x \in \mathbb{R}^n : x(N) = c(N),\ x(S) \leq c(S) \ \text{ for all } \ S \in \mathcal{E}\}.$$

Let (N, c) be a subadditive cost game. A coalition R is called *c-separable* if $c(R) = c(R_1) + c(R_2)$ for some partition in two nonempty sets. Then a coalition R is not essential if and only if R is c-separable. We note that the grand coalition N is not essential if and only if $c(N) = \sum_{i \in I} c(S_i)$ for some $\mathcal{E}$-partition of N, that is, for some partition of N such that $S_i \in \mathcal{E}$ for all $i \in I$.

Let $\mathcal{F}$ be a collection of subsets of a finite set N. The $N \times \mathcal{F}$ *incidence matrix* $M(\mathcal{F})$ is the $\{0,1\}$-matrix whose columns are the incidence vectors $\mathbf{1}_S$ of the sets S in $\mathcal{F}$. Then $M(\mathcal{F})_{i,S} = 1$ if $i \in S$, and 0 otherwise. Faigle and Kern [82] obtained the following characterization of subadditive cost games with nonempty core by using integer linear programming.

Theorem 2.4.2 *A subadditive cost game (N, c) is balanced if and only if the following linear program has an integral optimal solution*

$$\min\left\{\sum_{S\in\mathcal{E}} y_S\, c(S) : y \geq 0,\ M(\mathcal{E})\, y = \mathbf{1}_N\right\},$$

where $M(\mathcal{E})$ is the incidence matrix of the collection of its nonempty essential coalitions.

Proof. The Duality Theorem 2.2.4 implies

$$\begin{aligned}\max\left\{\mathbf{1}_N^T x : M(\mathcal{E})^T x \leq c\right\} &= \max\left\{x(N) : x(S) \leq c(S) \text{ for all } S \in \mathcal{E}\right\}\\ &= \min\left\{\sum_{S\in\mathcal{E}} y_S\, c(S) : y \geq 0,\ M(\mathcal{E})\, y = \mathbf{1}_N\right\}\\ &\leq \min_{y \text{ integral}}\left\{\sum_{S\in\mathcal{E}} y_S\, c(S) : y \geq 0,\ M(\mathcal{E})\, y = \mathbf{1}_N\right\}\\ &= \min\left\{\sum_{i\in I} c(S_i) : \{S_i\} \text{ is an } \mathcal{E}\text{-partition of } N\right\}\\ &= c(N).\end{aligned}$$

Since $N \in \mathcal{E}$ or $c(N) = \sum_{j\in J} c(T_j)$ for some $\mathcal{E}$-partition of N, it follows from the subadditivity of c that the last equation holds. Suppose first that there exists $x \in C(c)$. Then

$$c(N) \leq x(N) = \max\left\{\mathbf{1}_N^T x : M(\mathcal{E})^T x \leq c\right\},$$

so that the inequality holds with equality and therefore the linear program in the statement of the theorem has an integral solution.

Conversely, if the linear program has an optimal integer solution then the above equations imply that

$$\max\left\{\mathbf{1}_N^T x : M(\mathcal{E})^T x \leq c\right\} = c(N),$$

and hence every optimal solution belongs to $C(c)$. □

Corollary 2.4.3 *Let $c : 2^N \to \mathbb{Q}$ be a subadditive cost game (with rational utility) such that the system $x(S) \leq c(S)$, for all $S \in \mathcal{E}$ is totally dual integral. Then $C(c) \neq \emptyset$.*

Corollary 2.4.4 (Kuipers) *Let (N, c) be a cost game such that its incidence matrix $M(\mathcal{E})$ is perfect. Then (N, c) is balanced if and only if for each $\mathcal{E}$-partition of N, we have $c(N) \leq \sum_{j \in J} c(T_j)$.*

Proof. We observe that if the matrix $M(\mathcal{E})$ is perfect, then

$$\max\left\{\mathbf{1}_N^T x : M(\mathcal{E})^T x \leq c\right\} = \min\left\{\sum_{i \in I} c(S_i) : \{S_i\} \text{ is an } \mathcal{E}\text{-partition of } N\right\}.$$

□

Corollary 2.4.5 (Le Breton, Owen and Weber) *A subadditive cost game (N, c) is balanced if and only if the collection $\mathcal{E}$ of its nonempty essential coalitions is normal.*

Proof. Let $M(\mathcal{E})$ be the incidence matrix of the collection $\mathcal{E}$. Then $\mathcal{E}$ is normal if and only if the polyhedron $P = \{y : y \geq 0,\ M(\mathcal{E})\, y \leq \mathbf{1}_N\}$ is integral (see Duchet [60]). Le Breton, Owen and Weber [155, Lemma 6.4] proved that the above polyhedron P is integral if and only if the partition polyhedron $\overline{P} = \{y : y \geq 0,\ M(\mathcal{E})\, y = \mathbf{1}_N\}$ is integral. □

Proposition 2.4.6 *Let (N, c) be a subadditive cost game such that the incidence matrix $M(\mathcal{E})$ is balanced. Then (N, c) is totally balanced.*

Proof. Let $S \subseteq N$ be a nonempty coalition. To prove that the core of the subgame $c_S : 2^S \to \mathbb{R}$ is nonempty, we observe that if $S \in \mathcal{E}$ then the incidence matrix of $\mathcal{E}_S = \{T \subseteq S : T \in \mathcal{E}\}$ is balanced. Also, if $S \notin \mathcal{E}$ then there exists an $\mathcal{E}$-partition of S such that $c(S) = \sum_{j \in J} c(T_j)$ and the subadditivity of c implies that

$$\min\left\{\sum_{k \in K} c(T_k) : \{T_k\}_{k \in K} \text{ is an } \mathcal{E}_S\text{-partition of } S\right\} = c(S).$$

□

Remark 2.4.3 *Similar results holds for superadditive profit games (N, v) and the linear program*

$$\max\left\{\sum_{S \in \mathcal{E}'} y_S\, v(S) : y \geq 0,\ M(\mathcal{E}')\, y = \mathbf{1}_N\right\}.$$

Example: Component additive games (Potters and Reijnierse [201]).

In a communication structure given by the graph $G = (N, E)$, we consider the collection $\mathcal{F}_G$ of coalitions $S \subseteq N$ such that the subgraph induced by S is connected. For any $S \subseteq N$, the maximal feasible subsets of S are called components of S in G. The components of any S, denoted by $\Pi_S = \{T_1, \ldots, T_p\}$, form a partition of S.

Definition 2.4.3 *A cost game (N, c) is a G-component additive game if it is subadditive and*

$$c(S) = \sum \{c(T) : T \in \Pi_S\} \quad \textit{for all } S \subseteq N.$$

The collection of its nonempty essential coalitions is $\mathcal{F}_G \setminus \{\emptyset\}$. Le Breton, Owen and Weber [155, Proposition 3.1] proved the following characterization:

The collection $\mathcal{F}_G \setminus \{\emptyset\}$ is normal if and only if G is a forest.

Therefore, Corollary 2.4.5 implies that component additive forest games are balanced. We shall prove that these games are totally balanced as a consequence of the following proposition.

Proposition 2.4.7 *The incidence matrix $M(\mathcal{F}_G \setminus \{\emptyset\})$ is totally balanced if and only if G is a forest (acyclic graph).*

Proof. ($\Rightarrow$) Suppose that graph G contains a cycle $v_0, v_1, \ldots, v_k, v_0$. Hence the incidence matrix $M(\mathcal{F}_G \setminus \{\emptyset\})$ contains a square submatrix such that its columns are the incidence vectors of the edges $\{v_0, v_1\}, \{v_1, v_2\}, \ldots, \{v_k, v_0\}$. Then by permuting rows and columns we obtain a forbidden submatrix which contradicts the balancedness of $M(\mathcal{F}_G \setminus \{\emptyset\})$.

($\Leftarrow$) If the matrix $M(\mathcal{F}_G \setminus \{\emptyset\})$ is not balanced, then it contains a square submatrix such that its columns are the incidence vectors of a cycle in G. □

Proposition 2.4.8 *Let $G = (N, E)$ be a graph and let (N, c) be a G-component additive game. If G is a forest then the game (N, c) is totally balanced.*

Proof. It follows from Proposition 2.4.6. □

Example: Matching games (Shapley and Shubik [218]).

If $G = (N, E)$ is a graph and $w : E \to \mathbb{R}_+$ is a weight function, then the *w-matching game* $v : 2^N \to \mathbb{R}$ is defined by the optimal value of the following combinatorial optimization problem:

$$v(S) = \max \left\{ \sum_{e \in M} w(e) : M \text{ is a matching in the subgraph } G[S] \right\}.$$

Shapley and Shubik showed the following property of these games.

Proposition 2.4.9 *Let $G = (N, E)$ be a bipartite graph and let $w : E \to \mathbb{R}_+$ be a weight function. If (N, v) is a w-matching game, then it is totally balanced.*

Proof. Since any subgame of v is also a w-matching game, it is enough to prove that v has a nonempty core.

First, the definition implies that for all $S \subseteq N$,

$$v(S) = \max\left\{w^T x : x \in \{0,1\}^E,\ x(\delta(i)) \leq \mathbf{1}_S(i) \text{ for all } i \in N\right\}.$$

Next, since graph G is bipartite we obtain that Theorem 2.3.2 and the Duality Theorem 2.2.4 imply

$$\begin{aligned} v(S) &= \max\left\{w^T x : x \in \mathbb{R}_+^E,\ x(\delta(i)) \leq \mathbf{1}_S(i) \ \text{ for all } i \in N\right\} \\ &= \min\left\{y^T \mathbf{1}_S : y \in \mathbb{R}_+^n,\ y_i + y_j \geq w_{ij} \ \text{ for all } ij \in E\right\} \\ &= \min\left\{y(S) : y \in Q(w)\right\}, \end{aligned}$$

where

$$Q(w) = \left\{y \in \mathbb{R}_+^n : y_i + y_j \geq w_{ij} \ \text{ for all } ij \in E\right\}.$$

If y^* is an optimal solution for the above linear program taking $S = N$, then $y^*(N) = v(N)$ and y^* is a feasible vector for all $S \subseteq N$, i.e., $y^* \in Q(w)$. This implies that

$$y^*(S) \geq \min\{y(S) : y \in Q(w)\} = v(S)$$

for all $S \subseteq N$, and hence $y^* \in C(v)$. □

Remark 2.4.4 *Since the family of nonempty essential coalitions of a matching game (N, v) is*

$$\mathcal{E} = \{\{i\} : i \in N\} \cup \{\{i,j\} : ij \in E\},$$

we have

$$C(v) = \{y \in Q(w) : y(N) = v(N)\}.$$

Moreover, if the graph is bipartite, then the core of a w-matching game (N, v) is the set of all optimal solutions for the LP problem

$$\min\{y(N) : y \in Q(w)\}.$$

2.5 The greedy algorithm

Matroids are important in combinatorial optimization because one can find an independent set of maximum weight by using the *greedy algorithm.* This algorithm is an abstraction of Kruskal's algorithm [146] for finding a maximum (minimum) weighted spanning tree in a graph. We use the terminology introduced by Bixby and Cunningham [22].

Let $(N,\mathcal{I})$ be an independence system and let $c : N \to \mathbb{R}$ be a weight function. The problem is to find a feasible set of maximum weight, i.e.,

$$\max\left\{\sum_{i\in S} c(i) : S \in \mathcal{I}\right\}.$$

Greedy algorithm:

Order $N = \{e_1, \ldots, e_n\}$ so that $c(e_1) \geq \cdots \geq c(e_m) > 0 \geq \cdots \geq c(e_n)$;
$F \leftarrow \emptyset$;
For $i = 1$ to m do
If $F \cup \{e_i\} \in \mathcal{I}$ then $F \leftarrow F \cup \{e_i\}$.

This algorithm generates a feasible set for any independence system. For some families $\mathcal{I}$ the greedy algorithm works (produces an optimal solution) and for some it does not. The family of the forests of a graph has the property that the greedy algorithm finds a maximum-weight independent set. However, if we consider the problem $\max\left\{\sum_{e\in M} c(e) : M \text{ matching in } G\right\}$, then the greedy algorithm does not always work.

In Figure 2.4 we show an example for which the greedy algorithm produces the matching $\{a, c\}$, but the matching $\{b, d\}$ is optimal. Note that the set of matchings of this graph (which is an independence system) is not always a matroid.

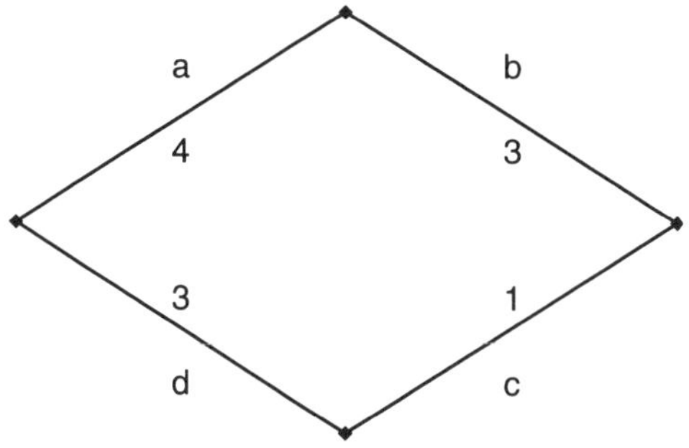

Figure 2.4 A edge-weighted cycle

Edmonds [71] showed that the greedy algorithm works for matroids.

Theorem 2.5.1 *Let $(N, \mathcal{M})$ be a matroid and let $c : N \to \mathbb{R}$ be a weight function. Then the greedy algorithm generates a maximum-weight independent set.*

Proof. Suppose on the contrary that the greedy algorithm produces an independent set $E = \{e_1, \ldots, e_r\}$ which is not optimal. Let $F = \{f_1, \ldots, f_t\}$ be an independent set such that

$$c(f_1) \geq \cdots \geq c(f_t) \text{ and } c(F) > c(E).$$

This implies that there exists an index i, $1 \leq i \leq r$ such that $c(f_i) > c(e_i)$ or $c(f_i) > 0$ and $i > r$. We claim that $\{e_1, \ldots, e_{i-1}\}$ is a basis of the set $H = \{e_1, \ldots, e_{i-1}, f_1, \ldots, f_i\}$. Otherwise $\{e_1, \ldots, e_{i-1}\} \cup \{f_j\} \in \mathcal{M}$ for some j, $1 \leq j \leq i$. Since

$$c(f_1) \geq \cdots \geq c(f_i) > c(e_i)$$

the greedy algorithm must be added one of $f_1, \ldots, f_i$, contrary to the choice of e_i. Therefore, we obtain a contradiction with property (M4) (see Theorem 1.5.1), since $\{f_1, \ldots, f_i\}$ is a larger independent subset of H. □

We will show that matroids are exactly the independence systems for which the greedy algorithm works.

Theorem 2.5.2 *An independence system $(N, \mathcal{I})$ is a matroid if and only if for all weight functions $c : N \to \mathbb{R}$ the greedy algorithm generates a maximum-weight independent set.*

Proof. It suffices to prove that $(N, \mathcal{I})$ satisfies property (M3). Let $T, S \in \mathcal{I}$ with $|T| = p$ and $|S| = p + 1$. We want to show that there exists $i \in S \setminus T$ such that $T \cup i \in \mathcal{I}$. We define the following weight function:

$$c(i) = \begin{cases} (p+1)/(p+2), & \text{if } i \in T \\ p/(p+1), & \text{if } i \in S \setminus T \\ 0, & \text{otherwise.} \end{cases}$$

Since $(p+1)^2 > p(p+2)$ the greedy algorithm generates T first. Moreover, the following inequalities

$$c(S) \geq (p+1)\left(\frac{p}{p+1}\right) = p > p\left(\frac{p+1}{p+2}\right) = c(T)$$

imply that the greedy algorithm finds $i \in S \setminus T$ such that $T \cup i \in \mathcal{I}$. □

In the following examples, we formulate the matroid optimization problem in terms of polyhedra. We know that the greedy algorithm gives a solution to this problem, but it will be useful to solve it as a linear programming problem.

Example: The matroid polytope (Edmonds [71] [72]).

For any $(N, \mathcal{M})$ matroid, the convex hull of the incidence vectors of the independent sets $S \in \mathcal{M}$ is called the *matroid polytope* of $\mathcal{M}$. Then the matroid optimization problem for $c \in \mathbb{R}^N$ is

$$\max \left\{ \sum_{i \in S} c(i) : S \in \mathcal{M} \right\} = \max \left\{ c^T x : x \in P(\mathcal{M}) \right\},$$

where $P(\mathcal{M}) = \operatorname{conv}\{\mathbf{1}_S : S \in \mathcal{M}\}$. The rank function r of a matroid is

$$r(A) := \max\{|S| : S \subseteq A,\ S \in \mathcal{M}\} \quad \text{for all } A \subseteq N.$$

Let $S \in \mathcal{M}$ be an independent set with incidence vector $\mathbf{1}_S \in \mathbb{R}^n$. Since $S \cap A \in \mathcal{M}$ for all $A \subseteq N$, we have $\mathbf{1}_S(A) = |S \cap A| \leq r(A)$. Therefore, the vectors of the matroid polytope satisfy the following inequalities:

$$\begin{aligned} x(i) &\geq 0 && \text{for all } i \in N, \\ x(A) &\leq r(A) && \text{for all } A \subseteq N. \end{aligned}$$

Edmonds [71] [72] proved that these inequalities are a complete description of the matroid polytope.

Theorem 2.5.3 *Let $(N, \mathcal{M})$ be a matroid with rank function r. Then we have*

$$P(\mathcal{M}) = \left\{ x \in \mathbb{R}^N : x \geq 0,\ x(A) \leq r(A) \ \textit{for all}\ \ A \subseteq N \right\}.$$

Proof. It suffices to prove that every extreme point x' of the polytope defined by the inequalities is the incidence vector of an independent set. If we take such x' then there exists $c \in \mathbb{R}^N$ such that x' is the unique optimal solution of the LP problem

$$\max \left\{ c^T x : x \geq 0,\ x(A) \leq r(A) \ \text{for all}\ \ A \subseteq N \right\}.$$

We will show that the greedy algorithm (for this c) generates an independent set such that its incidence vector is an optimal solution of the LP problem, and therefore, coincides with x'.

Let x^* be the incidence vector of the set F generated by the greedy algorithm for c. We construct a feasible solution y^* for the dual problem

$$\min \left\{ \sum_{A \subseteq N} r(A)\, y_A : y \geq 0,\ \sum_{A \ni e} y_A \geq c(e) \text{ for all } e \in N \right\}$$

that satisfies the complementary slackness conditions with x^*. These conditions are the following for (x, y)

$$x_e > 0 \text{ implies } \sum_{A \ni e} y_A = c(e) \text{ for each } e \in N,$$
$$y_A > 0 \text{ implies } x(A) = r(A) \text{ for each } A \subseteq N.$$

To compute the solution y^* we apply the following *Dual greedy algorithm*:

Order $N = \{e_1, \ldots, e_n\}$ so that $c(e_1) \geq \cdots \geq c(e_m) > 0 \geq \cdots \geq c(e_n)$;
For $i = 1$ to m define $T_i = \{e_1, \ldots, e_i\}$ and

$$y_A^* = \begin{cases} c(e_i) - c(e_{i+1}), & \text{if } A = T_i,\ 1 \leq i < m \\ c(e_m), & \text{if } A = T_m \\ 0, & \text{otherwise.} \end{cases}$$

By definition $y_A^* \geq 0$ for all $A \subseteq N$. To check that y^* is feasible for the dual problem, we consider any $e_j \in N$ and distinguish two cases:
If $j > m$ then

$$\sum_{A \ni e_j} y_A^* = 0 \geq c(e_j).$$

If $j \leq m$ then

$$\begin{aligned} \sum_{A \ni e_j} y_A^* &= \sum_{i=j}^{m} y_{T_i}^* \\ &= \sum_{i=j}^{m-1} (c(e_i) - c(e_{i+1})) + c(e_m) \\ &= c(e_j). \end{aligned}$$

We now prove that y^* satisfies the complementary slackness conditions with $x^* = \mathbf{1}_F$. First, if $x_e^* > 0$ then $e \in F \subseteq \{e_1, \ldots, e_m\}$, so that $\sum_{A \ni e} y_A^* = c(e)$. Next, we suppose that $y_A^* > 0$. Then $A = T_i$ for some i, $1 \leq i \leq m$ and it is enough to prove that $y_{T_i}^* > 0$ implies $x^*(T_i) = r(T_i)$ for each $1 \leq i \leq m$. Since $x^*(T_i) = \mathbf{1}_F(T_i) = |F \cap T_i|$ we must prove that $F \cap T_i$ is a basis of T_i for each $1 \leq i \leq m$. Suppose on the contrary that there is an element $e_j \in T_i \setminus F$ such that $(F \cap T_i) \cup \{e_j\} \in \mathcal{M}$. Then the greedy algorithm must be added e_j to F and we obtain a contradiction, since $e_j \notin F$. □

Corollary 2.5.4 *Let $(N, \mathcal{M})$ be a matroid with rank function r. Then the constraints system $x \geq 0$, $x(A) \leq r(A)$ for all $A \subseteq N$, is totally dual integral.*

Proof. It follows from the formula of the dual greedy algorithm that the dual optimal solution $(y^*_A)_{A \subseteq N}$ is integral, for each integral vector $c \in \mathbb{R}^N$. □

Let $(N, \mathcal{M})$ be a matroid with rank function r. Then we have proved that:

1. The greedy algorithm works for the LP problem

$$\max \left\{ c^T x : x \geq 0,\ x(A) \leq r(A) \ \text{ for all } \ A \subseteq N \right\}.$$

2. The dual greedy algorithm works for the dual problem

$$\min \left\{ \sum_{A \subseteq N} r(A)\, y_A : y \geq 0,\ \sum_{A \ni e} y_A \geq c(e) \text{ for all } e \in N \right\}.$$

Example: The forest polytope (Grötschel *et al.* [116, p. 217]).

Let $G = (V, E)$ be a graph and let $(E, \mathcal{M})$ be the forest matroid of G, i.e., the family that consists of those subsets of E that not contain a cycle of G. The convex hull of the incidence vectors of the independent sets, i.e., forests of G, is called the *forest polytope* of G. The rank function r of a set $A \subseteq E$ is given by

$$\begin{aligned} r(A) &= \max \{ |F| : F \subseteq A,\ F \in \mathcal{M} \} \\ &= \max \{ |F| : F \subseteq A,\ (V, F) \text{ is a forest} \} \\ &= |V| - \text{number of components of } (V, A). \end{aligned}$$

The following theorem gives a complete description of the forest polytope.

Theorem 2.5.5 *Let $G = (V, E)$ be a graph. The convex hull of the incidence vectors of all forests of G is given by the set*

$$\left\{ x \in \mathbb{R}^E : x \geq 0,\ x(\gamma(W)) \leq |W| - 1 \ \textit{for all} \ \emptyset \neq W \subseteq V \right\}.$$

Proof. It follows from Theorem 2.5.3 that the vectors $x \in \mathbb{R}^E$ of the forest polytope satisfy $x \geq 0$ and $x(\gamma(W)) \leq r(\gamma(W)) \leq |W| - 1$, for $\emptyset \neq W \subseteq V$.

Conversely, let x be a vector which satisfies the above inequalities and let $A \subseteq E$ be a set of edges. If $(V_1, A_1), \ldots, (V_t, A_t)$ are the components of (V, A) then

$$x(A) = \sum_{i=1}^{t} x(A_i) \leq \sum_{i=1}^{t} x(\gamma(V_i)) \leq \sum_{i=1}^{t} (|V_i| - 1) = |V| - t = r(A),$$

so that the vector x is an element of the forest polytope. □

2.6 Polymatroids

Edmonds [71] combined matroid theory with polyhedral combinatorics and contributed the concept of polyhedral matroid (*polymatroid*). These special kinds of polytopes play an important role in combinatorial optimization and game theory. For a discussion of polymatroids, see the survey of Girlich and Kovalev [107].

Definition 2.6.1 *We say that a function* $f : 2^N \to \mathbb{R}$ *is submodular if*

$$f(S \cup T) + f(S \cap T) \le f(S) + f(T)$$

for all $S, T \subseteq N$. *The polymatroid associated with* f *is the polytope*

$$P^+(f) = \left\{x \in \mathbb{R}^N_+ : x(A) \le f(A) \text{ for all } A \subseteq N\right\}.$$

The submodular polyhedron associated with f *is*

$$P(f) = \left\{x \in \mathbb{R}^N : x(A) \le f(A) \text{ for all } A \subseteq N\right\}.$$

The base polyhedron associated with f *is*

$$B(f) = P(f) \cap \left\{x \in \mathbb{R}^N : x(N) = f(N)\right\}.$$

Note that $P^+(f) \neq \emptyset$ if and only if $f \ge 0$, and $P(f) \neq \emptyset$ if and only if $f(\emptyset) \ge 0$. It follows from Theorem 2.5.3 that the polymatroid $P(r)$ defined by the rank function of a matroid $(N, \mathcal{M})$ is the convex hull of the incidence vectors of all independent sets of $(N, \mathcal{M})$.

Definition 2.6.2 *A function* $g : 2^N \to \mathbb{R}$ *is supermodular if*

$$g(S \cup T) + g(S \cap T) \ge g(S) + g(T)$$

for all $S, T \subseteq N$. *The supermodular polyhedron associated with* g *is*

$$P(g) = \left\{x \in \mathbb{R}^N : x(A) \ge g(A) \text{ for all } A \subseteq N\right\}.$$

The base polyhedron associated with g *is*

$$B(g) = \{x \in P(g) : x(N) = g(N)\}.$$

Remark 2.6.1 *The core of a submodular (concave) cost game* $C(c)$ *is the base polyhedron* $B(c)$ *and the core of a supermodular (convex) profit game* $C(v)$ *is the base polyhedron* $B(v)$.

Definition 2.6.3 *Let $P^+(f)$ be a polymatroid and let $x \in P^+(f)$. A set $T \subseteq N$ is tight (or x-tight) if $x(T) = f(T)$.*

Proposition 2.6.1 *For a vector $x \in P^+(f)$, the family of all x-tights is closed under union and intersection.*

Proof. Let x be a vector in the polymatroid $P^+(f)$. If S and T are x-tights, then the submodularity of f implies

$$\begin{aligned} x(S \cup T) &\leq f(S \cup T) \leq f(S) + f(T) - f(S \cap T) \\ &= x(S) + x(T) - f(S \cap T) \\ &\leq x(S) + x(T) - x(S \cap T) = x(S \cup T). \end{aligned}$$

Therefore, $x(S \cup T) = f(S \cup T)$ and $x(S \cap T) = f(S \cap T)$. □

The following result shows that for any vector $u \in \mathbb{R}_+^N$, all maximal vectors $x \in P^+(f)$ with $x \leq u$ have the same component-sum.

Theorem 2.6.2 *Let $f : 2^N \to \mathbb{R}_+$ be a submodular function and let $u \in \mathbb{R}_+^N$. Then the following max-min equation holds:*

$$\max\left\{x(N) : x \in P^+(f),\ x \leq u\right\} = \min\left\{f(A) + u(N \setminus A) \ :\ A \subseteq N\right\}.$$

Proof. We fix $u \in \mathbb{R}_+^N$. For each $A \subseteq N$ and for each $x \in P^+(f)$ such that $x \leq u$, we have

$$x(N) = x(A) + x(N \setminus A) \leq f(A) + u(N \setminus A),$$

and hence

$$\max\left\{x(N) : x \in P^+(f),\ x \leq u\right\} \leq \min\left\{f(A) + u(N \setminus A) \ :\ A \subseteq N\right\}.$$

To complete the proof it is enough to show that there exists $A \subseteq N$ such that $x(N) = f(A) + u(N \setminus A)$, where x is a solution of the above max problem, i.e., a maximal vector $x \in P^+(f)$ with $x \leq u$. For this x, we define the set

$$A = \bigcup \{T : T \text{ is } x\text{-tight}\}$$

and we claim that $x(N \setminus A) = u(N \setminus A)$. Otherwise $x(N \setminus A)$ can be increased, since $x(S) < f(S)$ for all $S \subseteq N \setminus A$.

By Proposition 2.6.1, the set A is x-tight and hence

$$x(N) = x(A) + x(N \setminus A) = f(A) + u(N \setminus A),$$

and this completes the proof of the max-min relation. □

Corollary 2.6.3 *Let $f : 2^N \to \mathbb{R}_+$ be a submodular function. Then, for each $S \neq \emptyset$, we have*

$$\max\left\{x(S) : x \in P^+(f)\right\} = \min\left\{f(T) : S \subseteq T \subseteq N\right\}.$$

Proof. For each nonempty S, we apply Theorem 2.6.2 to the vector $u_S \in \mathbb{R}^N_+$, defined by

$$u_S(i) = \begin{cases} \beta, & \text{if } i \in S \\ 0, & \text{if } i \in N \setminus S, \end{cases}$$

where β is sufficiently large. □

Remark 2.6.2 *For the submodular polyhedron $P(f)$ we can define tight sets and the max-min relation of Theorem 2.6.2 still holds for any vector $u \in \mathbb{R}^N$, with the same proof.*

Fujishige [97] obtained the following min-max relation which generalizes the max-min equation of Theorem 2.6.2. Let $c : 3^N \to \mathbb{R}$ be a bisubmodular game. Recall that the bisubmodular polyhedron associated with c is given by

$$P_*(c) = \left\{x \in \mathbb{R}^n : x(S) - x(T) \leq c(S,T) \text{ for all } (S,T) \in 3^N\right\}.$$

Theorem 2.6.4 *For any vector $u \in \mathbb{R}^n$,*

$$\min\left\{\sum_{i \in N} |x_i - u_i| : x \in P_*(c)\right\} = \max\left\{u(S,T) - c(S,T) : (S,T) \in 3^N\right\},$$

where $u(S,T) = u(S) - u(T)$. Moreover, if c is integer valued and the vector u is integral, then there exists an integral $x \in P_*(c)$ that attains the minimum in the left-hand side of the equation.

Edmonds [71] showed that the greedy algorithm, defined as follows, also works for the class of polymatroids.

Polymatroid greedy algorithm:

Order $N = \{e_1, \ldots, e_n\}$ so that $c(e_1) \geq \cdots \geq c(e_m) > 0 \geq \cdots \geq c(e_n)$;

For $i = 1$ to m do

$$x^*(e_i) = \max\{x \in \mathbb{R} : (x^*(e_1), \ldots, x^*(e_{i-1}), x, 0, \ldots, 0) \in P^+(f)\}.$$

Note that the output vector x^* satisfies $x^*(e_i) = 0$ for all i, $m < i \leq n$. We say that the polymatroid greedy algorithm works for $c \in \mathbb{R}^N$ if for every nonincreasing ordering of the components of c the above maxima exist and the output vector $x^* \in \mathbb{R}^N_+$ satisfies

$$c^T x^* = \max\left\{c^T x : x \in P^+(f)\right\}.$$

Theorem 2.6.5 *Let $f : 2^N \to \mathbb{R}_+$ be a submodular function and let $c \in \mathbb{R}^N$. Then the polymatroid greedy algorithm works for c. Moreover, if f is integer-valued, then its output vector x^* is integral.*

Proof. Let us order $N = \{e_1, \dots, e_n\}$ so that

$$c(e_1) \geq \cdots \geq c(e_m) > 0 \geq c(e_{m+1}) \geq \cdots \geq c(e_n).$$

Let $T_i = \{e_1, \dots, e_i\}$ for $1 \leq i \leq n$ and let $T_0 = \emptyset$. If $x^* \in \mathbb{R}_+^N$ is the vector generated by the polymatroid greedy algorithm, then Theorem 2.6.2 implies that $x^*(T_i) = \max\{x(T_i) : x \in P^+(f)\}$ for $1 \leq i \leq n$. Hence

$$\begin{aligned}
c^T x &= \sum_{i=1}^{m} c(e_i)\, x(e_i) + \sum_{i=m+1}^{n} c(e_i)\, x(e_i) \\
&= \sum_{i=1}^{m} c(e_i)\left(x(T_i) - x(T_{i-1})\right) + \sum_{i=m+1}^{n} c(e_i)\, x(e_i) \\
&= \sum_{i=1}^{m-1} \left(c(e_i) - c(e_{i+1})\right) x(T_i) + c(e_m)\, x(T_m) + \sum_{i=m+1}^{n} c(e_i)\, x(e_i) \\
&\leq \sum_{i=1}^{m-1} \left(c(e_i) - c(e_{i+1})\right) x^*(T_i) + c(e_m)\, x^*(T_m) \\
&= c^T x^*,
\end{aligned}$$

for all $x \in P^+(f)$. It follows from the proof of Theorem 2.6.2 that $x^*(T_i)$, for $1 \leq i \leq n$ are integers for integer-valued functions. Hence x^* is integral. □

For the submodular polyhedron $P(f)$, the (modified) polymatroid greedy algorithm choose $x^*(e_i)$ as large as possible so that $x \in P(f)$. We will show that this also works for any $c \in \mathbb{R}_+^N$.

Theorem 2.6.6 *Let $f : 2^N \to \mathbb{R}$ be a submodular function with $f(\emptyset) = 0$ and let $c \in \mathbb{R}_+^N$. Then the (modified) polymatroid greedy algorithm works for c. Moreover, the output vector x^* is an extreme point of the core $C(f)$ given by the formulas: $x^*(e_i) = f(T_i) - f(T_{i-1})$, where $T_i = \{e_1, \dots, e_i\}$ and $T_0 = \emptyset$, for all $1 \leq i \leq n$.*

Proof. Let us order N so that $c(e_1) \geq \cdots \geq c(e_n) \geq 0$. Let $T_i = \{e_1, \dots, e_i\}$ for $1 \leq i \leq n$ and let $T_0 = \emptyset$. The output vector x^* for the modified polymatroid greedy algorithm is given by

$$x^*(e_i) = \max\{x \in \mathbb{R} : (x^*(e_1), \dots, x^*(e_{i-1}), x, 0, \dots, 0) \in P(f)\},$$

for $1 \le i \le n$. These maxima can be computed by the following formulas:

$$\begin{aligned} x^*(e_1) &= f(T_1) \\ x^*(e_2) &= f(T_2) - f(T_1) \\ &\vdots \\ x^*(e_n) &= f(T_n) - f(T_{n-1}). \end{aligned}$$

To prove that $x^* \in P(f)$ is an extreme point of the core $C(f)$, we note that it is the unique solution of the following system of n equations:

$$\begin{aligned} x^*(T_i) &= f(T_i) \quad 1 \le i < n, \\ x^*(N) &= f(N). \end{aligned}$$

By using the same method of Theorem 2.6.5, we prove that x^* maximizes the linear function $c^T x$ over $P(f)$ and the maximum value is given by

$$\begin{aligned} c^T x^* &= \sum_{i=1}^{n-1} (c(e_i) - c(e_{i+1}))\, x^*(T_i) + c(e_n)\, x^*(T_n) \\ &= \sum_{i=1}^{n-1} (c(e_i) - c(e_{i+1}))\, f(T_i) + c(e_n)\, f(N). \end{aligned}$$

□

The following characterization of the core associated to a submodular function is due to Edmonds [71] and Shapley [217]. For results of adjacent extreme points in the core of a convex game, we refer the reader to Topkis [235].

Theorem 2.6.7 *Let $f : 2^N \to \mathbb{R}$ be a submodular function with $f(\emptyset) = 0$. A vector $x \in \mathbb{R}^N$ is an extreme point of the core $C(f)$ if and only if there exists an ordering of $N = \{e_1, \ldots, e_n\}$ such that the components of x are given by $x(e_i) = f(T_i) - f(T_{i-1})$, where $T_i = \{e_1, \ldots, e_i\}$ for $1 \le i \le n$, and $T_0 = \emptyset$.*

Proof. It suffices to prove the "only if" part. Suppose that $x \in C(f)$ is an extreme point. Then x is the unique solution of a system of equations $x(A) = f(A)$, $A \subseteq N$, where the sets A can be taken to be exactly the x-tights sets. Now, for each $e \in N$ define

$$t(e) = \bigcap \{A : e \in A,\ x(A) = f(A)\}.$$

Proposition 2.6.1 implies that $t(e)$ is the minimal x-tight which contains e. Bixby *et al.* [21] proposed an algorithm for finding $t(e)$ in a polymatroid.

This algorithm, adapted by Fujishige [96, p. 62] for submodular functions, is as follows:

Initialize: $T \leftarrow \emptyset$;

Iterate: For $i = 1$ to n do
If there exists $e \in N \setminus T$ with $x(T \cup e) = f(T \cup e)$ then
$T \leftarrow T \cup e$ and $e_i \leftarrow e$;
Otherwise, stop and $x \notin C(f)$.

The output of the algorithm is a maximal chain

$$\emptyset = T_0 \subset T_1 \subset \cdots \subset T_{n-1} \subset T_n = N,$$

where $T_i = \{e_1, \ldots, e_i\}$ for $1 \leq i \leq n$, are x-tights. We therefore obtain an ordering of N such that

$$x(e_i) = x(T_i) - x(T_{i-1}) = f(T_i) - f(T_{i-1}),$$

for $1 \leq i \leq n$. □

A real-valued function defined on subsets of N normalized ($f(\emptyset) = 0$), non-decreasing ($S \subseteq T \subseteq N$ implies $f(S) \leq f(T)$) and submodular is called a *polymatroid function.* Note that the rank function of a matroid satisfies these properties. If f is a polymatroid function, then f is monotone and $f \geq 0$, so that the polymatroid greedy algorithm generates $x^* \in \mathbb{R}_+^N$ such that

$$x^*(e_i) = \begin{cases} f(T_i) - f(T_{i-1}), & 1 \leq i \leq m \\ 0, & m < i \leq n. \end{cases}$$

For an arbitrary submodular function f, the function given by

$$f_{\text{mon}}(S) := \min\{f(T) : S \subseteq T \subseteq N\}$$

for all $S \subseteq N$, is monotone and submodular.

Further $f_{\text{mon}} \leq f$ and hence $P^+(f_{\text{mon}}) \subseteq P^+(f)$. To prove the reverse inclusion, for each $A \subseteq N$ we can find $A_* \supseteq A$ such that $f_{\text{mon}}(A) = f(A_*)$. Then for all $x \in P^+(f)$ we obtain

$$f_{\text{mon}}(A) \geq x(A_*) \geq x(A)$$

for all $A \subseteq N$, since $x \geq 0$. Hence $P^+(f_{\text{mon}}) = P^+(f)$.

This implies that every polymatroid can be defined by a monotone submodular function. Further, if the polymatroid $P^+(f) \neq \emptyset$ then this monotone submodular function is uniquely determined by $P^+(f)$, except on the emptyset, by the formula

$$f_{\text{mon}}(S) = \max\{x(S) : x \in P^+(f)\}$$

for $\emptyset \neq S \subseteq N$. Conversely, if f is a polymatroid function then $f = f_{\text{mon}}$ and $f(\emptyset) = 0$. We therefore deduce the following result.

Proposition 2.6.8 *Every nonempty polymatroid, i.e., defined by a nonnegative submodular function, is determined by a unique (except on the emptyset) polymatroid function.*

The submodular (supermodular) polyhedron and the base polyhedron also arise from more general functions. Fujishige [96, Section 3.4] considered intersecting and crossing submodular functions. Faigle and Kern [80] introduced the following weak submodularity. Let $P = (N, \leq)$ be a partial order on the finite set N, and let $J(P)$ be the lattice of order ideals of P. A function $f : J(P) \to \mathbb{R}$ is submodular if for all $A, B \in J(P)$, we have

$$f(A \cup B) + f(A \cap B) \leq f(A) + f(B).$$

Queyranne, Spieksma an Tardella [203] presented a common framework for the transportation problem satisfying a *Monge property* (see Burkard [32]) and the work of Edmonds on polymatroids. The greedy algorithm proposed by these authors also applies to model of Faigle and Kern [80] under submodular constraints relative to the lattice $J(P)$ of a *rooted forest* P.

Frank [91] obtained a two-phase greedy algorithm to increase, at a minimum cost, the rooted node-connectivity of a digraph by one. Frank writes:

> "I must admit I do not know any natural and/or interesting concrete application in which a more general decreasing intersecting supermodular function is involved. (A possible next research task –comparing the present algorithm to that of Faigle and Kern [80]– may reveal such applications)."

A more general model of submodular functions on algebraic posets has been introduced by Faigle and Kern [83]. They presented an algorithmic generalization of the basic matroid model that also shows the structural connections between solution concepts in cooperative game theory and combinatorial optimization problems. These authors proposed the following question:

> "Another interesting open problem concerns the primal-dual greedy algorithm of Frank [91] that seems to bear some formal resemblance with our present greedy algorithm in Section 6. Yet, neither of the two appears to imply the other. Does there exist a common algorithmic generalization?"

Chapter 3

Discrete convex analysis

In this chapter a theory of discrete optimization called *discrete convex analysis* will be introduced. For a comprehensive treatment we refer the reader to Stoer and Witzgall [227], Fujishige [96, Chapter IV], and Murota [179] [181]. Martínez-Legaz [165] seems to have been the first writer to apply the discrete convex analysis to the cooperative game theory.

3.1 Fenchel conjugation

In this section, we present a generalization to set functions of the Fenchel conjugation and his duality theory. This general duality framework, due to Fujishige [94], Martínez-Legaz and Singer [164], and Murota [180], motivate us to investigate solution concepts of cooperative games from the view point of convex analysis.

Let $N = \{1, \ldots, n\}$ be a finite set and let $\mathcal{F} \subseteq 2^N$ be a nonempty family. We make the following two definitions.

Definition 3.1.1 *For a profit game $v : \mathcal{F} \to \mathbb{R}$ we define $v^\circ : \mathbb{R}^n \to \mathbb{R}$ by*

$$v^\circ(x) = \min \{x(S) - v(S) : S \in \mathcal{F}\}.$$

We call the concave function v° the *concave conjugate function* of v.

Definition 3.1.2 *For a cost game $c : \mathcal{F} \to \mathbb{R}$ we define $c^\bullet : \mathbb{R}^n \to \mathbb{R}$ by*

$$c^\bullet(x) = \max \{x(S) - c(S) : S \in \mathcal{F}\}.$$

We call the convex function $c^\bullet$ the *convex conjugate function* of c. For the functions $v^\circ : \mathbb{R}^n \to \mathbb{R}$ and $c^\bullet : \mathbb{R}^n \to \mathbb{R}$, we define their concave and convex conjugate functions (see Stoer and Witzgall [227, Section 4.6]).

These conjugate functions $(v^\circ)^\circ : \mathbb{R}^n \to \mathbb{R}$ and $(c^\bullet)^\bullet : \mathbb{R}^n \to \mathbb{R}$ are given by

$$\begin{aligned} v^{\circ\circ}(y) &= \inf\left\{y^T x - v^\circ(x) : x \in \mathbb{R}^n\right\} \\ c^{\bullet\bullet}(y) &= \sup\left\{y^T x - c^\bullet(x) : x \in \mathbb{R}^n\right\}. \end{aligned}$$

We denote by $\overline{\mathcal{F}} = \operatorname{conv}(\mathcal{F})$ the convex hull of the collection of incidence vectors $\{\mathbf{1}_S : S \in \mathcal{F}\} \subseteq \mathbb{R}^n$, that is,

$$\overline{\mathcal{F}} = \left\{ y \in \mathbb{R}^n : y = \sum_{S \in \mathcal{F}} \lambda_S \mathbf{1}_S,\ \lambda \in \Lambda_{\mathcal{F}} \right\},$$

where

$$\Lambda_{\mathcal{F}} = \left\{ \lambda \in \mathbb{R}^{\mathcal{F}} : \sum_{S \in \mathcal{F}} \lambda_S = 1,\ \lambda_S \geq 0 \text{ for all } S \in \mathcal{F} \right\}.$$

Since $\mathcal{F}$ is finite, $v^\circ(x) = \min\left\{(\mathbf{1}_S)^T x - v(S) : S \in \mathcal{F}\right\}$ is a polyhedral concave function (minimum of a finite number of linear functions) and $c^\bullet$ is a polyhedral convex function (maximum of a finite number of linear functions). Stoer and Witzgall [227, Section 4.8] proved that the conjugates of a polyhedral function satisfy the following relations:

$$v^{\circ\circ}(y) = \begin{cases} \max\left\{\sum_{S \in \mathcal{F}} \lambda_S v(S) : y = \sum_{S \in \mathcal{F}} \lambda_S \mathbf{1}_S,\ \lambda \in \Lambda_{\mathcal{F}}\right\} & \text{if } y \in \overline{\mathcal{F}} \\ -\infty & \text{if } y \notin \overline{\mathcal{F}}. \end{cases}$$

$$c^{\bullet\bullet}(y) = \begin{cases} \min\left\{\sum_{S \in \mathcal{F}} \lambda_S c(S) : y = \sum_{S \in \mathcal{F}} \lambda_S \mathbf{1}_S,\ \lambda \in \Lambda_{\mathcal{F}}\right\} & \text{if } y \in \overline{\mathcal{F}} \\ +\infty & \text{if } y \notin \overline{\mathcal{F}}. \end{cases}$$

The following proposition shows that $v^{\circ\circ}$ (or $c^{\bullet\bullet}$) is an extension of v (or c) and also that the correspondence between a set function v (or c) and its concave (or convex) conjugate v° (or $c^\bullet$) is one to one.

Proposition 3.1.1 *For a profit game $v : \mathcal{F} \to \mathbb{R}$ and a cost game $c : \mathcal{F} \to \mathbb{R}$ we have that*

1. $v^{\circ\circ}(\mathbf{1}_S) = v(S)$ *for all* $S \in \mathcal{F}$.
2. $c^{\bullet\bullet}(\mathbf{1}_S) = c(S)$ *for all* $S \in \mathcal{F}$.
3. $\max\left\{v^{\circ\circ}(y) : y \in \overline{\mathcal{F}}\right\} = \max\{v(S) : S \in \mathcal{F}\}$.
4. $\min\left\{c^{\bullet\bullet}(y) : y \in \overline{\mathcal{F}}\right\} = \min\{c(S) : S \in \mathcal{F}\}$.

Proof. For all $x \in \mathbb{R}^n$ we have that $v^\circ(x) \le x(S) - v(S)$ for $S \in \mathcal{F}$. Hence

$$x(S) - v^\circ(x) = \max\{x(S) - x(S') + v(S') : S' \in \mathcal{F}\} \ge v(S).$$

Taking the vector $x = k\left(\mathbf{1}_{N\setminus S} - \mathbf{1}_S\right)$ we see that

$$x(S) - x(S') = -k(|S \cup S'| - |S \cap S'|)$$

so that for sufficiently large positive k, the maximum is attained by $S' = S$ and its value is $v(S)$. We have thus shown that

$$v^{\circ\circ}(\mathbf{1}_S) = \min\{x(S) - v^\circ(x) : x \in \mathbb{R}^n\} = v(S).$$

Assertion 2 follows from the following relations

$$(-c)^\circ(x) = -c^\bullet(-x) \quad \text{and} \quad (-c)^{\circ\circ}(y) = -c^{\bullet\bullet}(y).$$

Proposition 2.1.3 and the above relations imply 3 and 4. □

Note that assertions 1 and 2 of this proposition show that the correspondence between games $v : \mathcal{F} \to \mathbb{R}$, $c : \mathcal{F} \to \mathbb{R}$, and their conjugates is, for any $S \in \mathcal{F}$,

$$\begin{aligned} v(S) &= \min\{x(S) - v^\circ(x) : x \in \mathbb{R}^n\}, \\ c(S) &= \max\{x(S) - c^\bullet(x) : x \in \mathbb{R}^n\}. \end{aligned}$$

Let $\mathcal{F}$ and $\mathcal{G}$ be nonempty families of subsets of N. We shall introduce a Fenchel type theorem into a form of primal and dual program. Given a cost game $c : \mathcal{F} \to \mathbb{R}$, a profit game $v : \mathcal{G} \to \mathbb{R}$, and their conjugate functions $c^\bullet$ and v°, respectively, we consider

Primal problem: *Find* $\min\{c(S) - v(S) : S \in \mathcal{F} \cap \mathcal{G}\}$,

Dual problem: *Find* $\max\{v^\circ(x) - c^\bullet(x) : x \in \mathbb{R}^n\}$.

Murota [180] showed the following weak duality proposition by using the Fenchel's duality theorem in the standard convex analysis.

Proposition 3.1.2 *For a cost game $c : \mathcal{F} \to \mathbb{R}$, a profit game $v : \mathcal{G} \to \mathbb{R}$, and their conjugate functions $c^\bullet$ and v°, we have*

$$\begin{aligned} \min\{c(S) - v(S) : S \in \mathcal{F} \cap \mathcal{G}\} &\ge \min\left\{c^{\bullet\bullet}(y) - v^{\circ\circ}(y) : y \in \overline{\mathcal{F}} \cap \overline{\mathcal{G}}\right\} \\ &= \max\{v^\circ(x) - c^\bullet(x) : x \in \mathbb{R}^n\} \\ &\ge \max\{v^\circ(x) - c^\bullet(x) : x \in \mathbb{Z}^n\}. \end{aligned}$$

Proof. Proposition 3.1.1 implies that $c(S) - v(S) = c^{\bullet\bullet}(\mathbf{1}_S) - v^{\circ\circ}(\mathbf{1}_S)$ for all $S \in \mathcal{F} \cap \mathcal{G}$ and this gives the first inequality. The next equality is the duality theorem of Fenchel [227, Section 5.1] and the last inequality is obvious. □

Definition 3.1.3 *The vector rank function* $r_c : \mathbb{R}^n \to \mathbb{R}$ *of a submodular cost game* $c : 2^N \to \mathbb{R}$, *is defined by* $r_c(u) = \min\{c(S) + u(N \setminus S) : S \subseteq N\}$.

Theorem 2.6.2 implies the following min-max equation for all $u \in \mathbb{R}^n$,

$$r_c(u) = \max\{x(N) : x \in P(c),\ x \leq u\}. \tag{3.1}$$

The following intersection theorem extends equation (3.1) and generalizes the intersection theorem for polymatroids due to Edmonds [71].

Theorem 3.1.3 *For distributive lattices* $\mathcal{F}_1, \mathcal{F}_2 \subseteq 2^N$, *let* $c_1 : \mathcal{F}_1 \to \mathbb{R}$ *and* $c_2 : \mathcal{F}_2 \to \mathbb{R}$ *be submodular cost games. If there exists a set* $S \in \mathcal{F}_1$ *such that* $N \setminus S \in \mathcal{F}_2$, *then we have*

$$\begin{aligned} & \min\{c_1(S) + c_2(N \setminus S) : S \in \mathcal{F}_1,\ N \setminus S \in \mathcal{F}_2\} \\ = & \max\{x(N) : x \in P(c_1) \cap P(c_2)\} \end{aligned}$$

Moreover, if c_1 *and* c_2 *are integer valued, then the maximum is attained by an integral vector.*

Proof. See Fujishige [96, Theorem 4.9]. □

Fujishige [96, Theorem 4.12] proved the following Frank's discrete separation theorem [90] by using Theorem 3.1.3.

Theorem 3.1.4 *For distributive lattices* $\mathcal{F}, \mathcal{G} \subseteq 2^N$ *with* $\emptyset, N \in \mathcal{F} \cap \mathcal{G}$, *let* $c : \mathcal{F} \to \mathbb{R}$ *be a submodular cost game and let* $v : \mathcal{G} \to \mathbb{R}$ *be a supermodular profit game. If* $v(S) \leq c(S)$ *for all* $S \in \mathcal{F} \cap \mathcal{G}$, *there exists* $x \in \mathbb{R}^n$ *such that* $v(S) \leq x(S)$ *for all* $S \in \mathcal{G}$ *and* $x(S) \leq c(S)$ *for all* $S \in \mathcal{F}$.

Proof. We consider the dual game of v defined by the submodular cost game $v^*(N \setminus S) = v(N) - v(S)$, for all $S \in \mathcal{G}$. From Theorem 3.1.3, we deduce that if $v \leq c$ then

$$\begin{aligned} \max\{x(N) : x \in P(c) \cap P(v^*)\} &= \min\{c(S) + v^*(N \setminus S) : S \in \mathcal{F} \cap \mathcal{G}\} \\ &= \min\{c(S) + v(N) - v(S) : S \in \mathcal{F} \cap \mathcal{G}\} \\ &= v(N). \end{aligned}$$

Then there exists $x \in P(c) \cap P(v^*)$ such that $x(N) = v(N)$ and hence $x \in P(c) \cap C(v^*) = P(c) \cap C(v)$. It therefore follows from the definitions that x satisfies $v \leq x \leq c$. □

We end this section with a Fenchel-type min-max duality theorem for supermodular and submodular functions showed by Fujishige [94, Theorem 6.3].

Theorem 3.1.5 *For distributive lattices $\mathcal{F}, \mathcal{G} \subseteq 2^N$ with $\emptyset, N \in \mathcal{F} \cap \mathcal{G}$, let $c : \mathcal{F} \to \mathbb{R}$ be a submodular cost game and let $v : \mathcal{G} \to \mathbb{R}$ be a supermodular profit game. Then we have*

$$\min \{c(S) - v(S) : S \in \mathcal{F} \cap \mathcal{G}\} = \max \{v^\circ (x) - c^\bullet (x) : x \in \mathbb{R}^n\}. \tag{3.2}$$

Moreover, if c and v are integer valued, then the maximum is attained by a vector $x^* \in \mathbb{Z}^n$.

Proof. Let v^* be dual of v. Then v^* is a submodular cost game defined by $v^*(N \setminus S) = v(N) - v(S)$, for all $S \in \mathcal{G}$. We observe now that the min-max equation (3.2) is equivalent to

$$\min \{c(S) + v^*(N \setminus S) : S \in \mathcal{F} \cap \mathcal{G}\} = \max \{v^\circ (x) + v(N) - c^\bullet (x) : x \in \mathbb{R}^n\}.$$

If r_c and r_{v^*} are the vector rank functions of the submodular games c and v^*, respectively, then for all $x \in \mathbb{R}^n$,

$$\begin{aligned}
c^\bullet (x) &= \max \{x(S) - c(S) : S \in \mathcal{F}\} = -\min \{c(S) - x(S) : S \in \mathcal{F}\} \\
&= x(N) - \min \{c(S) + x(N \setminus S) : S \in \mathcal{F}\} = x(N) - r_c(x). \\
v^\circ (x) &= \min \{x(S) + v(N) - v(S) : S \in \mathcal{G}\} - v(N) \\
&= \min \{v^*(N \setminus S) + x(S) : S \in \mathcal{G}\} - v(N) = r_{v^*}(x) - v(N).
\end{aligned}$$

It therefore follows that the min-max equation (3.2) is equivalent to

$$\min \{c(S) + v^*(N \setminus S) : S \in \mathcal{F} \cap \mathcal{G}\} = \max \{r_c(x) + r_{v^*}(x) - x(N) : x \in \mathbb{R}^n\}.$$

Let $x \in \mathbb{R}^n$. Then, equation (3.1) implies that

$$\begin{aligned}
r_c(x) &= \max \{y(N) : y \in P(c),\ y \leq x\}, \\
r_{v^*}(x) &= \max \{z(N) : z \in P(v^*),\ z \leq x\},
\end{aligned}$$

and there exist vectors $y, z \in \mathbb{R}^n$ such that $r_c(x) = y(N),\ y \in P(c),\ y \leq x$ and $r_{v^*}(x) = z(N),\ z \in P(v^*),\ z \leq x$. By taking $u = \min\{y, z\}$, we deduce that $u \in P(c) \cap P(v^*)$. From the above relations we obtain

$$\begin{aligned}
r_c(x) + r_{v^*}(x) - x(N) &= y(N) + z(N) - x(N) \\
&= u(N) + \max\{y, z\}(N) - x(N) \\
&\leq u(N) = r_c(u) + r_{v^*}(u) - u(N).
\end{aligned}$$

Therefore, the min-max equation (3.2) is equivalent to

$$\min \{c(S) + v^*(N \setminus S) : S \in \mathcal{F} \cap \mathcal{G}\} = \max \{u(N) : u \in P(c) \cap P(v^*)\},$$

which is true due to Theorem 3.1.3. We may apply this theorem and the above equation to obtain the integrality property. □

3.2 Subgradients and subdifferentials

Fujishige [94] [95] introduced the concepts of subgradient and subdifferential for submodular functions and Murota [180] applied these concepts to set functions. We begin with the following two definitions for cooperative games.

Definition 3.2.1 *Let $\mathcal{F} \subseteq 2^N$ be a nonempty family and let $v : \mathcal{F} \to \mathbb{R}$ be a profit game. We call the vector $x \in \mathbb{R}^n$ a subgradient of v at $S \in \mathcal{F}$ if*

$$v(T) - v(S) \leq x(T) - x(S) \quad \textit{for all } T \in \mathcal{F}.$$

For a cost game $c : \mathcal{F} \to \mathbb{R}$ the vector $x \in \mathbb{R}^n$ is a subgradient of c at $S \in \mathcal{F}$ if

$$c(T) - c(S) \geq x(T) - x(S) \quad \textit{for all } T \in \mathcal{F}.$$

Definition 3.2.2 *Let $g : \mathcal{F} \to \mathbb{R}$ be a game. The subdifferential $\partial g(S)$ of the game g at the coalition $S \in \mathcal{F}$ is the set of all the subgradients of g at S.*

Note that $\mathbb{R}^n$ is divided into $|\mathcal{F}|$ unbounded polyhedra $\partial g(S)$, for $S \in \mathcal{F}$. Moreover, if $S, T \in \mathcal{F}$ and $S \neq T$ then the subdifferentials $\partial g(S)$ and $\partial g(T)$ may have common faces but not common interior vectors.

Proposition 3.2.1 *Let $v : 2^N \to \mathbb{R}$ be a profit game and let $c : 2^N \to \mathbb{R}$ be a cost game with dual games v^* and c^*. Then*

(i) $\partial v(\emptyset) = P(v)$ *and* $\partial v(N) = P(v^*)$,

(ii) $\partial c(\emptyset) = P(c)$ *and* $\partial c(N) = P(c^*)$,

(iii) $\partial v(\emptyset) \cap \partial v(N) = C(v) = C(v^*)$,

(iv) $\partial c(\emptyset) \cap \partial c(N) = C(c) = C(c^*)$.

Proof. (i) Notice that $v(\emptyset) = x(\emptyset) = 0$ imply that

$$\partial v(\emptyset) = \{x \in \mathbb{R}^n : v(T) \leq x(T) \text{ for all } T \subseteq N\} = P(v).$$

The dual game v^* is a cost game given by $v^*(S) = v(N) - v(N \setminus S)$ for all $S \subseteq N$. Thus

$$\begin{aligned} \partial v(N) &= \{x \in \mathbb{R}^n : v(T) - v(N) \leq x(T) - x(N) \text{ for all } T \subseteq N\} \\ &= \{x \in \mathbb{R}^n : x(N \setminus T) \leq v(N) - v(T) \text{ for all } T \subseteq N\} \\ &= \{x \in \mathbb{R}^n : x(S) \leq v^*(S) \text{ for all } S \subseteq N\} = P(v^*). \end{aligned}$$

(ii) Follows from the definitions of the subdifferentials and the polyhedra.

Properties (i), (ii) and Proposition 1.1.1 together give (iii) and (iv). □

Example: We consider the profit game (N, v), where $N = \{1, 2\}$ and the characteristic function is defined by $v(S) = 0$ if $|S| \leq 1$, and $v(N) = 1$. Then Figure 3.1 shows the four subdifferentials of the game.

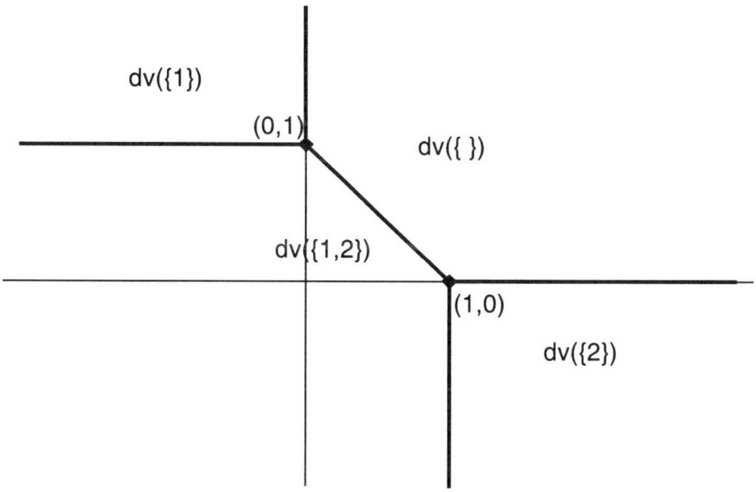

Figure 3.1 The subdifferentials $\partial v(S)$

Using Theorem 2.6.7, Fujishige [96] proved the following characterization of the extreme points of the subdifferentials for submodular cost games.

Theorem 3.2.2 *Let $c : \mathcal{F} \to \mathbb{R}$ be a submodular cost game on a distributive lattice $\mathcal{F} \subseteq 2^N$. For each $S \in \mathcal{F}$, a vector $x \in \mathbb{R}^n$ is an extreme point of $\partial c(S)$ if and only if there exists an ordering of $N = \{e_1, \ldots, e_n\}$ such that the components of x are $x(e_i) = c(S_i) - c(S_{i-1})$, where $S_i = \{e_1, \ldots, e_i\}$ for $1 \leq i \leq n$, $S_0 = \emptyset$ and $S = S_j$ for some $1 \leq j \leq n$.*

Proof. See [96, Theorem 6.11]. □

Proposition 3.2.3 *Let $v : \mathcal{F} \to \mathbb{R}$ be a profit game on $\mathcal{F} \subseteq 2^N$, let $x \in \mathbb{R}^n$ and let $S \in \mathcal{F}$. Then the following statements are equivalent:*

(i) $x \in \partial v(S)$,

(ii) $\max\{v(T) - x(T) : T \in \mathcal{F}\} = v(S) - x(S)$,

(iii) $v(S) + v^\circ(x) = x(S)$.

Proof. The result follows from the definitions. □

Proposition 3.2.4 *Let $c : \mathcal{F} \to \mathbb{R}$ be a cost game on $\mathcal{F} \subseteq 2^N$, let $x \in \mathbb{R}^n$ and let $S \in \mathcal{F}$. Then the following statements are equivalent:*

(i) $x \in \partial c(S)$,

(ii) $\min\{c(T) - x(T) : T \in \mathcal{F}\} = c(S) - x(S)$,

(iii) $c(S) + c^\bullet(x) = x(S)$.

Proof. The result follows from the definitions. □

It follows from the definition of subdifferential the following optimization property.

Proposition 3.2.5 *Let $\mathcal{F} \subseteq 2^N$ be a nonempty family, let $v : \mathcal{F} \to \mathbb{R}$ be a profit game and let $c : \mathcal{F} \to \mathbb{R}$ be a cost game. Then*

$$\begin{aligned}\max\{v(T) : T \in \mathcal{F}\} &= v(S) \;\; \textit{if and only if the vector } \mathbf{0} \in \partial v(S), \\ \min\{c(T) : T \in \mathcal{F}\} &= c(S) \;\; \textit{if and only if the vector } \mathbf{0} \in \partial c(S).\end{aligned}$$

For the concave conjugate function $v^\circ : \mathbb{R}^n \to \mathbb{R}$ and $x \in \mathbb{R}^n$ we define

$$\partial_2 v^\circ(x) = \{S \in \mathcal{F} : v^\circ(u) - v^\circ(x) \le u(S) - x(S) \text{ for all } u \in \mathbb{R}^n\}.$$

For the convex conjugate function $c^\bullet : \mathbb{R}^n \to \mathbb{R}$ and $x \in \mathbb{R}^n$ we define

$$\partial_2 c^\bullet(x) = \{S \in \mathcal{F} : c^\bullet(u) - c^\bullet(x) \ge u(S) - x(S) \text{ for all } u \in \mathbb{R}^n\}.$$

The collections of coalitions $\partial_2 v^\circ(x) \subseteq \mathcal{F}$ and $\partial_2 c^\bullet(x) \subseteq \mathcal{F}$ are called the binary subdifferentials of v° and $c^\bullet$ at x.

Proposition 3.2.6 *Let $\mathcal{F} \subseteq 2^N$ be a nonempty family, let $v : \mathcal{F} \to \mathbb{R}$ be a profit game and let $c : \mathcal{F} \to \mathbb{R}$ be a cost game. Then*

(i) *The vector $x \in \partial v(S)$ if and only if the set $S \in \partial_2 v^\circ(x)$.*

(ii) *The vector $x \in \partial c(S)$ if and only if the set $S \in \partial_2 c^\bullet(x)$.*

Proof. (i) From Proposition 3.1.1 (1) we obtain for all $S \in \mathcal{F}$,

$$v(S) = v^{\circ\circ}(\mathbf{1}_S) = \min\{u(S) - v^\circ(u) : u \in \mathbb{R}^n\}. \tag{3.3}$$

Proposition 3.2.3, equation (3.3) and the respective definitions together give the following equivalences

$$\begin{aligned} x \in \partial v(S) &\iff x(S) - v^\circ(x) = v(S) \\ &\iff x(S) - v^\circ(x) = \min\{u(S) - v^\circ(u) : u \in \mathbb{R}^n\} \\ &\iff x(S) - v^\circ(x) \le u(S) - v^\circ(u) \quad \text{for all } u \in \mathbb{R}^n \\ &\iff S \in \partial_2 v^\circ(x). \end{aligned}$$

The statement (ii) follows similarly from Propositions 3.1.1 (2) and 3.2.4. □

We recall that $v^\circ : \mathbb{R}^n \to \mathbb{R}$ and $v^{\circ\circ} : \overline{\mathcal{F}} \to \mathbb{R}$ are concave functions and then we can define the following subdifferentials in the sense of convex analysis (see Rockafellar [207]):

$$\begin{aligned} \partial v^\circ(x) &= \left\{y \in \mathbb{R}^n : v^\circ(u) - v^\circ(x) \le y^T(u-x) \quad \text{for all } u \in \mathbb{R}^n\right\}, \\ \partial v^{\circ\circ}(y) &= \left\{x \in \mathbb{R}^n : v^{\circ\circ}(z) - v^{\circ\circ}(y) \le x^T(z-y) \quad \text{for all } z \in \overline{\mathcal{F}}\right\}. \end{aligned}$$

To characterize these subdifferentials we prove the following lemma.

Lemma 3.2.7 *Let $\mathcal{F} \subseteq 2^N$ be a nonempty family, let $v : \mathcal{F} \to \mathbb{R}$ be a profit game and let $u \in \mathbb{R}^n$. Then*

1. $(v-u)^\circ(x) = v^\circ(x+u)$ *for all* $x \in \mathbb{R}^n$.
2. $(v-u)^{\circ\circ}(y) = v^{\circ\circ}(y) - y^T u$ *for all* $y \in \overline{\mathcal{F}}$.
3. $\max\left\{v^{\circ\circ}(y) - y^T u : y \in \overline{\mathcal{F}}\right\} = \max\{v(S) - u(S) : T \in \mathcal{F}\}$.

Proof. 1. For each $x \in \mathbb{R}^n$ we obtain

$$\begin{aligned} (v-u)^\circ(x) &= \min\{x(S) - (v-u)(S) : S \in \mathcal{F}\} \\ &- \min\{(x+u)(S) - v(S) : S \subset \mathcal{F}\} \\ &= v^\circ(x+u). \end{aligned}$$

2. For each $y \in \overline{\mathcal{F}}$ we have

$$\begin{aligned} (v-u)^{\circ\circ}(y) &= \min\left\{y^T x - (v-u)^\circ(x) : x \in \mathbb{R}^n\right\} \\ &= \min\left\{y^T(x+u) - v^\circ(x+u) : x \in \mathbb{R}^n\right\} - y^T u \\ &= v^{\circ\circ}(y) - y^T u. \end{aligned}$$

3. From Proposition 3.1.1 and 2 we obtain

$$\begin{aligned} \max\left\{v^{\circ\circ}(y) - y^T u : y \in \overline{\mathcal{F}}\right\} &= \max\left\{(v-u)^{\circ\circ}(y) : y \in \overline{\mathcal{F}}\right\} \\ &= \max\left\{(v-u)(S) : S \in \mathcal{F}\right\}. \end{aligned}$$

□

Theorem 3.2.8 *Let $v : \mathcal{F} \to \mathbb{R}$ be a profit game on a nonempty $\mathcal{F} \subseteq 2^N$. Then*

1. $\partial v^\circ(x) = \left\{y \in \overline{\mathcal{F}} : v^{\circ\circ}(y) + v^\circ(x) = y^T x\right\}$ *for all* $x \in \mathbb{R}^n$.

2. $\partial v^{\circ\circ}(y) = \left\{x \in \mathbb{R}^n : v^\circ(x) + v^{\circ\circ}(y) = x^T y\right\}$ *for all* $y \in \overline{\mathcal{F}}$.

3. *If* $(x, y) \in \mathbb{R}^n \times \overline{\mathcal{F}}$ *then* $y \in \partial v^\circ(x) \Longleftrightarrow x \in \partial v^{\circ\circ}(y)$.

4. $\partial v^{\circ\circ}(\mathbf{1}_S) = \partial v(S)$ *for all* $S \in \mathcal{F}$.

Proof. 1. For each $x \in \mathbb{R}^n$ we have

$$\begin{aligned}
\partial v^\circ(x) &= \left\{y \in \mathbb{R}^n : v^\circ(u) - y^T u \le v^\circ(x) - y^T x \text{ for all } u \in \mathbb{R}^n\right\} \\
&= \left\{y \in \mathbb{R}^n : \max\left\{v^\circ(u) - y^T u : u \in \mathbb{R}^n\right\} = v^\circ(x) - y^T x\right\} \\
&= \left\{y \in \mathbb{R}^n : \min\left\{y^T u - v^\circ(u) : u \in \mathbb{R}^n\right\} = y^T x - v^\circ(x)\right\} \\
&= \left\{y \in \overline{\mathcal{F}} : v^{\circ\circ}(y) + v^\circ(x) = y^T x\right\}.
\end{aligned}$$

2. Let $y \in \overline{\mathcal{F}}$. Using Lemma 3.2.7, we can write

$$\begin{aligned}
\partial v^{\circ\circ}(y) &= \left\{x \in \mathbb{R}^n : v^{\circ\circ}(z) - x^T z \le v^{\circ\circ}(y) - x^T y \text{ for all } z \in \overline{\mathcal{F}}\right\} \\
&= \left\{x \in \mathbb{R}^n : \max\left\{v^{\circ\circ}(z) - x^T z : z \in \overline{\mathcal{F}}\right\} = v^{\circ\circ}(y) - x^T y\right\} \\
&= \left\{x \in \mathbb{R}^n : \max\left\{v(T) - x(T) : T \in \mathcal{F}\right\} = v^{\circ\circ}(y) - x^T y\right\} \\
&= \left\{x \in \mathbb{R}^n : \min\left\{x(T) - v(T) : T \in \mathcal{F}\right\} = x^T y - v^{\circ\circ}(y)\right\} \\
&= \left\{x \in \mathbb{R}^n : v^\circ(x) + v^{\circ\circ}(y) = x^T y\right\}.
\end{aligned}$$

3. If $(x, y) \in \mathbb{R}^n \times \overline{\mathcal{F}}$ then

$$y \in \partial v^\circ(x) \Longleftrightarrow v^{\circ\circ}(y) + v^\circ(x) = y^T x \Longleftrightarrow x \in \partial v^{\circ\circ}(y).$$

4. From Propositions 3.1.1 and 3.2.3 we obtain the following equivalences

$$\begin{aligned}
x \in \partial v^{\circ\circ}(\mathbf{1}_S) &\Longleftrightarrow v^{\circ\circ}(\mathbf{1}_S) + v^\circ(x) = x^T \mathbf{1}_S \\
&\Longleftrightarrow v(S) + v^\circ(x) = x(S) \\
&\Longleftrightarrow x \in \partial v(S),
\end{aligned}$$

for all $S \in \mathcal{F}$ and $x \in \mathbb{R}^n$. □

Remark 3.2.1 *The above results for the subdifferentials of v° and $v^{\circ\circ}$ can be obtained similarly for $c^\bullet$ and $c^{\bullet\bullet}$.*

3.3 Indirect functions of cooperative games

The *indirect function* of a cooperative game (N, v) with transferable utility given by $v : 2^N \to \mathbb{R}$, was introduced by Martínez-Legaz [165]. It is a convex function defined on $\mathbb{R}^n$, where $n = |N|$, which provides an alternative definition of the game as it contains exactly the same information as the characteristic function.

Definition 3.3.1 *Let $v : 2^N \to \mathbb{R}$ be a game. We define its indirect function $\pi : \mathbb{R}^n \to \mathbb{R}$ by $\pi(w) = \max\{v(S) - w(S) : S \subseteq N\}$, for all $w \in \mathbb{R}^n$.*

The indirect function admits an economic interpretation. Suppose an employer has to choose among the players those who will produce the maximum profit for him. If the subset S is selected then the total amount of output that S will produce, using the resources owned by the employer, is $v(S)$. If the (possibly negative) salary demanded by player i is w_i then $\pi(w)$ represents the maximum net profit the employer can obtain under the vector of salaries w.

Indirect functions of cooperative games are characterized by Martínez-Legaz [165] according to the three following properties:

(i) $\partial\pi(w) \bigcap \{0, -1\}^n \neq \emptyset$, for all $w \in \mathbb{R}^n$,

(ii) $\{0, -1\}^n \subset \bigcup \{\partial\pi(w) : w \in \mathbb{R}^n\}$,

(iii) $\min\{\pi(w) : w \in \mathbb{R}^n\} = 0$,

where the symbol ∂ stands for subdifferential in the sense of convex analysis.

The following alternative characterization of indirect functions, in terms of gradients, is also given by Martínez-Legaz [167].

Proposition 3.3.1 *Let $\pi : \mathbb{R}^n \to \mathbb{R}$ be a function and let $\mathrm{dom}(\nabla\pi)$ be the set of vectors such that the gradient $\nabla\pi$ exists. Then π is an indirect function if and only if it is convex and satisfies properties (iii) and*

(iv) $\nabla\pi(\mathrm{dom}(\nabla\pi)) = \{0, -1\}^n$.

Moreover, it is proved by Martínez-Legaz [165] that the characteristic function v can be obtained from its indirect function π, by using the formula

$$v(S) = \min\{\pi(w) + w(S) : w \in \mathbb{R}^n\} \quad \text{for all } S \subseteq N.$$

The following characterization of totally balanced games given by Martínez-Legaz [166] will be based on the fact, proved by Kalai and Zemel [136], that these games can be expressed as minima of finite collections of additive games.

Theorem 3.3.2 *Let (N, v) be a game and let $\pi : \mathbb{R}^n \to \mathbb{R}$ be its indirect function. Then the following statements are equivalent:*

1. *The game (N, v) is totally balanced.*
2. *The indirect function satisfies*
$$\pi(w) = \max_{S \subseteq N} \left\{ \min_{\bar{w} \in \pi^{-1}(0)} \sum_{i \in S} (\bar{w}_i - w_i) \right\},$$
for all $w \in \mathbb{R}^n$.
3. $\{0, -1\}^n \subset \bigcup \{\partial \pi(w) : w \in \pi^{-1}(0)\}$.

Proof. 1 ⇒ 2. Since (N, v) is the minimum game of a finite collection of additive games, for every $S \subseteq N$ we have

$$v(S) = \min \left\{ \sum_{i \in S} \bar{w}_i : \sum_{i \in S'} \bar{w}_i \geq v(S') \text{ for all } S' \subseteq N \right\}.$$

The condition $\sum_{i \in S'} \bar{w}_i \geq v(S')$ for all $S' \subseteq N$ can be equivalently written as

$$\pi(\bar{w}) = \max \left\{ v(S') - \sum_{i \in S'} \bar{w}_i : S' \subseteq N \right\} \leq 0.$$

We note that if $S' = \emptyset$ then the maximum is 0, and the above relation is an equality. Therefore, we obtain

$$v(S) = \min \left\{ \sum_{i \in S} \bar{w}_i : \bar{w} \in \pi^{-1}(0) \right\},$$

for all $S \subseteq N$, and this, together with the definition of π, implies assertion 2.

2 ⇒ 3. Let $-\mathbf{1}_S$ be any point in $\{0, -1\}^n$, with $S \in 2^N$. According to property (ii), there exists $\tilde{w} \in \mathbb{R}^n$ such that $-\mathbf{1}_S \in \partial \pi(\tilde{w})$. By statement 2, we have

$$\pi(\tilde{w}) \geq \min \left\{ \sum_{i \in S} (\bar{w}_i - \tilde{w}_i) : \bar{w} \in \pi^{-1}(0) \right\},$$

whence there exists $\bar{w} \in \pi^{-1}(0)$ such that

$$\pi(\tilde{w}) \geq \sum_{i \in S} (\bar{w}_i - \tilde{w}_i) = \langle \tilde{w} - \bar{w}, -\mathbf{1}_S \rangle .$$

Then, for any $w \in \mathbb{R}^n$,

$$\begin{aligned} \pi(w) &\geq \pi(\tilde{w}) + \langle w - \tilde{w}, -\mathbf{1}_S \rangle \\ &\geq \langle \tilde{w} - \bar{w}, -\mathbf{1}_S \rangle + \langle w - \tilde{w}, -\mathbf{1}_S \rangle \\ &= \langle w - \bar{w}, -\mathbf{1}_S \rangle \\ &= \pi(\bar{w}) + \langle w - \bar{w}, -\mathbf{1}_S \rangle , \end{aligned}$$

and hence $-\mathbf{1}_S \in \partial\pi(\bar{w})$. This proves statement 3.

$3 \Rightarrow 1$. For each coalition $S \in 2^N$ wc can find a vector $w^S \in \pi^{-1}(0)$ such that $-\mathbf{1}_S \in \partial\pi\left(w^S\right)$. Given any coalition S and any vector $w \in \mathbb{R}^n$, we have

$$\pi(w) \geq \pi\left(w^S\right) + \left\langle w - w^S, -\mathbf{1}_S \right\rangle$$

or equivalently,

$$\pi(w) + \sum_{i \in S} w_i \geq \pi\left(w^S\right) + \sum_{i \in S} w_i^S .$$

Hence also $\pi\left(w^S\right) = 0$, so that $v(S) = v^S(S)$, where v^S is the additive game defined by $v^S(i) = w_i^S$. On the other hand, for any $S' \in 2^N$,

$$v(S) \leq \pi\left(w^{S'}\right) + \sum_{i \in S} w_i^{S'} = v^{S'}(S).$$

It follows that $v = \min\left\{v^{S'} : S' \subseteq N\right\}$, and this implies that (N, v) is the minimum game of a finite collection of additive games. □

The characterization of indirect functions obtained in Theorem 3.3.2 implies the following corollary.

Corollary 3.3.3 *A function $\pi : \mathbb{R}^n \to \mathbb{R}$ is the indirect function of a (unique) totally balanced game if and only if it satisfies properties (i), (iii) and*

(ii)* $\{0, -1\}^n \subset \bigcup \left\{\partial\pi(w) : w \in \pi^{-1}(0)\right\}$.

Let us mention that balanced games, i.e., games with a nonempty core, also admit a simple characterization in terms of indirect functions. Martínez-Legaz [165] obtained the following representation of the core of (N, v) in terms of its indirect function π.

Theorem 3.3.4 *Let (N, v) be a balanced game and let $\pi : \mathbb{R}^n \to \mathbb{R}$ be its indirect function. Then the core of v satisfies*

$$C(v) = \left\{x \in \mathbb{R}^n : (-t, \ldots, -t) \in \partial\pi(x) \ \ \textit{for all}\, t \in [0, 1]\right\}.$$

From this and the fact that the indirect function π^S of the subgame (S, v_S) induced by the coalition S is given by

$$\pi^S(w_S) = \min_{z_{N\setminus S}} \pi\left(w_S,\, z_{N\setminus S}\right) = \lim_{z_{N\setminus S}\to+\infty} \pi\left(w_S,\, z_{N\setminus S}\right),$$

it is possible to derive a (somewhat complicated) alternative characterization of totally balanced games in terms of indirect functions.

Another consequence of the Kalai and Zemel [136] characterization of totally balanced games underlying our results is that, for an arbitrary game (N, v), there exists a least upper bound $\bar{v}$ of v such that the game $(N, \bar{v})$ is totally balanced. Indeed, one can define $\bar{v}$ as the pointwise infimum of the collection of additive upper bounds of v, namely,

$$\bar{v}(S) = \inf\left\{\sum_{i\in S} w_i : \sum_{i\in S'} w_i \geq v\left(S'\right) \quad \forall\, S' \subseteq N\right\},$$

for all $S \in 2^N$. Being a linear programming problem with a nonempty feasible set and a bounded from below (by $v(S)$) objective function, the minimization problem defined by the right hand side of this equality has an optimal solution $w^S = (w_1^S, \ldots, w_n^S)$. One can easily prove that $\bar{v} = \min\left\{v^S : S \subseteq N\right\}$, with v^S denoting the additive game given by $v^S(i) = w_i^S$. Hence, $\bar{v}$ is totally balanced. By the Duality Theorem 2.2.4 of linear programming, an equivalent definition of $\bar{v}$ is

$$\begin{aligned}\bar{v}(S) &= \max\left\{\sum_{S'\subseteq N} \gamma_{S'} v\left(S'\right) : \sum_{S'\subseteq N} \gamma_{S'} \mathbf{1}_{S'} = \mathbf{1}_S,\ \{\gamma_{S'}\}_{S'\subseteq N} \geq 0\right\} \\ &= \max\left\{\sum_{S'\subseteq S} \gamma_{S'} v\left(S'\right) : \sum_{S'\subseteq S} \gamma_{S'} \mathbf{1}_{S'} = \mathbf{1}_S,\ \{\gamma_{S'}\}_{S'\subseteq S} \geq 0\right\},\end{aligned}$$

which coincides with the definition of the cover of a game given by Shapley and Shubik [216]. Although these authors proved that $\bar{v} \geq v$ with equality if and only if the game is totally balanced, they did not observe that $\bar{v}$ is just the smallest majorant of v that defines a totally balanced game.

The equivalence between totally balanced games and market games was proved by Shapley and Shubik [216] by showing that the cover of a game (N, v) is generated by the direct market induced by the game defined as follows. The traders of the direct market are the players of (N, v), the commodity space is $\mathbb{R}_+^n$, the initial endowments are the unit vectors, that is, each trader possesses one unit of *himself*, and all traders share the same utility function $u : \mathbb{R}_+^n \to \mathbb{R}$, given by

$$u(x) = \max\left\{ \sum_{S\subseteq N} \gamma_S v(S) : \sum_{S\subseteq N} \gamma_S \mathbf{1}_S = x, \;\; \gamma_S \geq 0 \text{ for all } S \subseteq N \right\}.$$

This utility function can be explained by supposing that one tries to distribute in an optimal way the existing x_i *units of trader* i among all possible coalitions. Using the Duality theorem 2.2.4 of linear programming, the utility function can be written as

$$u(x) = \min\left\{ \sum_{i=1}^{n} x_i w_i : \sum_{i\in S} w_i \geq v(S) \quad \text{for all } S \subseteq N \right\}.$$

But, as observed in the proof of Theorem 3.3.2, the inequality system appearing in the latter expression can be simply written as $\pi(w) = 0$. Thus, we obtain

$$u(x) = \min\left\{ \langle w,\, x\rangle : w \in \pi^{-1}(0) \right\},$$

which gives the utility function of the direct market in terms of the indirect function of the game. In the converse direction, the indirect function of the market game generated by a direct market can be obtained from the utility function by means of

$$\pi(w) = \max\left\{ u\left(\mathbf{1}_S\right) - \sum_{i\in S} w_i : S \subseteq N \right\},$$

since $v(S) = u\left(\mathbf{1}_S\right)$ for all $S \subseteq N$ (see Shapley and Shubik [216]).

A game (N, v) is said to be *monotone* if v is nondecreasing with respect to inclusion, i.e., if $v(S) \leq v(T)$ whenever $S \subseteq T$. Martínez-Legaz [164] proved that, for monotone games, it is enough to know the restriction of the indirect function to the nonnegative orthant in order to recover the characteristic function. This restriction, denoted by $\pi^+ : \mathbb{R}^n_+ \to \mathbb{R}$, is called the restricted indirect function of (N, v). Then the following formula holds

$$v(S) = \min_{w_{N\setminus S}} \pi^+\left(\mathbf{0}_S, w_{N\setminus S}\right) \quad \text{for all } S \subseteq N.$$

Also, only two properties characterize restricted indirect functions, namely (i) and (iii) with π and $\mathbb{R}^n$ replaced by π^+ and $\mathbb{R}^n_+$, respectively.

3.4 The least increment function

The notion of least increment function was introduced by Bilbao and Martínez-Legaz [13]. We recall that for a cost game (N, c), we consider the polyhedron

$$P(c) = \{x \in \mathbb{R}^n : x(S) \leq c(S) \text{ for all } S \subseteq N\},$$

and for a profit game (N, v), we define the polyhedron

$$P(v) = \{x \in \mathbb{R}^n : x(S) \geq v(S) \text{ for all } S \subseteq N\}.$$

Definition 3.4.1 *Let (N, c) be a cost game. We define its least increment function $\epsilon_c : \mathbb{R}^n \to \mathbb{R}$ by*

$$\epsilon_c(x) = \min\{x(N) - y(N) : y \in P(c),\ y \leq x\}.$$

For a profit game (N, v) we define its least increment function $\epsilon_v : \mathbb{R}^n \to \mathbb{R}$ by

$$\epsilon_v(x) = \max\{x(N) - y(N) : y \in P(v),\ y \geq x\}.$$

Example: Let $(\{1\}, c)$ be a cost game defined by $c(\emptyset) = 0$ and $c(1) = c$. The least increment function for this game is

$$\begin{aligned}
\epsilon_c(x) &= \min\{x - y : y \leq c,\ y \leq x\} \\
&= \min\{x - y : y \leq \min\{c, x\}\} \\
&= x - \max\{y : y \leq \min\{c, x\}\} \\
&= x - \min\{c, x\} \\
&= \begin{cases} x - c & \text{if } x \geq c \\ 0 & \text{if } x \leq c. \end{cases}
\end{aligned}$$

Example: Let $(\{1\}, v)$ be a profit game defined by $v(\emptyset) = 0$ and $v(1) = v$. The least increment function for this game is

$$\begin{aligned}
\epsilon_c(x) &= \max\{x - y : y \geq v,\ y \geq x\} \\
&= \max\{x - y : y \geq \max\{v, x\}\} \\
&= x - \min\{y : y \geq \max\{v, x\}\} \\
&= x - \max\{v, x\} \\
&= \begin{cases} x - v & \text{if } x \leq v \\ 0 & \text{if } x \geq v. \end{cases}
\end{aligned}$$

Proposition 3.4.1 *If (N, c) is a cost game then its least increment function satisfies $\epsilon_c(x) \geq \max\{x(S) - c(S) : S \subseteq N\}$ for all $x \in \mathbb{R}^n$. If (N, v) is a profit game then $\epsilon_v(x) \leq \min\{x(S) - v(S) : S \subseteq N\}$ for all $x \in \mathbb{R}^n$.*

Proof. The least increment function of (N, c) satisfies

$$\begin{aligned}\epsilon_c(x) &= \min\{x(N) - y(N) : y \in P(c),\ \ y \leq x\} \\ &= x(N) - \max\{y(N) : y \in P(c),\ \ y \leq x\}.\end{aligned}$$

We fix $x \in \mathbb{R}^n$. For each $S \subseteq N$ and for each $y \in P(c)$ such that $y \leq x$, we have that

$$y(N) = y(S) + y(N \setminus S) \leq c(S) + x(N \setminus S),$$

and therefore,

$$\max\{y(N) : y \in P(c),\ y \leq x\} \leq \min\left\{c(S) + x(N \setminus S) : S \subseteq N\right\}.$$

It follows now that $\epsilon_c(x) - x(N) \geq -\min_{S \subseteq N}\{c(S) - x(S)\} - x(N)$ and this gives the first result. The inequality for profit games follows similarly. □

We shall next prove a result in which equality holds and in which least increment functions and the convex (or concave) conjugate function of the cost (or profit) game coincides.

Proposition 3.4.2 *If (N, c) is a submodular cost game then, for all $x \in \mathbb{R}^n$,*

$$\epsilon_c(x) = \max\{x(S) - c(S) : S \subseteq N\} = c^\bullet(x).$$

If (N, v) is a supermodular profit game then, for all $x \in \mathbb{R}^n$,

$$\epsilon_v(x) = \min\{x(S) - v(S) : S \subseteq N\} = v^\circ(x).$$

Proof. The result for submodular games follows from the max-min equation proved in Theorem 2.6.2, that is,

$$\begin{aligned}\epsilon_c(x) &= \min\{x(N) - y(N) : y \in P(c),\ y \leq x\} \\ &= x(N) - \max\{y(N) : y \in P(c),\ \ y \leq x\} \\ &= x(N) - \min\left\{c(S) + x(N \setminus S) : S \subseteq N\right\} \\ &= \max\{x(S) - c(S) : S \subseteq N\}.\end{aligned}$$

By the equation for submodular cost games applied to $-v$, we deduce the equation for supermodular profit games. □

3.5 The Lovász extension of games

Let $N = \{1, 2, \ldots, n\}$ be a finite set of players and $v : 2^N \to \mathbb{R}$ be a cooperative game with $v(\emptyset) = 0$. To every coalition $S \subseteq N$ is associated its *characteristic vector* $\mathbf{1}_S \in \{0,1\}^n$, where $\mathbf{1}_S(i) = 1$ if $i \in S$, and $\mathbf{1}_S(i) = 0$ if $i \notin S$. This is the natural correspondence between 2^N and $\{0,1\}^n$. Through this identification, a cooperative game is a function $v : \{0,1\}^n \to \mathbb{R}$, with $v(\mathbf{0}) = 0$. Any cooperative game has a unique expression as a multilinear polynomial in n discrete variables (Hammer and Rudeanu [118]):

$$v(x) = \sum_{S \subseteq N} \left(d_S(v) \prod_{i \in S} x_i \right), \quad \text{for all } x \in \{0,1\}^n,$$

where the coefficients $d_S(v) \in \mathbb{R}$ are the dividends of the coalitions S in the cooperative game v (see Section 1.1).

We now consider an extension of a cooperative game (N, v) to a function on $\mathbb{R}^n$. We say that a non-negative function $f : 2^N \to \mathbb{R}_+$ is a *weighted chain* if the family $\mathcal{F} = \{S \subseteq N : f(S) > 0\}$ is a chain, i.e. $S \subseteq T$ or $T \subseteq S$, for every pair $S, T \in \mathcal{F}$. To every weighted chain f, we associate a non-negative vector

$$x = \sum_{S \subseteq N} f(S) \mathbf{1}_S \in \mathbb{R}^n_+,$$

called the *depth vector* of f.

This is a one-to-one correspondence: for a non-negative vector $x \in \mathbb{R}^n_+$ let $0 \leq x^1 < x^2 < \cdots < x^k$ be the components of x with different values and let $S_p = \{i \in N : x_i \geq x^p\}$. Then we define

$$f_x(S) = \begin{cases} x^p - x^{p-1}, & \text{if } S = S_p \\ 0, & \text{otherwise,} \end{cases}$$

where $x^0 = 0$. Obviously f_x is a weighted chain and its depth vector is x.

Let $v : 2^N \to \mathbb{R}$ be a cooperative game. There is a natural way of extending v to all non-negative vectors.

Definition 3.5.1 *Let $x \in \mathbb{R}^n_+$ be a non-negative vector and f_x its weighted chain, the Lovász extension of the game v is*

$$\widehat{v}(x) = \sum_{S \subseteq N} f_x(S) v(S).$$

The function $\widehat{v}$ is an extension of v because $\widehat{v}(\mathbf{1}_S) = v(S)$, for all $S \in 2^N$ and it has the following properties:

1. $\widehat{v}$ is positively homogeneous, i.e., $\widehat{v}(\lambda x) = \lambda \widehat{v}(x)$ for all $\lambda \geq 0$.

2. $\widehat{v_1 + v_2} = \widehat{v_1} + \widehat{v_2}$.

3. $\widehat{\lambda v} = \lambda \widehat{v}$ for all $\lambda \in \mathbb{R}$.

Shapley [217] introduced convex games, that is, supermodular functions such that $v(\emptyset) = 0$.

Definition 3.5.2 *A game (N, v) is called convex if for all $S, T \subseteq N$,*

$$v(S \cup T) + v(S \cap T) \geq v(S) + v(T).$$

A game (N, c) is concave if the reverse inequality holds.

The Lovász extension of a supermodular function satisfies (see Fujishige [96, Section 6.3]) the next optimization property:

$$\widehat{v}(x) = \min \{\langle x, y \rangle : y \in P(v)\},$$

where

$$P(v) = \left\{ y \in \mathbb{R}^n : \sum_{i \in S} y_i \geq v(S), \quad \text{for all } S \subseteq N \right\}.$$

The following theorem is the supermodular version given by Lovász [159] for submodular functions (see also [96, Theorem 6.13]).

Theorem 3.5.1 *A function $v : 2^N \to \mathbb{R}$ is supermodular if and only if the Lovász extension $\widehat{v}$ of v is concave.*

Let us assume a total ordering of the elements of N, defined by

$$i_1 < i_2 < \cdots < i_n.$$

Given the previous ordering C, consider the following chain of coalitions,

$$C_0 \subset C_1 \subset \cdots \subset C_{n-1} \subset C_n,$$

where $C_0 = \emptyset$ and $C_k = \{i_1, i_2, \ldots, i_k\}$, $k = 1, \ldots, n$.

Definition 3.5.3 *The marginal worth vector $a^C(v) \in \mathbb{R}^n$ with respect to the ordering C in the game v is given by $a^C_{i_k}(v) = v(C_k) - v(C_{k-1})$, $k = 1, \ldots, n$. The Weber set of the game v is the convex hull of the marginal worth vectors,*

$$Weber(v) = \text{conv}\left\{a^C(v) : C \text{ is an ordering of } N\right\}.$$

It is easy to prove that

$$\sum_{i \in C_k} a_i^C(v) = v(C_k) \quad \text{for } k = 1, \ldots, n.$$

Weber [244] showed that any game satisfies $C(v) \subseteq Weber(v)$ and Ichiishi [129] proved that if $Weber(v) \subseteq C(v)$ then v is a convex game. These results imply the following characterization of convex games.

Theorem 3.5.2 *The game* $v : 2^N \rightarrow \mathbb{R}$ *is convex if and only if we have* $C(v) = Weber(v)$.

The Lovász extension is strongly related to the greedy algorithm. We say that an ordering $i_1 < i_2 < \cdots < i_n$ is *compatible* with the vector $x \in \mathbb{R}_+^n$ if

$$x_{i_1} \geq x_{i_2} \geq \cdots \geq x_{i_n} \geq 0.$$

Theorem 3.5.3 *Let* (N, v) *be a convex game and let* $x \in \mathbb{R}_+^n$. *The marginal worth vector* $a^C(v)$ *with respect to an* x*-compatible ordering* C *satisfies*

$$\widehat{v}(x) = \min\{\langle x, y\rangle : y \in C(v)\} = \left\langle x, a^C(v)\right\rangle.$$

Proof. If v is a convex game then $a^C(v) \in C(v) \subset P(v)$ and hence we obtain the result if we prove $\langle x, y\rangle \geq \left\langle x, a^C(v)\right\rangle$, for all $y \in P(v)$. Let $y \in P(v)$, the summation by parts implies that

$$\begin{aligned}
\langle x, y\rangle &= \sum_{k=1}^{n} x_{i_k} y_{i_k} \\
&= \sum_{k=1}^{n-1} \left[\left(x_{i_k} - x_{i_{k+1}}\right) \sum_{j=1}^{k} y_{i_j}\right] + x_{i_n} \sum_{j=1}^{n} y_{i_j} \\
&\geq \sum_{k=1}^{n-1} \left(x_{i_k} - x_{i_{k+1}}\right) v(C_k) + x_{i_n} v(C_n) \\
&= \sum_{k=1}^{n} x_{i_k} \left[v(C_k) - v(C_{k-1})\right] \\
&= \sum_{k=1}^{n} x_{i_k} a_{i_k}^C(v) = \left\langle x, a^C(v)\right\rangle.
\end{aligned}$$

□

Unanimity games are considered in order to obtain a formula for the Lovász extension of v, in terms of its dividends. Recall that for any $S \subseteq N$, $S \neq \emptyset$,

$$u_S(T) = \begin{cases} 1, & \text{if } T \supseteq S \\ 0, & \text{otherwise,} \end{cases}$$

is called the S-unanimity game. Every game is a linear combination of unanimity games, $v = \sum_{S \subseteq N} d_S(v) u_S$, where the coefficients $\{d_S(v) : S \subseteq N,\ S \neq \emptyset\}$ are the dividends of the coalition S in the game v.

Thus, we have that $v(T) = \sum_{S \subseteq T} d_S(v)$ and the Möbius inversion formula for the lattice 2^N implies (see [224, p. 116]) that

$$d_S(v) = \sum_{T \subseteq S} \mu(T,S) v(T) = \sum_{T \subseteq S} (-1)^{|S|-|T|} v(T).$$

Furthermore, the dividends satisfy the following recursive formula:

$$d_S(v) = \begin{cases} 0, & \text{if } S = \emptyset \\ v(S) - \sum_{T \subset S} d_T(v), & \text{if } S \neq \emptyset. \end{cases}$$

Theorem 3.5.4 *Let (N,v) be a game with dividends $\{d_S(v) : S \subseteq N\}$. Then the Lovász extension of v satisfies, for each $x \in \mathbb{R}^n_+$,*

$$\widehat{v}(x) = \sum_{S \subseteq N} \left(d_S(v) \min_{i \in S} x_i \right).$$

Proof. The properties 2 and 3 of the Lovász extension imply that

$$v = \sum_{S \subseteq N} d_S(v) \widehat{u_S}.$$

Every unanimity game u_S is a supermodular function and we can use the optimization property showed in Theorem 3.5.3. Let $\{e_1, \ldots, e_n\}$ be the natural basis in $\mathbb{R}^n$. Thus,

$$\begin{aligned} \widehat{u_S}(x) &= \min\{\langle x, y \rangle : y \in C(u_S)\} \\ &= \min\{\langle x, e_i \rangle : i \in S\} \\ &= \min_{i \in S} x_i, \end{aligned}$$

where the second equation follows from the characterization of the core for unanimity games (see Einy and Wettstein [74]):

$$C(u_S) = Weber(u_S) = \text{conv}\{e_i : i \in S\}.$$

□

Remark 3.5.1 *Driessen and Rafels [58] have studied several formulas for the Lovász extension of k-convex games.*

We will now consider a game model of a pure exchange economy. Let N be a set of n traders and let us suppose that they participate in a market encompassing trade in m commodities. The space $\mathbb{R}^m_+$ is considered as the commodity space. Every trader $i \in N$ is characterized by means of an initial endowment vector $w^{(i)} \in \mathbb{R}^m_+$ and by a utility function $u_i : \mathbb{R}^m_+ \to \mathbb{R}$ which measures the worth, for him/her, of any bundle of commodities. The individual utility functions u_i are continuous and concave. The triple

$$\mathcal{M} = \left(N, m, \left\{w^{(i)}\right\}_{i\in N}, \{u_i\}_{i\in N}\right),$$

is called a *market.* We denote the aggregate endowment of the coalition of traders S by $w(S) = \sum_{i\in S} w^{(i)} \in \mathbb{R}^m_+$.

Then, $w(S)$ can be reallocated as a collection $\{a^{(i)} : i \in S\}$ of bundles such that each $a^{(i)} \in \mathbb{R}^m_+$ and $a(S) = \sum_{i\in S} a^{(i)} = w(S)$. We denote the set of these collections by $A(S)$.

Since the individual utility functions are continuous and $A(S)$ is a compact set, we can define a cooperative game $(N, v_{\mathcal{M}})$ as

$$v_{\mathcal{M}}(S) = \max\left\{\sum_{i\in S} u_i\left(a^{(i)}\right) \;:\; \{a^{(i)} : i \in S\} \in A(S)\right\},$$

for all $S \subseteq N$. This model is called a *market game* (see Kannai [139]), and it corresponds to the original market in a natural way.

Remark 3.5.2 *Note that a game $v : 2^N \to \mathbb{R}$ is totally balanced if for all $S \in 2^N$ and all $\mathbf{1}_S$-balanced collection, i.e., $\{\lambda_T\}_{T\subseteq N}$ with $\sum_{T\subseteq N} \lambda_T \mathbf{1}_T = \mathbf{1}_S$ and $\lambda_T \geq 0$, satisfies $\sum\limits_{T\subseteq N} \lambda_T\, v(T) \leq v(S)$.*

The following result is due to Shapley and Shubik [216].

Theorem 3.5.5 *A game is a market game if and only if it is totally balanced.*

For any totally balanced game (N, v) we define a market, called *direct market*, $\mathcal{M}_0 = \left(N, n, \left\{\mathbf{1}_{\{i\}}\right\}_{i\in N}, u\right)$, where u is the same utility function for all traders, defined by

$$u(x) = \max\left\{\sum_{T\subseteq N} \lambda_T\, v(T) \;:\; \{\lambda_T\}_{T\subseteq N} \text{ is an } x\text{-balanced collection}\right\}.$$

Every convex game is totally balanced. In this case we obtain the following property of the Lovász extension.

Theorem 3.5.6 *Let (N, v) be a totally balanced game with common utility function $u : \mathbb{R}^n_+ \to \mathbb{R}$. Then (N, v) is a convex game if and only if $u = \widehat{v}$, where $\widehat{v} : \mathbb{R}^n_+ \to \mathbb{R}$ is the Lovász extension of v.*

Proof. The utility function is the solution of the next linear programming problem,

$$u(x) = \max \left\{ \sum_{T \subseteq N} \lambda_T \, v(T) : A\lambda = x, \quad \lambda \geq 0 \right\},$$

where $\lambda = (\lambda_T)_{T \subseteq N}$, $v = (v(T))_{T \subseteq N}$ and the matrix $A = (\mathbf{1}_T(i))_{i \in N, T \subseteq N}$. The Duality Theorem 2.2.4 of linear programming implies that

$$\begin{aligned} u(x) &= \min \left\{ y^T x \, : \, y^T A \geq v^T \right\} \\ &= \min \left\{ y^T x \, : \, y(T) \geq v(T) \ \text{ for all } T \subseteq N \right\} \\ &= \min \left\{ y^T x : \, y \in P(v) \right\}. \end{aligned}$$

If the game (N, v) is convex then v is supermodular and Theorem 3.5.3 implies that $\widehat{v} = u$. Conversely, if the utility function satisfies $u = \widehat{v}$ then the Lovász extension $\widehat{v}$ is concave and Theorem 3.5.1 implies that (N, v) is a convex game. □

Chapter 4

Computational complexity

This chapter contains some basic notions of the theory of computational complexity. In this theory, *computational problems* are mathematical objects defined by parameters and properties of a solution. An *instance* of the problem is obtained when all the parameters are a set of input data. An *algorithm* describes a sequence of operations for solving a computational problem. There are often several algorithms for solving a problem and we are interested in to analyze the computational resources required to solve a certain problem. The complexity of a problem is the order of computational resources which are necessary and sufficient to solve the problem.

For a model of computation, the *computational complexity* of a problem is the minimal cost of solving the problem over all possible algorithms and the *algorithm complexity* is the cost of a particular algorithm. We say that a problem has *polynomial complexity* if its computational complexity is a polynomial in the measure of input size. The idea that polynomial time is a desirable property of computational problems was mentioned implicitly by von Neumann [243] and explicitly by Cobham [41] and Edmonds [69].

For a more detailed exposition, see the books of Bovet and Crescenzi [30], Cormen, Leiserson and Rivest [46], Gács and Lovász [101], Grötschel, Lovász and Schrijver [116, Chapter 1], Kronsjö [145], Papadimitriou [198], and Schrijver [210, Chapter 2]. A concise survey of the results that have been obtained in the last years is due to Shmoys and Tardos [220]. Blum, Cucker, Shub and Smale [23] developed a theory of real computation and complexity over the ring of integers $\mathbb{Z}$; the field of the integers mod 2, $\mathbb{Z}_2$; the rationals $\mathbb{Q}$; the reals $\mathbb{R}$; and the complex numbers $\mathbb{C}$. This theory over noncountable structures includes the classical theory of computational complexity over $\mathbb{Z}_2$ and provides a bridge between complexity theory and numerical analysis.

4.1 Models of computation

We begin the development of the complexity theory with the definition of a *model of computation,* which includes a number system, arithmetic and costs associated with arithmetic. The classical model of computation is the *binary* or *Turing machine* model. In this model, the *size* of an integer or rational instance (input) of the problem is the number of symbols needed to encode the instance and the cost of an algorithm is the number of bit operations performed by a Turing machine for obtaining a solution (output).

In the *real number* model of computation, the number system is $\mathbb{R}$ and the cost is the number of basic algebraic operations from input to output. We call the former measure *logarithmic* or *bit cost* and the latter *algebraic* or *unit cost.*

Number system	Arithmetic	Cost
$\mathbb{Z}$ or $\mathbb{Q}$	exact or approximate	logarithmic (bit)
$\mathbb{R}$	exact	algebraic (unit)

The main difference between the above models is that in the integer model the cost of an arithmetic operation depends of the size of the inputs, while in the real number model the cost of an arithmetic operation is one.

In the binary model, a positive integer p, $2^n \leq p < 2^{n+1}$, is represented by the $\{0,1\}$-vector $(\varepsilon_0, \varepsilon_1, \ldots, \varepsilon_n)$ where $p = \sum_{i=0}^{n} \varepsilon_i 2^i$ and $n \leq \log_2 p < n+1$. If $p \neq 0$ is an integer then we need one bit for the sign and $\lceil \log_2 (|p|+1) \rceil$ bits for the $\{0,1\}$-vector of the binary representation of its absolute value. Hence, the *sizes* of a rational number $r = p/q$ (with $q > 0$ and p and q coprime integers), of a rational vector $x = (x_1, \ldots, x_n)$ and of a rational $m \times n$ matrix $A=(a_{ij})$ are:

$$\begin{aligned}
\text{size}\,(r) &= 1 + \lceil \log_2 (|p|+1) \rceil + \lceil \log_2 (|q|+1) \rceil\,, \\
\text{size}\,(x) &= n + \sum_{j=1}^{n} \text{size}\,(x_j)\,, \\
\text{size}\,(A) &= mn + \sum_{i=1}^{m} \sum_{j=1}^{n} \text{size}\,(a_{ij})\,.
\end{aligned}$$

Proposition 4.1.1 *If* $A \in \mathbb{Q}^{n\times n}$ *then* $\text{size}\,(\det A) < 2\,\text{size}\,(A)\,.$

Proof. See Schrijver [210, Theorem 3.2]. □

Nemhauser and Wosley [186, Proposition 3.1] proved that the number of binary digits needed to represent extreme points of the integral polyhedron $P=\{x: \ Ax\leq b\}$ is bounded by $2n^2\log_2(nK)$, where $K=\max_{i,j}\{|a_{ij}|,|b_j|\}$. For the core of a integer valued game we have the following proposition.

Proposition 4.1.2 *Let $v: 2^N \to \mathbb{Z}$ be a balanced game and let $\overline{x}$ be an extreme point of its core $C(v)$. Then the coordinates of $\overline{x}$ satisfy, for any $j=1,\ldots,n$,*

$$\overline{x}_j=\frac{p_j}{q},\ \ 0\leq p_j\leq Kn^n,\ \ 1\leq q\leq n^n,\ \ p_j,q\in\mathbb{Z},$$

where $K=\max\{|v(S)|: S\subseteq N\}$. Moreover, $\overline{x}$ has size polynomially bounded by n and $\operatorname{size}(K)$.

Proof. Theorem 2.1.5 implies that $\overline{x}\in C(v)$ is an extreme point of $C(v)$ if and only if $x(S')=v(S')$ for some n-subsystem of the system of inequalities which describes $C(v)$. By Cramer's rule we obtain that

$$\overline{x}_j=\frac{\det M_j}{\det M}=\frac{p_j}{q}\quad\text{for}\quad j=1,\ldots,n$$

where M is the $\{0,1\}$-matrix which columns are the incidence vectors $\mathbf{1}_{S'}$ of the sets S' in the n-subsystem and M_j results from M by replacing its j-th column by $(v(S'))$. The entries of M and M_j are $0,1$ or integers (exactly in one column) bounded in absolute value by K, and the determinant is a sum of $n!<n^n$ factors. Hence $1\leq q\leq n^n$ and $0\leq p_j\leq Kn^n$, for all $j=1,\ldots,n$.

By Proposition 4.1.1, $\det M$ has size less than $4n^2$ and $\det M_j$ has size less than $2n\operatorname{size}(K)+4\left(n^2-n\right)$ for $j=1,\ldots,n$. □

4.2 The complexity classes P and NP

Throughout the remainder of this chapter, we will assume the bit cost model. Karp in the book of Blum *et al.* [23] writes:

> "Complexity theory groups problems into *complexity classes* such as P, the class of problems that can be solved in polynomial time (i.e., in a number of steps bounded by a polynomial function of the length of the input), and NP, the class of problems for which solutions can be verified in polynomial time. Solvability in polynomial time can be equated roughly with computational tractability, and accordingly there is great interest in determining whether particular problems lie in P."

Suppose $\Sigma = \{0,1\}$ and let Σ^∞ the set of strings (finite, but unbounded sequences) over Σ. The size of a string is the number of its components. If we have an procedure to encode each input and output of a problem as strings of Σ^∞, then a problem is a subset $\Pi \subseteq \Sigma^\infty \times \Sigma^\infty$.

A *search problem* Π is: For each instance (input) $x \in \Sigma^\infty$, find a solution (output) $y \in \Sigma^\infty$ such that $(x,y) \in \Pi$, or to declare that no such solution y exists. A *decision problem* Π is: For each instance $x \in \Sigma^\infty$, exactly one of $(x,0) \in \Pi$ or $(x,1) \in \Pi$ is true, and if $(x,y) \in \Pi$ then $y \in \{0,1\}$. In this context, 0 stands for `no`, and 1 for `yes`.

An *algorithm* is a finite list of instructions to solve a problem in a finite number of steps. An algorithm can be encoded as a string $A \in \Sigma^\infty$. We say that A solves problem Π if, for any instance x of Π, when we give the string (A,x) to a universal Turing machine, then, after a finite number of iterations, the machine halts and we have a string y which is a solution of the problem, that is $(x,y) \in \Pi$.

The time complexity function $f : \mathbb{Z}_+ \to \mathbb{Z}_+$ of an algorithm A is the maximal number $f(n)$ of iterations of a universal Turing machine makes before halting, taken over all inputs of size n.

Let f and g be functions from $\mathbb{Z}_+$ to $\mathbb{Z}_+$. We write $f(n) = \mathrm{O}(g(n))$, in words *f is of the order of g*, if there are positive integers c and n_0 such that $f(n) \leq cg(n)$ for all $n \geq n_0$. We write $f(n) = \Omega(g(n))$ if the opposite happens, that is, $g(n) = \mathrm{O}(f(n))$. If f and g have exactly the same rate of growth, then we write $f(n) = \Theta(g(n))$. For instance, if $p(n)$ is a polynomial of degree d, then $p(n) = \Theta(n^d)$. The above $\mathrm{O}\Omega\Theta$-notation was proposed by Knuth [143].

Definition 4.2.1 *An algorithm is called polynomial-time if its time complexity function f satisfies $f(n) = O(p(n))$, for some polynomial p.*

Definition 4.2.2 *The class of decision problems that can be solved in polynomial time is called the class P.*

The Euclidean algorithm for finding the greatest common divisor of two integers and the Gaussian elimination for solving linear equations, are polynomial-time algorithms. The Fourier-Motzkin elimination algorithm for linear inequalities is not polynomial-time (see [210, page 156]).

Example: Linear algebra.

Consider the linear system $Ax = b$, where A is an $n \times n$ nonsingular matrix. This problem has $n^2 + n$ input data and n output data. In the real number model, the cost of the *Gaussian elimination* algorithm is $\mathrm{O}(n^3)$ arithmetic operations. Therefore, the computational complexity of the linear system problem is at most $\mathrm{O}(n^3)$. For the integer model, let L be the total number of digits used to represent the input data. Edmonds [70] described an algorithm so that the numerators and the denominators of the solutions can be computed with polynomial cost in n and L. The above results imply the polynomial complexity of the following problems in linear algebra:

- Computing the rank of a matrix.
- Computing the determinant of a matrix.
- Computing the inverse of a nonsingular matrix.
- Testing whether a finite set of vectors are linearly independent.

Example: Linear inequalities.

Linear programming can be reduced to the solution of m linear inequalities in n variables $Ax \leq b$. A classical algorithm for this computational problem, the *simplex method*, is not a polynomial algorithm, but this method has been strongly successful in the solution of real-world instances. In contrast, the first polynomial algorithm discovered for this problem, the *Kachiyan ellipsoide algorithm*, appears to be impracticality slow. In the integer model, Khachiyan [141] proved that the complexity of the linear inequalities problem is a polynomial in n, m, and L.

The linear programming feasibility and optimization problems over $\mathbb{Q}$ can be solved in polynomial time. Some *interior point methods* (see Megiddo [175]) have been proven to have polynomial complexity for rational linear programming. Blum *et al.* [23, Chapter 15] proved that this problem is in P over $\mathbb{Q}$ using the *barrier method* (see also Gonzaga [108]).

In the real number model, Traub and Wozniakowski [236] established that the ellipsoide algorithm is not a polynomial algorithm:

> "The fact that the number of steps in the ellipsoide algorithm increases as L (which of course is true in either model) causes the difficulty when the algorithm is used in numerical computations."

In fact, it is an open problem whether the linear inequalities problem has polynomial complexity in the real number model (Blum *et al.* [23, page 275]).

Example: Cost allocation games.

Let $N = \{1, \ldots, n\}$ be a set of users, let $T = (N \cup \{0\}, A)$ an arborescence with root 0 and let $w : A \rightarrow \mathbb{R}_+$ a nonnegative arc-weight function. Megiddo [173] introduced the following *cost allocation tree game* which generalizes the *airport game* (tree which consists of a single path) analyzed by Littlechild and Owen [156], and Littlechild [157]. For every $S \subseteq N$ the cost $c(S)$ of any coalition S is the total cost of arcs that belongs to some path from the root 0 to a node $i \in S$, that is

$$c(S) = \sum_{a \in P_S} w(a),$$

where P_S is a path from 0 to some $i \in S$.

Megiddo showed that the Shapley value can be found in $O(n)$ operations and the nucleolus is computed within $O(n^3)$ operations. This cost allocation problem is a special case of the min-cost spanning tree game defined in page 10. Granot and Granot [110] generalized the Megiddo's tree game with the *fixed cost spanning forest game.* They proved that, in some instances of the problem, core points can be generated in strongly polynomial time, i.e., its computational complexity is polynomial in the real number and Turing machine models. Furthermore, the nucleolus can be computed, in some instances, by a strongly polynomial algorithm.

Example: The minimum spanning tree problem.

Given a complete graph $G = (V, E)$ with order $|V| = n$, size $|E| = m$, and a weigth function $w : E \to \mathbb{Q}$, find a spanning tree of minimum weigth. The greedy algorithm is a polynomial-time algorithm for this problem, since spanning trees form the bases of the graphic matroid of a graph. We note that since weigths may be negative, the maximum spanning tree problem is included in this problem. Kruskal [146] and Prim [202] designed the classical polynomial-time algorithms of complexity $O(mn)$ and $O(n^2)$ respectively. Gabow, Galil, Spencer and Tarjan [99] described a fastest method to solve this problem. Fredman and Tarjan [93] showed that the complexity of Prim's algorithm, using a data structure called a *Fibonacci heap*, is $O(m + n \log_2 n)$. Kruskal's algorithms requires a *union-find* data structure studied in Lovász, Shmoys and Tardos [160, Section 3] and implemented in Skiena [221, Section 6.2].

Example: The weighted matching problem.

Given a graph $G = (V, E)$ and a weigth function $w : E \to \mathbb{Q}_+$, find a matching of maximum weight. Recall that the matching polytope P is determined by the following inequalities:

$$\begin{aligned} x(e) &\geq 0, \quad \text{for all } e \in E, \\ x(\delta(v)) &\leq 1, \quad \text{for all } v \in V, \\ x(\gamma(S)) &\leq \frac{|S|-1}{2}, \quad \text{for all } S \subseteq V,\ |S| \geq 3 \text{ and } |S| \text{ odd.} \end{aligned}$$

This problem can be described as the linear program

$$\max\left\{w^T \mathbf{1}_M : M \text{ matching in } G\right\} = \max\left\{w^T x : x \in P\right\}.$$

Although this linear programming formulation has a number of inequalities that are exponential in the size of the data, an optimal dual solution can be described by at most $|E|$ inequalities and this problem is in P.

Example: The symmetric traveling salesman problem (Lawler *et al.* [154]).

Given n cities $1, \ldots, n$, and a nonnegative rational distance d_{ij} between any two cities i and j (we assume that $d_{ij} = d_{ji}$ for all i and j), find a tour of the cities as small as possible. We can formulate this problem as follows. Given a complete graph $G = (V, E)$ with vertex set $V = \{1, \ldots, n\}$ and a symmetric distance matrix $(d_{ij}) \in \mathbb{Q}_+^{n \times n}$, find a Hamiltonian cycle in G with minimum length. We can solve this problem by enumeration of the $\frac{1}{2}(n-1)!$ tours, computing the distances of each, and taking the best. However, the number of tours grows exponentially with the size of the graph and the algorithm is not polynomial-time.

There are no polynomial-time algorithms known for the traveling salesman problem and we might suppose that no polynomial-time algorithm exists for this problem (unless P = NP).

The class of decision problems solvable in polynomial time is denoted by P. Another complexity class, denoted by NP, includes most combinatorial optimization problems. The class NP can be described as the class of those decision problems Π with the property that for any instance x of Π that has a positive answer, there is a *certificate* from which the correctness of this answer has a proof of length polynomially bounded by the size of x.

The letters NP stand for *nondeterministic polynomial time* and *not* for non-polynomial time. We will now make the following formal definition.

Definition 4.2.3 *The class NP consists of those decision problems* Π *for which there exists a decision problem* Π' *in P and a polynomial* p*, such that for each input* $x \in \Sigma^\infty$:

$$(x, 1) \in \Pi \iff \exists y \in \Sigma^\infty \text{ such that } ((x, y), 1) \in \Pi' \text{ and } \operatorname{size}(y) \leq p(\operatorname{size}(x)).$$

Note that the only requirement is the existence of y with the mentioned properties, but it is not required that we are able to find y in polynomial time. The string y can be seen as a certificate for x to show that $(x, 1) \in \Pi$.

The decision problem obtained by negating the question of a decision problem Π is the complementary problem $\overline{\Pi}$, defined by $\overline{\Pi} = \{(x, 1 - y) : (x, y) \in \Pi\}$.

Definition 4.2.4 *For any class* C *of decision problems, let co*C *denote the class of decision problems that are complementary to problems in* C.

The definition of the class P is symmetric with respect to the `yes/no` question $(x, 1) \in \Pi$ or $(x, 0) \in \Pi$, so that P = coP. In this respect, NP is very different and the new class coNP can be used with the class NP to classify decision problems with `yes/no` answers.

Example: Hamiltonian graph problem (Garey and Johnson [102]).

Given a graph $G = (V, E)$, we ask whether G contains a Hamiltonian cycle. The decision problem $\Pi = \{G : G \text{ is Hamiltonian}\}$ belongs to the class NP, since the problem

$$\Pi' = \{(G, H) : G \text{ is a graph and } H \text{ is a Hamiltonian cycle}\}$$

is in P. For this we use that H is a Hamiltonian cycle if it has n vertices of degree 2 and H is connected, and each of these properties is proved in polynomial time. In fact, it is unknown whether the Hamiltonian graph problem is in coNP.

Note that coP $\subseteq$ coNP implies P $\subseteq$ NP $\cap$ coNP. The class NP $\cap$ coNP consists of those decision problems for which both a positive answer and a negative answer to an instance x have a proof of length polynomially bounded by the size of x. That is, the class NP $\cap$ coNP consists of all decision problems Π for which there exist decision problems Π', Π'' in P and a polynomial p, such that for each $x \in \Sigma^\infty$:

$$\begin{aligned}(x,1) &\in \Pi \Longleftrightarrow \exists\, y \in \Sigma^\infty \text{ such that } ((x,y),1) \in \Pi' \text{ and } \operatorname{size}(y) \le p(\operatorname{size}(x))\\ (x,0) &\in \Pi \Longleftrightarrow \exists\, z \in \Sigma^\infty \text{ such that } ((x,z),1) \in \Pi'' \text{ and } \operatorname{size}(z) \le p(\operatorname{size}(x))\end{aligned}$$

Edmonds [69] called problems in the class NP $\cap$ coNP *well-characterized.* A good characterization for a well-characterized problem is the following type of theorem:

$$\exists\, y : ((x,y),1) \in \Pi' \Longleftrightarrow \forall z : ((x,z),0) \in \Pi'', \tag{4.1}$$

where Π' and Π'' satisfy, for a certain polynomial p :

$$\begin{aligned}\text{If } ((x,y),1) &\in \Pi' \text{ then } ((x,y'),1) \in \Pi' \text{ for some } y' \text{ with } \operatorname{size}(y') \le p(\operatorname{size} x)\\ \text{If } ((x,z),1) &\in \Pi'' \text{ then } ((x,z'),1) \in \Pi'' \text{ for some } z' \text{ with } \operatorname{size}(z') \le p(\operatorname{size} x)\end{aligned}$$

Many deep theorems in combinatorics (Kuratowski's, Menger's, König's, Tutte's Matching) are good characterizations.

Example: Fundamental Theorem of Linear Algebra.

An example of a good characterization is the *first Fredholm theorem*:

Let A be an $m \times n$ matrix and let $b \in \mathbb{R}^m$. Then the system $Ay = b$ has a solution if and only if $z^T b = 0$ holds for each vector $z \in \mathbb{R}^m$ with $z^T A = 0$.

The problem $\Pi = \{(A, b) : Ax = b \text{ has a solution}\}$ is in NP $\cap$ coNP, i.e., it is well-characterized, since the associated problems

$$\begin{aligned}\Pi' &= \{((A,b),y) : Ay = b\},\\ \Pi'' &= \{((A,b),z) : z^T A = 0 \text{ and } z^T b \neq 0\},\end{aligned}$$

satisfy the above conditions and Fredholm theorem is equivalent to formula 4.1.

Most well-characterized problems have been proved to be solvable in polynomial time. A interesting exception is the PRIME NUMBER: *Given an integer n, is n prime?*

Theorem 4.2.1 PRIME NUMBER *belongs to class NP ∩ coNP.*

Proof. See *Testing primality* in Bovet and Crescenzi [30, pages 135–139]. □

At the present, it is not known whether there is a polynomial algorithm for PRIME NUMBER in the bit cost model. If a problem Π is in P, then we can take the empty string as a certificate for any instance x to show that Π is in NP. The most important open problem in computer science today is whether the classes P and NP are equal. In fact, we have the following three open problems:

1. Is P = NP?
2. Is NP = coNP?
3. Is P = NP ∩ coNP?

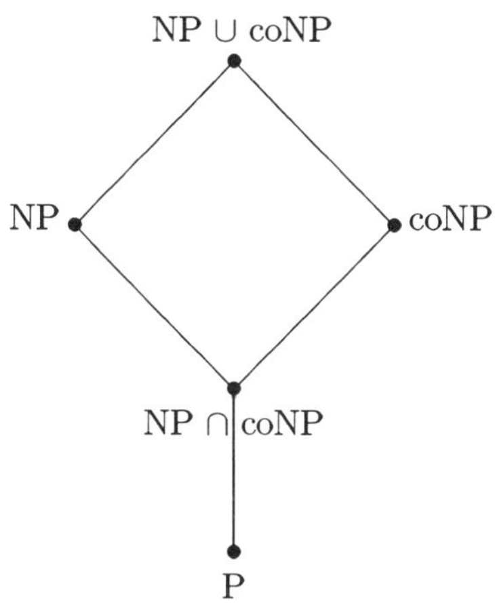

Figure 4.1 Inclusions between complexity classes

Note that if P = NP, then P = coNP, and hence all classes drawn coincide. According to the following maxim handed by Trefethen [237]:

> "The two most important unsolved problems in mathematics are the Riemann hypothesis and P $\stackrel{?}{=}$ NP. Of the two, it is the latter whose solution will have the greater impact."

4.3 NP-completeness and #P-completeness

The notion of the NP-complete problem was introduced by S. A. Cook [43] who proved the NP-completeness of the SATISFIABILITY problem of propositional calculus (see W. J. Cook *et al.* [45, Section 9.7]). Karp [140] proved that 21 decision problems from several areas of discrete mathematics were also NP-complete. The most important encyclopedia on NP-completeness is *Computers and Intractability: A Guide to the Theory of NP-Completeness* by Garey and Johnson [102]. The basic idea of the NP-completeness concept and the P versus NP question are well summarized by Karp [23, Foreword]:

> "A major open question is whether the classes P and NP are equal; it would be surprising if they were, since our experience suggests that checking a solution is much easier than finding one. Further evidence against the equality of P and NP is the plethora of problems that lie in NP but are not known to lie in P. A problem in NP is called NP-*complete* if every problem in NP can be reduced to it in polynomial-time. The classes P and NP are equal if and only if the NP-complete problems lie in P. This fact is taken as evidence that the NP-complete problems are intractable."

Definition 4.3.1 *A decision problem Π is called polynomial-time reducible to decision problem Π_1 if there is a polynomial-time algorithm A such that for any input $x \in \Sigma^\infty$ we obtain an output $y = A(x)$ and $(x, 1) \in \Pi \Longleftrightarrow (y, 1) \in \Pi_1$.*

We shall denote this reduction by $\Pi \propto \Pi_1$. Clearly, $\propto$ is both reflexive and transitive. If $\Pi \propto \Pi_1$ then Π_1 is at least as *hard* as Π. Note that any decision procedure for Π_1 can be easily converted into a decision composite procedure for Π. For example, if Π_1 is in P then also Π is in P, since the composition procedure is polynomial time. Similarly, if Π_1 belongs to NP, and $\Pi \propto \Pi_1$, then also Π belongs to NP.

Definition 4.3.2 *A decision problem Π_1 is NP-complete if Π_1 is in NP and for all Π in NP, $\Pi \propto \Pi_1$. A decision problem Π_1 is coNP-complete if Π_1 is in coNP and for all Π in coNP, $\Pi \propto \Pi_1$.*

The following results follow from the transitive property of the reduction $\propto$.

1. P = NP if and only if there exists a NP-complete problem Π in P.

2. If Π is NP-complete, Π_1 is in NP and $\Pi \propto \Pi_1$, then Π_1 is NP-complete.

The first result says that any NP-complete problem characterizes the P versus NP question. The second result gives a strategy to prove the NP-completeness of a problem Π_1, provides that a first NP-complete problem is known. This way is:

(i) Show that Π_1 belongs to NP.

(ii) Find a problem Π already known to be NP-complete such that $\Pi \propto \Pi_1$.

Example: The k-minimum spanning tree problem.
Given a complete graph $G = (V, E)$ and a weigth function $w : E \to \mathbb{Q}_+$, and a positive integer $k \leq n$, find a tree that spans at least k vertices of G and has minimum weigth. Since all weigths are nonnegative, there exists an optimal tree which spans exactly k vertices. This problem is NP-complete if $d : E \to \{1, 2, 3\}$, but there is a polynomial-time algorithm if $d : E \to \{1, 2\}$ (see Ravi *et al.* [205]). Dudás, Klinz and Woeginger [61] studied easy and difficult cases of the k-minimum spanning tree problem in graded matrices.

Starting with SATISFIABILITY and using this strategy, thousands of decision problems have been shown to be NP-complete. The HAMILTONIAN GRAPH problem, the TRAVELING SALESMAN problem, the STABLE SET problem (given a graph G and a bound k, decide if there exist k pairwise non-adjacent vertices in G), the CLIQUE problem (decide if there exists a k-clique in G), and the NODE COVER problem (decide if there exists a set of k vertices such that every edge is incident on a vertex in the set) are NP-complete.

Proposition 4.3.1 *Let Π_1 be a decision problem that belongs to NP $\cap$ coNP. If Π_1 is NP-complete or coNP-complete, then NP = coNP.*

Proof. Suppose that Π_1 is NP-complete. Let us consider any problem Π in NP. Since Π_1 is NP-complete, we have $\Pi \propto \Pi_1$. The same reduction shows that the complementary problems satisfy $\overline{\Pi} \propto \overline{\Pi_1}$. By hypothesis $\overline{\Pi_1}$ is in NP and the composition procedure implies that $\overline{\Pi}$ is in NP and hence Π is in coNP.

Conversely, for any Π in coNP we have that $\overline{\Pi}$ is in NP and then $\overline{\Pi} \propto \Pi_1$. This reduction implies $\Pi \propto \overline{\Pi_1}$ and since $\overline{\Pi_1}$ is in NP, we obtain that Π is in NP. We can use a similar argument if Π_1 is coNP-complete. □

An *optimization problem* A is given by the set of instances or input objects, the set of feasible solutions or output objects, and the measure defined for any feasible solution. This means that for each instance x we have a set $F(x)$ of feasible solutions, and for each such solution $s \in F(x)$ we have a positive integer measure $m(s)$. The function opt mapping an instance x to the measure of an optimal solution, that is, $\text{opt}(x) = \min\{m(s) : s \in F(x)\}$ or $\text{opt}(x) = \max\{m(s) : s \in F(x)\}$.

Let A be an optimization problem whose corresponding decision problem is an NP-complete problem. Then A is not solvable in polynomial time (unless P = NP). In this case, we are looking for approximate solutions computable in polynomial time.

Definition 4.3.3 *Let T be an algorithm that, for any instance x of A, returns a feasible solution $T(x) \in F(x)$. Given $\varepsilon \geq 0$, we say that T is an ε-approximate algorithm for A if, for any instance x, we have*

$$\frac{|m(T(x)) - \mathrm{opt}(x)|}{\max\{\mathrm{opt}(x), m(T(x))\}} \leq \varepsilon.$$

An algorithm is ε-approximate if the relative error of the approximate solution is bounded by ε, which satisfies $0 \leq \varepsilon \leq 1$. For maximization problems, any ε-approximate solution $T(x)$ satisfies $(1-\varepsilon)\,\mathrm{opt}(x) \leq m(T(x))$. For minimization problems, $(1-\varepsilon)\, m(T(x)) \leq \mathrm{opt}(x)$.

If A is a NP-complete optimization problem, then we shall be interested in obtaining the infimum of all $\varepsilon > 0$ such that there is a polynomial time ε-approximate algorithm for A. See Papadimitiou [198, Chapter 13] for examples and results for NP-complete optimization problems. An extensive list of optimization problems and their approximability status is collected by Crescenzi and Kann [47].

Let $G = (V \cup W, E)$ be a bipartite graph, with vertex sets $V = \{v_1, \ldots, v_n\}$ and $W = \{w_1, \ldots, w_n\}$. Recall that a set of edges $M \subseteq E$ is a matching, if every vertex of G is incident with at most one edge from M. A matching is called perfect if every vertex is of G is incident with exactly one edge from M, that is $|M| = n$. The corresponding decision problem asks whether a bipartite graph has a perfect matching and belongs to P. We now consider the *counting problem* of computing the number of different perfect matchings in G. The adjacency matrix $A = (a_{ij})$ of the bipartite graph G is the $n \times n$ matrix defined by $a_{ij} = 1$ if $v_i w_j \in E$, and 0 otherwise. The permanent of A is defined by $\mathrm{perm}(A) = \sum_\pi a_{1\pi(1)} a_{2\pi(2)} \cdots a_{n\pi(n)}$, and this function is precisely the number of perfect matchings in G, counting problem which is known as PERMANENT.

A counting problem usually arises from a search problem and asks how many solutions exist, that is, given an instance of the search problem, what is the cardinality of the solution set? We define a new complexity class by using the number of certificates for the instances of a problem.

Definition 4.3.4 *The class #P consists of those problems for which there exists a decision problem Π' in P and a polynomial p, such that for each input $x \in \Sigma^\infty$, the only acceptable output is*

$$z = |y \in \Sigma^\infty : ((x,y), 1) \in \Pi' \text{ and } \mathrm{size}(y) \leq p(\mathrm{size}(x))|.$$

Note that the problem of counting perfect matchings in a bipartite graph is in #P. The notion of *#P-completeness* was introduced by Valiant [238] who proved that the PERMANENT problem for $\{0,1\}$-matrices is #P-complete. This made it possible to prove more #P-completeness results. An important example is the problem of computing the volume of a polytope by Dyer and Frieze [64].

4.4 Complexity in cooperative game theory

Let $v : 2^N \to \mathbb{Q}$ be a cooperative game with rational worths. The coalition description of v is given by its 2^n coalitional values $\{v(S) : S \subseteq N\}$. If the complexity is measured in the size of these values, then the input is exponential in n, and thus, the most of the complexity questions would become easy. Therefore, let us start with games (N, v) with a description polynomial in n, where n is the number of players.

Example: A *weighted voting game* $[q;\ w_1, \ldots, w_n]$, where w_i is the number of votes of each player i, is a description polynomial in n of the game v, defined by

$$v(S) = \begin{cases} 1, & \text{if } w(S) \geq q \\ 0, & \text{if } w(S) < q, \end{cases}$$

where $w(S) = \sum_{i \in S} w_i$. We will denote these games by wVG.

Example: Deng and Papadimitriou [53] studied computational complexity of cooperative solution concepts for the following game. Given a graph $G = (N, E)$ with weight $w : E \to \mathbb{Z}$, the *weighted graph game* $v_G : 2^N \to \mathbb{Z}$ is given by

$$v_G(S) = \sum_{e \in G[S]} w(e),$$

and the class of these games is denoted by wGG. The game v_G evaluates the profit $v_G(S)$ of any coalition S as the profit of the subgraph of G induced by the vertex set S. We can use it for the fair division between n cities of the income from a communication network connecting them. Since $|E| \leq \frac{1}{2}n(n-1)$, this game has a description polynomial in n.

Example: Nagamochi, Zeng, Kabutoya and Ibaraki [185] consider a matroid $(E, \mathcal{M})$ with a weight function $w : E \to \mathbb{Q}$. The *minimum base game* $c : 2^E \to \mathbb{Q}$ is given by

$$c(S) = \min \left\{ \sum_{e \in B(S)} w(e) : B(S) \text{ is a basis of } S \right\},$$

and we denoted by MBG the class of these games. If the matroid $(E, \mathcal{M})$ is the forest matroid of a graph $G = (N, E)$, then the *minimum forest game* is

$$c(S) = \min \left\{ \sum_{e \in F_S} w(e) : F_S \text{ is a maximal forest on } S \right\},$$

for all $S \subseteq E$ and the class is MFG.

For the last game, we define the encoding length or size of a matroid $(E, \mathcal{M})$ as the number of its independents sets $|\mathcal{M}|$ and we assume that the matroid is given by an *oracle* that testing if a given set $S \subseteq E$ is independent or not. In this context, a matroid on $|E|$ elements can be represented by a string of length $2^{|E|}$, where each character indicates whether a subset $S \subseteq E$ is independent.

In this model, a minimum base $B(S)$ on S can be obtained, in polynomial time, by the greedy algorithm as follows:

Order $S = \{e_1, \ldots, e_s\}$ so that $c(e_1) \leq \cdots \leq c(e_t) < 0 \leq \cdots \leq c(e_s)$;
$B(S) \leftarrow \emptyset$;
For $i = 1$ to s do
If $B(S) \cup \{e_i\} \in \mathcal{M}$ then $B(S) \leftarrow B(S) \cup \{e_i\}$.

In order to study the complexity of *cooperative game problems*, we assume that a cooperative game $v : 2^N \to \mathbb{Q}$ is given by an oracle returning $v(S)$ for each query $S \subseteq N$. For this oracle we have a polynomial p_v such that for every input of size at most n, the answer of the oracle has size at most $p_v(n)$. Therefore, we know an upper bound $\beta \geq \max\{\text{size}(f(S)) : S \subseteq N\}$.

Definition 4.4.1 *Let $v : 2^N \to \mathbb{Q}$ be a game given by an oracle with upper bound β. The size of v is defined by* $\text{size}(v) = |N| + \beta$.

Definition 4.4.2 *We say that a problem for a game $v : 2^N \to \mathbb{Q}$ given by an oracle, is solvable in oracle-polynomial time if the number of computational steps, counting each call to the oracle as one step, is polynomially bounded in* $\text{size}(v)$.

Example: A *min-cost spanning tree game* is defined by a set $N = \{1, \ldots, n\}$ of players, a supply vertex 0, a complete graph $G = K_{n+1} = (N \cup \{0\}, E)$ and a nonnegative edge-weight function $w : E \to \mathbb{Q}_+$. The cost $c(S)$ of any coalition $S \subseteq N$, is defined by

$$c(S) = \min\left\{\sum_{e \in T_S} w(e) : T_S \text{ is a spanning tree of } G[S \cup \{0\}]\right\}.$$

The class of these games is denoted by *MSTG.*

Remark 4.4.1 *We know that the greedy algorithm is a polynomial-time algorithm for the min-cost spanning tree problem. Therefore, the oracle is any algorithm that computes $c(S)$ in polynomial time with respect to $|N| + \text{size}(w)$, for all $S \subseteq N$.*

We consider now several cooperative game problems and we give a list of complexity results.

Problem: MEMBER CORE.

- Instance: A game $v : 2^N \to \mathbb{Q}$ with size (v), and $x \in I(v) \cap \mathbb{Q}^N$.
- Question: Is x an element of the core of v?

Good News: For the class wGG, Deng and Papadimitriou [53] proved that it is in P if the weight function w is nonnegative. If v is convex (or concave) then it can be solved in oracle-polynomial time by the greedy algorithm (Grötschel, Lovász and Schrijver [116, Section 10.2]).

Bad News: coNP-complete for the classes wGG and $MSTG$ (Deng and Papadimitriou [53] and Faigle, Kern, Fekete and Hochstättler [81]).

Comments: Deng and Papadimitriou used reduction from MAX-CUT and Faigle *et al.* [81] used reduction from EXACT COVER BY 3-SETS (Papadimitriou [198, page 201]). It is still open the complexity for the classes MBG andMFG.

Problem: CORE.

- Instance: A game $v : 2^N \to \mathbb{Q}$ with size (v).
- Question: Is the core of v empty?

Good News: For wGG with nonnegative weights and MFG with no all-negative circuit is in P. For MBG such that the underlying matroid has no all-negative circuits it is solvable in oracle-polynomial time (Deng and Papadimitriou [53] and Nagamochi *et al.* [185]).

Bad News: NP-complete for wGG (Deng and Papadimitriou [53]).

Problem: SUBADDITIVITY.

- Instance: A game $c : 2^N \to \mathbb{Q}$ with size (v).
- Question: Is the game c subadditive?

Good News: Nagamochi *et al.* [185] obtained an $O(\text{size}(v))$ time algorithm for the class MFG and proved that it is solvable in oracle linear time for MBG.

Problem: SUBMODULARITY.

- Instance: A game $c : 2^N \to \mathbb{Q}$ with size (v).
- Question: Is the game c submodular?

Good News: It is solvable in oracle-polynomial time for the class MBG and in $O(\text{size}(v))$ time for MFG (Nagamochi *et al.* [185]).

Problem: SHAPLEY VALUE.

- Instance: A game $v : 2^N \to \mathbb{Q}$ with size (v).
- Solution: The Shapley value $\Phi(v)$.

Good News: Megiddo [173] showed that it can be found in $O(n)$ time for *MSTG*, in the special case where the underlying graph is a tree. For *wGG*, Deng and Papadimitriou [53] proved that it can be computed in $O(n^2)$.

Bad News: For the class *wVG*, Deng and Papadimitriou [53] proved that is #P-complete, that is, as difficult as any counting problem in NP. Nagamochi *et al.* [185] proved that is #P-complete for the class *MFG* and that there is no oracle-polynomial time algorithm for the class *MBG*.

Comments: Deng and Papadimitriou [53] used the counting problem associated with KNAPSACK (Papadimitriou [198, page 202]). Nagamochi *et al.* [185] used the #P-completeness of GRAPH RELIABILITY (Valiant [239]).

Problem: TIJS VALUE.

- Instance: A game $v : 2^N \to \mathbb{Q}$ with size (v).
- Solution: The Tijs value $\tau(v)$.

Good News: For the class *MFG*, Nagamochi *et al.* [185] proved that it can be computed in $O(m^2 + mn \log_2 n)$ time, where $n = |N|$ and $m = |E|$.

Bad News: For the class *MBG*, there is no oracle-polynomial time algorithm.

Comments: Nagamochi *et al.* [185] used the Dijkstra's algorithm implemented by Fredman and Tarjan [93] (see also Lovász *et al.* [160, Theorem 3.3]).

Problem: NUCLEOLUS.

- Instance: A game $v : 2^N \to \mathbb{Q}$ with size (v).
- Solution: The nucleolus $\eta(v)$.

Good News: Granot and Granot [110] and Granot, Maschler, Owen and Zhu [111] showed that it can be found in strongly polynomial time for a subclass of *MSTG*. Solymosi and Raghavan [222] provides an $O(n^4)$ algorithm for the class of assignments games. For the class *wGG*, Deng and Papadimitriou [53] proved that it can be computed in $O(n^2)$. If v is convex (or concave) then it can be solved in oracle-polynomial time (Kuipers [150] and Faigle, Kern and Kuipers [84]). The $\mathcal{B}$-nucleolus can be calculated in strongly polynomial time for games with size $(\mathcal{B})$ polynomially bounded in n (Granot, Granot and Zhu [112], Kuipers, Solymosi and Aarts [151], Solymosi, Aarts and Driessen [223]).

Bad News: NP-hard, that is, at least as difficult as any optimization problem in NP, for the class *MSTG* (Faigle, Kern and Kuipers [85]).

4.5 Project: Locally concave (convex) games

Megiddo [174] was the first to show that linear programming problems can be solved in linear time when the number of variables d is fixed. The complexity of Megiddo's LP algorithm, in the unit cost model, is $O\left(2^{2^d} n\right)$, where n is the number of constraints. Clarkson [38] and Dyer [63] presented algorithms with complexity $O\left(3^{d^2} n\right)$ and Dyer [63] showed that Megiddo's approach is $\Omega(d!\,n)$. Clarkson [39] improves these bounds using *Las Vegas algorithms* (always compute the correct answer, but its time complexity is only *expected* time complexity) for solving LP and ILP problems.

Several other problems in computational geometry, such as the distance between convex polyhedra, can be efficiently solvable by a randomized algorithm. To describe this type of problems, Amenta [4], Gärtner [105], Matoušek, Sharir and Welzl [171], Sharir and Welzl [219] have formulated a theory of LP-type problems, also known as *Generalized Linear Programming.*

Definition 4.5.1 *A generalized linear program (GLP) is a pair* (H, w), *where* H *is a finite set of constraints, and* $w : 2^H \to (P, \leq)$ *is an objective function valued on a totally ordered set, with the following properties:*

1. $F \subseteq G \subseteq H$ *implies* $w(F) \leq w(G)$.
2. *For all* $F \subset H$ *and all* $p, q \in H$,
$$w(F) = w(F \cup p) = w(F \cup q) \;\; \textit{implies} \;\; w(F) = w(F \cup \{p, q\}).$$

Property 2 is known as *local submodularity.* Another example is the rank function of a greedoid. This function is nondecreasing and locally submodular as the following theorem states.

Theorem 4.5.1 *A function* $r : 2^N \to \mathbb{Z}_+$ *is the rank function of a greedoid if and only if for all* $X, Y \subseteq N$ *and all* $x, y \in N$ *the following conditions hold:*

1. $r(\emptyset) = 0$,
2. $r(X) \leq |X|$,
3. $X \subseteq Y$ *implies* $r(X) \leq r(Y)$,
4. $r(X) = r(X \cup x) = r(X \cup y)$ *implies* $r(X) = r(X \cup \{x, y\})$.

Proof. See Korte, Lovász and Schrader [144, Chapter V, Theorem 1.1]. □

We will now define a new class of cooperative games and investigate its properties.

Definition 4.5.2 *A game (N,c) is locally concave if the function c is locally submodular, i.e., for all coalitions $S \subset N$ and all players $i,j \in N \setminus S$,*

$$c(S) = c(S \cup i) = c(S \cup j) \text{ implies } c(S) = c(S \cup \{i,j\}).$$

A game (N,v) is locally convex if the function v is locally supermodular, i.e., for all nonempty coalitions $S \subseteq N$ and all players $i,j \in S$,

$$v(S) = v(S \setminus i) = v(S \setminus j) \text{ implies } v(S) = v(S \setminus \{i,j\}).$$

Recall that if (N,c) is a game then its dual game (N,c^*) is defined by $c^*(S) = c(N) - c(N \setminus S)$ for all $S \subseteq N$. The following results are immediate:

(i) c is nondecreasing if and only if c^* is nondecreasing.

(ii) (N,c) is concave (convex) if and only if (N,c^*) is convex (concave).

Proposition 4.5.2 *(N,c) is locally concave (convex) if and only if its dual game (N,c^*) is locally convex (concave).*

Proof. Suppose first that the function c is locally submodular, and that

$$c^*(S) = c^*(S \setminus i) = c^*(S \setminus j)$$

for all nonempty $S \subseteq N$ and all $i,j \in S$. By definition, we have

$$c(N) - c(N \setminus S) = c(N) - c(N \setminus (S \setminus i)) = c(N) - c(N \setminus (S \setminus j)).$$

If we take $T = N \setminus S$, then $T \cup i = N \setminus (S \setminus i)$ and $T \cup j = N \setminus (S \setminus j)$. Hence $c(T) = c(T \cup i) = c(T \cup j)$ holds and the local submodularity of c implies $c(T) = c(T \cup \{i,j\})$. Further, since $c(N \setminus S) = c(N \setminus (S \setminus \{i,j\}))$ we obtain $c^*(S) = c^*(S \setminus \{i,j\})$. An argument similar shows the converse. The result for convexity follows from $(c^*)^* = c$. □

Proposition 4.5.3 *Let (N,c) be a nondecreasing concave (convex) game. Then (N,c) is locally concave (convex).*

Proof. Let $S \subseteq N$ and $i,j \in N \setminus S$ such that $c(S) = c(S \cup i) = c(S \cup j)$. Since c is concave it follows that

$$c(S \cup \{i,j\}) \leq c(S \cup i) + c(S \cup j) - c(S) = c(S).$$

The game c is nondecreasing, so that $c(S) = c(S \cup \{i,j\})$.

Suppose that $v : 2^N \to \mathbb{R}$ is a nondecreasing convex game. Let $S \subseteq N$ and $i,j \in S$ such that $v(S) = v(S \setminus i) = v(S \setminus j)$. The convexity of v implies

$$v(S \setminus \{i,j\}) \geq v(S \setminus i) + v(S \setminus j) - v(S) = v(S). \tag{4.2}$$

Since v is nondecreasing, we obtain $v(S) = v(S \setminus \{i,j\})$. □

There is a stronger local condition which we describe in the next result.

Proposition 4.5.4 *Let (N,c) and (N,v) be nondecreasing games. Then*

1. *(N,c) is locally concave if and only if for all $S,T \subseteq N$*

$$c(S) = c(T) = c(S \cap T) \text{ implies } c(S) = c(T) = c(S \cup T).$$

2. *(N,v) is locally convex if and only if for all $S,T \subseteq N$,*

$$v(S) = v(T) = v(S \cup T) \text{ implies } v(S) = v(T) = v(S \cap T).$$

Proof. 1. Clearly the local condition implies local submodularity. Conversely, let $S,T \subseteq N$ such that $c(S) = c(T) = c(S \cap T)$. Since c is nondecreasing, the above condition gives

$$\begin{aligned} c(S \cap T) &= c((S \cap T) \cup i) = c(S) \text{ for all } i \in S \setminus T, \\ c(S \cap T) &= c((S \cap T) \cup j) = c(T) \text{ for all } j \in T \setminus S. \end{aligned}$$

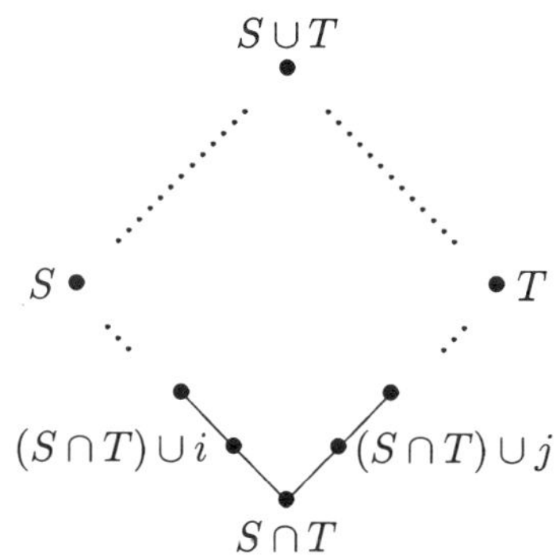

By repeatedly applying the local submodularity of c we deduce

$$c(S \cap T) = c((S \cap T) \cup \{i,j\}) = \cdots = c(S \cup T).$$

2. The proof is the same as 1. □

Definition 4.5.3 *A game (N,v) is zero-monotone if $v(S) - \sum_{i \in S} v(i)$ is nondecreasing, i.e., for all $T \subseteq S \subseteq N$,*

$$v(T) - \sum_{i \in T} v(i) \le v(S) - \sum_{i \in S} v(i).$$

The zero-monotonicity is equivalent to

$$v(T) + \sum_{i \in S \setminus T} v(i) \leq v(S)$$

for all $T \subseteq S \subseteq N$. Note that superadditivity implies zero-monotonicity.

Proposition 4.5.5 *Let (N, v) be a convex game such that $v(i) \geq 0$ for all $i \in N$. Then v is locally convex.*

Proof. Consider $S \subseteq N$ and $i, j \in S$ with $v(S) = v(S \setminus i) = v(S \setminus j)$. From (4.2) we deduce that $v(S) \leq v(S \setminus \{i, j\})$. Since convexity implies zero-monotonicity, we have

$$\begin{aligned} v(S \setminus i) + v(i) &\leq v(S) = v(S \setminus i), \\ v(S \setminus j) + v(j) &\leq v(S) = v(S \setminus j), \end{aligned}$$

and therefore $v(i) = v(j) = 0$. Further, zero-monotonicity implies

$$v(S \setminus \{i, j\}) = v(S \setminus \{i, j\}) + v(i) + v(j) \leq v(S),$$

and hence $v(S \setminus \{i, j\}) = v(S)$. □

Every rank function of a greedoid which is not a matroid is an example of locally concave game which is not concave. We introduce a class of games in order to study the equivalence of these concepts.

Definition 4.5.4 *An integer game $c : 2^N \to \mathbb{Z}$ has the unit marginal worth property if for all $S \subset N$ and $i \in N$, $c(S \cup i) - c(S) \in \{0, 1\}$.*

Theorem 4.5.6 *Let $c : 2^N \to \mathbb{Z}$ be an integer game which satisfies the unit marginal worth property. Then the following statements are equivalent:*

1. *(N, c) is locally concave.*

2. *c is the rank function of the matroid $\mathcal{M} = \{S \subseteq N : c(S) = |S|\}$.*

3. *(N, c) is nondecreasing and concave.*

Proof. $1 \Rightarrow 2$. The function $c : 2^N \to \mathbb{Z}$ satisfies the following three conditions:

(i) $c(\emptyset) = 0$,

(ii) For all $S \subset N$ and $i \in N$, the unit marginal worth property implies

$$0 \leq c(S \cup i) - c(S) \leq 1 \Longleftrightarrow c(S) \leq c(S \cup i) \leq c(S) + 1,$$

(iii) c is locally submodular.

Therefore, the statement 2 follows from the axiomatization given by Korte *et al.* [144, Chapter II, Theorem 1.3].

$2 \Rightarrow 3$. Follows from Korte *et al.* [144, Chapter II, Theorem 1.4].

$3 \Rightarrow 1$. Proposition 4.5.3 implies the result. □

Corollary 4.5.7 *Let $v : 2^N \to \mathbb{Z}$ be an integer game which satisfies the unit marginal worth property. Then the following statements are equivalent:*

1. *(N, v) is locally convex.*
2. *The function $r^d(S) = |S| - v(S)$ is the rank function of the matroid $\mathcal{M}^d = \{S \subseteq N : v(S) = 0\}$.*
3. *(N, v) is nondecreasing and convex.*

Proof. $1 \Rightarrow 2$. The dual game v^* is locally concave, so that v^* is the rank function of the matroid $\mathcal{M} = \{S \subseteq N : v^*(S) = |S|\}$. The rank function r^d of the dual matroid $\mathcal{M}^d$ (see Welsh [245, Section 5]) is given by

$$r^d(S) = |S| - v^*(N) + v^*(N \setminus S) = |S| - v(S).$$

Hence, the dual matroid is

$$\mathcal{M}^d = \left\{S \subseteq N : r^d(S) = |S|\right\} = \{S \subseteq N : v(S) = 0\}.$$

$2 \Rightarrow 3$ and $3 \Rightarrow 1$ are immediate consequences of the equivalences of the main theorem and the identity $(v^*)^* = v$. □

Corollary 4.5.8 *Let $v : 2^N \to \{0, 1\}$ be a game which is nondecreasing (simple game). Then the following statements are equivalent:*

1. *(N, v) is convex.*
2. *(N, v) is locally convex.*
3. *The collection of winning coalitions $\mathcal{W} = \{S \subseteq N : v(S) = 1\}$ is closed under intersection.*
4. *There exists a unique circuit (minimal dependent set) of the matroid $\mathcal{M}^d$.*
5. *$v = u_T$, where T is the unique circuit of the matroid $\mathcal{M}^d$.*

Note that the intersection of the winning coalitions is the minimal winning coalition of the convex simple game v, i.e., the unique circuit of the matroid $\mathcal{M}^d$. A simple game with these properties is called an *oligarchic game* (see van Deemen [240, Theorem 5.4]).

Chapter 5

Restricted games by partition systems

In this chapter, *partition systems* and their corresponding *restricted games* are introduced. We obtain the relationship between the core of a cooperative game and the core of the restricted game. The convexity is not transmitted, in a general way, to the corresponding restricted game. *Partition convex geometries* are defined as a special class of partition systems, in which the convexity is hereditary.

There have been two main lines in the literature on restricted games: the first line was started by Myerson [182] who studied graph-restricted games, and the second one was initiated by Kaneko and Wooders [138], Le Breton, Owen and Weber [155] and Faigle [77]. We provide a unified way of looking at the literature and establish connections between the two different lines on restricted games. The strength and advantages of this unified approach becomes clear in the study of the inheritance of the convexity from the game to the restricted game where an interesting result by van den Nouweland and Borm [187] on the convexity of graph-restricted games is turned into a direct consequence of the corresponding result by Faigle [77].

In section 5.3, we present several results of the core of restricted games, obtained by Algaba in her Ph. D. Thesis [1]. Next, the recursive procedure for computing the Hart and Mas-Colell potential of these games and the relation between the dividends of Harsanyi in the restricted game and the worths in the original game were proved by López [158].

The properties of partition convex geometries are used to obtain formulas for the Shapley and Banzhaf values of the players in the restricted game $v^{\mathcal{L}}$, in terms of the original game v (see Bilbao [10]). Finally, we consider the Owen multilinear extension for the restricted game.

5.1 Partition systems

In this section, a general cooperation structure is considered, which is an extension of the graph-communication structure. Recall that a set system on a finite ground set N is a pair $(N, \mathcal{F})$, with $\mathcal{F} \subseteq 2^N$. The sets belonging to $\mathcal{F}$ are called *feasible coalitions*. For any $S \subseteq N$, maximal feasible subsets of S are called *components* of S.

Definition 5.1.1 *A partition system is a set system $(N, \mathcal{F})$ satisfying:*

(P1) *$\emptyset \in \mathcal{F}$, and $\{i\} \in \mathcal{F}$ for every $i \in N$, and*

(P2) *for all $S \subseteq N$, the components of S, denoted by $\Pi(S) = \{T_1, \ldots, T_p\}$, form a partition of S.*

Proposition 5.1.1 *A set system $(N, \mathcal{F})$ which satisfies property (P1) is a partition system if and only if $F \in \mathcal{F}$, $G \in \mathcal{F}$ and $F \cap G \neq \emptyset$ imply $F \cup G \in \mathcal{F}$.*

Proof. Suppose $F \cup G \notin \mathcal{F}$ for some pair $\{F, G\} \subset \mathcal{F}$ with $F \cap G \neq \emptyset$. Then, there exist two components $\{T_1, T_2\}$ of $F \cup G$, such that $T_1 \supseteq F$ and $T_2 \supseteq G$. Hence, $T_1 \cap T_2 \supseteq F \cap G \neq \emptyset$, which contradicts property (P2).

Conversely, if $(N, \mathcal{F})$ satisfies property (P1), then $S = \bigcup_{i \in S}\{i\}$ for every $S \subseteq N$. Let $\Pi(S) = \{T_1, \ldots, T_p\}$ be the family of components of S. If $\Pi(S)$ is not a partition of S, then $T_i \cap T_j \neq \emptyset$, and hence $T_i \cup T_j \in \mathcal{F}$, which contradicts the maximality of T_i and T_j. □

Example: The following collections of subsets of N, given by $\mathcal{F} = 2^N$ and $\mathcal{F} = \{\emptyset, \{1\}, \ldots, \{n\}\}$, are the maximal and minimal partition systems.

Example: In a sequencing situation there is a queue, consisting of n customers waiting to be served at a counter. Curiel, Pederzoli and Tijs [49] introduced *sequencing games* (N, v), defined by $v(S) := \sum \{v(T) : T \in \Pi(S)\}$, where $v(T)$ is equal to the maximal cost savings the coalition can obtain by rearranging their positions in the queue. The components of $\Pi(S)$ are the maximal intervals of S in a total order on N. If (N, v) is a sequencing game then $(N, \leq)$ is a chain and the collection $\mathcal{F} = \{T \subseteq N : T \text{ is an interval of } N\}$, is a partition system.

Example: In a communication structure given by a graph $G = (N, E)$, the set system $(N, \mathcal{F})$ with $\mathcal{F} = \{S \subseteq N : (S, E(S)) \text{ is a connected subgraph of } G\}$, is a partition system.

Example: A hypergraph communication structure is a triple $(N, \mathcal{H}, v)$, where (N, v) is a game and $\mathcal{H} \subseteq 2^N$. The idea of modelling communication by means of *conferences* $H \in \mathcal{H}$ is due to Myerson [183] and van den Nouweland, Borm and Tijs [188].

5.2 $\mathcal{F}$-restricted games

Definition 5.2.1 *Let (N,v) be a game and let $(N,\mathcal{F})$ be a partition system. The $\mathcal{F}$-restricted game $\left(N,v^{\mathcal{F}}\right)$, is defined by $v^{\mathcal{F}}(S)=\sum\{v(T):T\in\Pi(S)\}$, where $\Pi(S)$ is the collection of the components of $S\subseteq N$.*

If $S\in\mathcal{F}$ then $v^{\mathcal{F}}(S)=v(S)$. Note that if the partition system $(N,\mathcal{F})$ is defined by a communication structure, the corresponding restricted game is called a *graph-restricted* game. The map $L_{\mathcal{F}}:\Gamma^N\to\Gamma^N$, $v\mapsto v^{\mathcal{F}}$, is a linear operator on the vector space of all cooperative games (N,v).

Remark 5.2.1 *If $(N,\mathcal{F})$ is the partition system given by the connected subgraphs of the graph G, then the game $v^{\mathcal{F}}$ is a G-component additive game which are studied by Owen [195] and Potters and Reijnierse [201].*

Every game is a linear combination of unanimity games,

$$v=\sum_{T\subseteq N}d_T(v)u_T,\quad d_T(v)=\sum_{S\subseteq T}(-1)^{|T|-|S|}v(S),$$

where $d_T(v)$ is the *dividend* of T in game v. The linearity implies that

$$v^{\mathcal{F}}=\sum_{T\subseteq N}d_T(v)u_T^{\mathcal{F}},\tag{5.1}$$

where game $u_T^{\mathcal{F}}$ satisfies

$$u_T^{\mathcal{F}}(S)=\sum_{F\in\Pi(S)}u_T(F)=\begin{cases}1, & \text{if there exists } F\in\mathcal{F} \text{ such that } T\subseteq F\subseteq S\\ 0, & \text{otherwise.}\end{cases}$$

Owen [195, Theorems 2 and 3] gave the following result: *The unanimity games u_T, where T is connected in the graph G, form a basis of the graph-restricted games.* We shall obtain a similar property for partition systems.

Theorem 5.2.1 *If $(N,\mathcal{F})$ is a partition system, then the unanimity games $\{u_T:T\in\mathcal{F},T\neq\emptyset\}$ form a basis of the vector space $\left\{\left(N,v^{\mathcal{F}}\right):v\in\Gamma^N\right\}$, i.e. $v^{\mathcal{F}}=\sum_{T\in\mathcal{F}}d_T\left(v^{\mathcal{F}}\right)u_T$ where $d_{\emptyset}\left(v^{\mathcal{F}}\right)=0$.*

Proof. If $\left(N,v^{\mathcal{F}}\right)$ is a restricted game, then it is uniquely determined by the values $\{v(S):S\in\mathcal{F},\ S\neq\emptyset\}$. Thus, the vector space of these games will be identified with $\mathbb{R}^{|\mathcal{F}|-1}$. The unanimity game $u_T^{\mathcal{F}}=u_T$ if and only if $T\in\mathcal{F}$, hence the subset $\{u_T:T\in\mathcal{F},\ T\neq\emptyset\}$ contains $|\mathcal{F}|-1$ games of the type $v^{\mathcal{F}}$ and it is linearly independent. Therefore, it is a basis. □

5.3 The core of restricted games

We observe that if the set system $(N, \mathcal{F})$ satisfies that $N \in \mathcal{F}$, then for every game (N, v) we have that $v(N) = v^{\mathcal{F}}(N)$.

Theorem 5.3.1 *Let $(N, \mathcal{F})$ be a partition system and let (N, v) be a profit game such that $v(N) = v^{\mathcal{F}}(N)$. Then the following statements hold:*

(a) $C\left(v^{\mathcal{F}}\right) = \{x \in \mathbb{R}^n : x(N) = v(N),\ \ x(S) \geq v(S) \text{ for all } S \in \mathcal{F}\}.$

(b) $C(v) \subseteq C\left(v^{\mathcal{F}}\right).$

(c) *If game (N, v) is balanced then so is the restricted game $\left(N, v^{\mathcal{F}}\right)$.*

Proof. (a) Let $x \in C\left(v^{\mathcal{F}}\right)$, then $x(N) = v^{\mathcal{F}}(N) = v(N)$ and $x(S) \geq v^{\mathcal{F}}(S)$, for all $S \subseteq N$. If $S \in \mathcal{F}$ then $x(S) \geq v^{\mathcal{F}}(S) = v(S)$. It remains to prove the inverse inclusion. Indeed, let $S \subseteq N$, and $\{S_k\}_{k \in K}$ be the collection of its components. Therefore,

$$x(S) = \sum_{i \in S} x_i = \sum_{k \in K}\left(\sum_{i \in S_k} x_i\right) = \sum_{k \in K} x(S_k) \geq \sum_{k \in K} v(S_k) = v^{\mathcal{F}}(S).$$

The statements (b) and (c) follow from (a). □

In general, $C(v) \subset C\left(v^{\mathcal{F}}\right)$. The condition $v(N) = v^{\mathcal{F}}(N)$ is also necessary in order to establish the above inclusion, because otherwise $C(v) \cap C\left(\tilde{v}^{\mathcal{F}}\right) = \emptyset$.

Theorem 5.3.2 *Let $(N, \mathcal{F})$ be a partition system. If (N, v) is totally balanced then so is $\left(N, v^{\mathcal{F}}\right)$.*

Proof. We have to show that for all nonempty $S \subseteq N$, the induced subgames $\left(S, v_S^{\mathcal{F}}\right)$ are balanced. If $S \in \mathcal{F}$, then the subgame (S, v_S) is balanced and $v(S) = v^{\mathcal{F}}(S)$. Theorem 5.3.1 implies that the game $\left(S, v_S^{\mathcal{F}}\right)$ is balanced. If $S \notin \mathcal{F}$, let $\Pi(S) = \{S_1, S_2, \ldots, S_k\}$ be the partition of S in their components, then by hypothesis $C\left(v_{S_t}\right) \neq \emptyset$, for all $t = 1, \ldots, k$.

To prove that the subgame $\left(S, v_S^{\mathcal{F}}\right)$ is balanced, it is sufficient to show that the core is nonempty. Indeed, if we consider $x^{S_1} \in C\left(v_{S_1}\right), \ldots, x^{S_k} \in C\left(v_{S_k}\right)$, then we have that every element $i \in S$ belongs to a unique component of S. Hence, if $i \in S_p \subset S$, $S_p \in \Pi(S)$, we can associate $i \mapsto x_i^{S_p}$, where $x_i^{S_p}$ is the component of the player i in the vector $x^{S_p} \in C\left(v_{S_p}\right)$. We take the vector $y \in \mathbb{R}^{|S|}$ such that $y_i = x_i^{S_p}$. This vector belongs to the core of the induced subgame $\left(S, v_S^{\mathcal{F}}\right)$ since

$$\begin{aligned} y(S) &= \sum_{i\in S} y_i = \sum_{i\in S} x_i^{S_p} = \sum_{S_p\in\Pi(S)} \left(\sum_{i\in S_p} x_i^{S_p}\right) \\ &= \sum_{S_p\in\Pi(S)} x^{S_p}(S_p) = \sum_{S_p\in\Pi(S)} v(S_p) = v_S^{\mathcal{F}}(S). \end{aligned}$$

If $T \subseteq S$ then $v_S^{\mathcal{F}}(T) = \sum_{k=1}^{h} v(T_k)$, where $\Pi(T) = \{T_1, T_2, \ldots, T_h\}$. Given $T_k \in \Pi(T)$ there exists a unique j such that $T_k \subseteq S_j$ and $x^{S_j} \in C\left(v_{S_j}\right)$. Therefore,

$$v(T_k) = v_{S_j}(T_k) \le x^{S_j}(T_k) = \sum_{i\in T_k} x_i^{S_j}.$$

From this, we conclude that

$$v_S^{\mathcal{F}}(T) = \sum_{k=1}^{h} v(T_k) \le \sum_{k=1}^{h} \left(\sum_{i\in T_k} x_i^{S_j}\right) = \sum_{i\in T} x_i^{S_j} = \sum_{i\in T} y_i = y(T).$$

□

Let $\mathcal{F} \subseteq 2^N$ be a family of coalitions and let (N, v) be a game. Faigle [77] introduced the family $\widetilde{\mathcal{F}}$ of the coalitions that can be expressed as disjoint unions of elements of $\mathcal{F}$ and the game $\tilde{v}^{\mathcal{F}}$. That is, if $A \in \widetilde{\mathcal{F}}$ then there exist $A_i \in \mathcal{F}$, $i = 1, \ldots, p$ such that $A = A_1 \cup \ldots \cup A_p$, and $A_i \cap A_j = \emptyset$ for $i \neq j$. The game $\tilde{v}^{\mathcal{F}} : \widetilde{\mathcal{F}} \to \mathbb{R}$ is given by

$$\tilde{v}^{\mathcal{F}}(A) = \max\left\{\sum_{i\in I} v(A_i) : \{A_i\}_{i\in I} \text{ is a } \mathcal{F}\text{-partition of } A\right\}.$$

The main difference between the definition given by Faigle and the Kaneko and Wooders' partitioning games is that in general $\widetilde{\mathcal{F}} \neq 2^N$. If the family $(N, \mathcal{F})$ is a partition system then $\widetilde{\mathcal{F}} = 2^N$. As we already have noticed, if the game (N, v) is not superadditive, in general, $v^{\mathcal{F}} \neq \tilde{v}^{\mathcal{F}}$. However, the equality between the core of the game $\left(N, \tilde{v}^{\mathcal{F}}\right)$ and $\left(N, v^{\mathcal{F}}\right)$ can be stated.

Proposition 5.3.3 *Let $(N, \mathcal{F})$ be a partition system and let (N, v) be a profit game. If $v(N) = v^{\mathcal{F}}(N) = \tilde{v}^{\mathcal{F}}(N)$, then*

$$C\left(v^{\mathcal{F}}\right) = C\left(\tilde{v}^{\mathcal{F}}\right) = \{x \in \mathbb{R}^n : x(N) = v(N),\ \ x(S) \ge v(S) \ \text{ for all } S \in \mathcal{F}\}.$$

Proof. Let $x \in C\left(\tilde{v}^{\mathcal{F}}\right)$, then $x(N) = \tilde{v}^{\mathcal{F}}(N) = v(N)$ and $x(S) \geq \tilde{v}^{\mathcal{F}}(S)$, for all $S \subseteq N$. If $S \in \mathcal{F}$ then $\{S\}$ is a $\mathcal{F}$-partition of S, hence $x(S) \geq \tilde{v}^{\mathcal{F}}(S) \geq v(S)$. It remains to prove the inverse inclusion. Indeed, let $S \subseteq N$, and let $\{S_j\}_{j \in J}$ be a $\mathcal{F}$-partition of S. Therefore,

$$x(S) = \sum_{i \in S} x_i = \sum_{j \in J} \left(\sum_{i \in S_j} x_i \right) = \sum_{j \in J} x(S_j) \geq \sum_{j \in J} v(S_j),$$

and it follows that

$$x(S) \geq \max \left\{ \sum_{i \in I} v(T_i) : \{T_i\}_{i \in I} \text{ is a } \mathcal{F}\text{-partition of } S \right\} = \tilde{v}^{\mathcal{F}}(S).$$

Finally, we apply Theorem 5.3.1 (a). □

Moreover, Kaneko and Wooders [138] and Faigle and Kern [82] obtained the following characterization by using integer linear programming.

Theorem 5.3.4 *Let $(N, \mathcal{F})$ be a partition system and let (N, v) be a profit game such that $v(N) = v^{\mathcal{F}}(N) = \tilde{v}^{\mathcal{F}}(N)$. The game $\left(N, \tilde{v}^{\mathcal{F}}\right)$ is balanced if and only if the following linear program has an integral optimal solution*

$$\max \left\{ \sum_{S \in \mathcal{F} \setminus \{\emptyset\}} y_S \, v(S) : y \geq 0,\ My = \mathbf{1}_N \right\},$$

where M is the incidence matrix of the collection $\mathcal{F} \setminus \{\emptyset\}$.

Proof. The Duality Theorem 2.2.4 implies

$$\begin{aligned}
\min \left\{ \mathbf{1}_N^T x : M^T x \geq v \right\} &= \min \{ x(N) : x(S) \geq v(S) \text{ for all } S \in \mathcal{F} \setminus \{\emptyset\} \} \\
&= \max \left\{ \sum_{S \in \mathcal{F} \setminus \{\emptyset\}} y_S \, v(S) : y \geq 0,\ My = \mathbf{1}_N \right\} \\
&\geq \max_{y \text{ integral}} \left\{ \sum_{S \in \mathcal{F} \setminus \{\emptyset\}} y_S \, v(S) : y \geq 0,\ My = \mathbf{1}_N \right\} \\
&= \max \left\{ \sum_{i \in I} v(S_i) : \{S_i\}_{i \in I} \text{ is a } \mathcal{F}\text{-partition of } N \right\} \\
&= \tilde{v}^{\mathcal{F}}(N).
\end{aligned}$$

If $x \in C\left(\tilde{v}^{\mathcal{F}}\right)$ then $\tilde{v}^{\mathcal{F}}(N) = v(N) \geq x(N) = \max \left\{ \mathbf{1}_N^T x : M^T x \geq v \right\}$, so that the inequality holds with equality and therefore the linear program in the statement of the theorem has an integral solution.

Conversely, if the linear program has an optimal integer solution then the above equations imply that $\max\{\mathbf{1}_N^T x : M^T x \geq v\} = \tilde{v}^{\mathcal{F}}(N) = v(N)$, and hence every optimal solution belongs to $C(\tilde{v}^{\mathcal{F}})$. □

Corollary 5.3.5 (Kuipers) *Let $(N, \mathcal{F})$ be a partition system and let (N, v) be a game such that $v(N) = v^{\mathcal{F}}(N) = \tilde{v}^{\mathcal{F}}(N)$. If the incidence matrix of $\mathcal{F} \setminus \{\emptyset\}$ is perfect, then $C(\tilde{v}^{\mathcal{F}}) \neq \emptyset$.*

Proof. Since the family of the nonempty essential coalitions is $\mathcal{F} \setminus \{\emptyset\}$, it follows from Corollary 2.4.4. □

Corollary 5.3.6 *Let $(N, \mathcal{F})$ be a partition system and let (N, v) be a superadditive game such that $v(N) = v^{\mathcal{F}}(N) = \tilde{v}^{\mathcal{F}}(N)$. If the incidence matrix of $\mathcal{F} \setminus \{\emptyset\}$ is balanced, then the game $(N, \tilde{v}^{\mathcal{F}})$ is totally balanced.*

Proof. Proposition 2.4.6 implies that every subgame has a nonempty core. □

Example: Let (N, v) be a game where the set of players is $N = \{1, 2, 3, 4\}$ and the worth function is $v(S) = |S| - 1$ if $S \neq \emptyset$, $v(\emptyset) = 0$. Note that v is a supermodular function. Let G be the graph diagrammed in Figure 5.1 and let $\mathcal{F}$ be the family of connected subgraphs of G. Then $(N, \mathcal{F})$ is a partition system. Notice that v is convex while $v^{\mathcal{F}}$ is not convex. If we take $A = \{1, 2, 3\}$ and $B = \{2, 3, 4\}$ then $v^{\mathcal{F}}(A \cup B) + v^{\mathcal{F}}(A \cap B) = 3 \not\geq v^{\mathcal{F}}(A) + v^{\mathcal{F}}(B) = 4$.

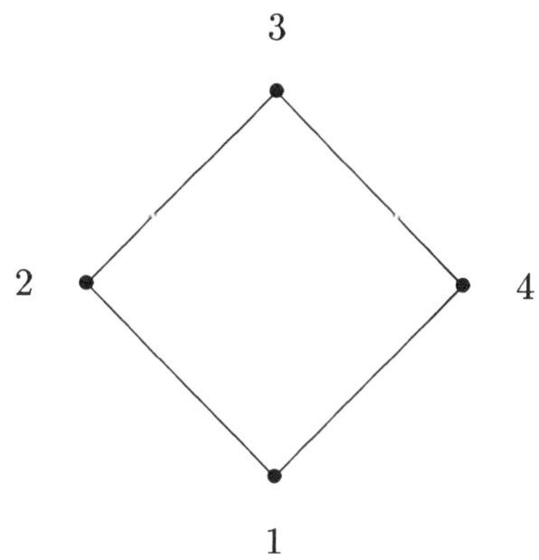

Figure 5.1 The four cycle

Van den Nouweland and Borm [187, Theorem 1], showed that if (N, v) is a convex game and graph G is cycle-complete (block graph), then the corresponding restricted game is convex. We will show that the convexity of the game (N, v) is transmitted to the restricted game when we consider a special kind of partition system.

Grötschel *et al.* [116] introduced the following concept.

Definition 5.3.1 *A set system $(N, \mathcal{F})$ is called an intersecting family if for all $S, T \in \mathcal{F}$ with $S \cap T \neq \emptyset$ we have $S \cap T \in \mathcal{F}$ and $S \cup T \in \mathcal{F}$.*

Proposition 5.1.1 implies that if an intersecting family $\mathcal{F}$ is *atomic*, that is $\{i\} \in \mathcal{F}$, for all $i \in N$ and $\emptyset \in \mathcal{F}$, then $(N, \mathcal{F})$ is a partition system which is stable for any intersections of the coalitions.

Faigle [77, Lemma 11] proved that if $\mathcal{F}$ is an intersecting family and the game v satisfies $v(A) + v(B) \leq v(A \cup B) + v(A \cap B)$, for all $A \in \mathcal{F},\ B \in \mathcal{F}$, then $\tilde{v}^{\mathcal{F}}(A) + \tilde{v}^{\mathcal{F}}(B) \leq \tilde{v}^{\mathcal{F}}(A \cup B) + \tilde{v}^{\mathcal{F}}(A \cap B)$, for all $A \in \widetilde{\mathcal{F}},\ B \in \widetilde{\mathcal{F}}$.

In the following theorem, we consider a special partition system, which is stable for the intersection in order to obtain a result very closed to the one in Faigle [77].

Theorem 5.3.7 *Let $(N, \mathcal{F})$ be an atomic intersecting family such that $\emptyset \in \mathcal{F}$. If (N, v) is convex then the restricted game $\left(N, v^{\mathcal{F}}\right)$ is convex.*

Proof. The family $(N, \mathcal{F})$ is a partition system which satisfies $\widetilde{\mathcal{F}} = 2^N$. We deduce that for all $S \subseteq N$,

$$\begin{aligned} v^{\mathcal{F}}(S) &= \sum_{T \in \Pi(S)} v(T) = \max\left\{\sum_{i \in I} v(S_i) : \{S_i\}_{i \in I} \text{ is a partition of } S\right\} \\ &= \tilde{v}^{\mathcal{F}}(S), \end{aligned}$$

since v is superadditive. The result follows from Faigle's Lemma. □

Definition 5.3.2 *A partition convex geometry is a convex geometry $(N, \mathcal{L})$ which satisfies properties (P1) and (P2).*

If $(N, \mathcal{L})$ is a partition convex geometry then it is an intersecting family and a partition system. Therefore, it is possible to consider the corresponding restricted game $v^{\mathcal{L}}(S) = \sum_{T \in \Pi(S)} v(T)$, where $\Pi(S)$ is the partition of S in maximal convex sets (we call convex components).

Recall that a graph G is a *block graph* if every block is a complete graph. Every block graph is a finite union of connected block graphs. Therefore, it is a finite union of atomic intersecting families that contain the empty set and, hence, an atomic intersecting system that contains the empty set. Thus, if (N, v) is a convex game and G is a block graph, then Theorem 5.3.7 implies that the graph-restricted game is convex.

5.4 Hart and Mas-Colell potential

The potential function for cooperative games was defined by Hart and Mas-Colell [123] [124]. Dragan [56] applied this approach to obtain a recursive formula for the Banzhaf value, and Ortmann [193] introduced the concept of *conservative value* by using properties of the conservative vector fields. Given a game (N, v) and a coalition $S \subseteq N$, the *subgame* (S, v) is obtained by restricting v to 2^S. Let Γ denote the set of all games. The *potential* is a function $P : \Gamma \to \mathbb{R}$ which assigns to each game (N, v) a real number $P(N, v)$ and satisfies the following equations

$$P(\emptyset, v) = 0, \quad P(S, v) = \frac{1}{|S|}\left[v(S) + \sum_{i \in S} P(S \setminus i, v)\right], \ S \in 2^N \setminus \{\emptyset\}. \quad (5.2)$$

Then the *marginal contribution* of i coincides with its Shapley value:

$$P(N, v) - P(N \setminus i, v) = \Phi_i(N, v) \text{ for all } i \in N.$$

Moreover, there are two explicit formulas for the potential (see [123]),

$$\begin{aligned} P(N, v) &= \sum_{S \subseteq N} \frac{d_S(v)}{|S|}, \\ P(N, v) &= \sum_{S \subseteq N} \frac{(s-1)!(n-s)!}{n!} v(S), \end{aligned}$$

where $n = |N|$ and $s = |S|$.

Definition 5.4.1 *Let $(N, \mathcal{F})$ be a partition system. The $\mathcal{F}$-restricted potential of the game (N, v) is defined by $P\left(N, v^{\mathcal{F}}\right)$.*

The recursive procedure defined by the formula (5.2) implies an algorithm for computing $P\left(N, v^{\mathcal{F}}\right)$. A new algorithm, in terms of v, is stated in the next theorem.

Theorem 5.4.1 *Let $(N, \mathcal{F})$ be a partition system. Then $P\left(N, v^{\mathcal{F}}\right)$ satisfies:*

$$\begin{aligned} P\left(S, v^{\mathcal{F}}\right) &= \frac{1}{|S|}\left[v(S) + \sum_{i \in S} P\left(S \setminus i, v^{\mathcal{F}}\right)\right], \quad \textit{for all } S \in \mathcal{F}. \\ P\left(S, v^{\mathcal{F}}\right) &= \sum \left\{P\left(S_k, v^{\mathcal{F}}\right) : S_k \in \Pi(S)\right\}, \quad \textit{for all } S \notin \mathcal{F}. \end{aligned}$$

Proof. If $S \in \mathcal{F}$, then $v^{\mathcal{F}}(S) = v(S)$. Let $S \notin \mathcal{F}$. It follows from [123, Proposition 1] that

$$P\left(S, v^{\mathcal{F}}\right) = \sum_{T \subseteq S} \frac{d_T\left(v^{\mathcal{F}}\right)}{|T|}.$$

By Theorem 5.2.1 we know that $d_T\left(v^{\mathcal{F}}\right) = 0$ unless $T \in \mathcal{F}$, hence

$$P\left(S, v^{\mathcal{F}}\right) = \sum_{\{T \in \mathcal{F} : T \subseteq S\}} \frac{d_T\left(v^{\mathcal{F}}\right)}{|T|}.$$

Since $S \notin \mathcal{F}$ and $(N, \mathcal{F})$ is a partition system, property (P2) implies that $S = \bigcup_{k=1}^{p} S_k$, where $\Pi(S) = \{S_1, \ldots, S_p\}$ is the collection of components of S. Then, we have the partition

$$\{T \in \mathcal{F} : T \subseteq S\} = \bigcup_{k=1}^{p} \{T \in \mathcal{F} : T \subseteq S_k\}.$$

This implies that

$$P\left(S, v^{\mathcal{F}}\right) = \sum_{k=1}^{p} \left[\sum_{\{T \in \mathcal{F} : T \subseteq S_k\}} \frac{d_T\left(v^{\mathcal{F}}\right)}{|T|} \right] = \sum_{k=1}^{p} P\left(S_k, v^{\mathcal{F}}\right).$$

□

Let $(N, \mathcal{L})$ be the partition convex geometry of subsets of vertices which induce connected subgraphs of the connected block graph $G = (N, E)$. If G is a tree, then Owen [195, Theorems 6 and 7] gave the following formula for computing the dividends in game $v^{\mathcal{L}}$,

$$d_S\left(v^{\mathcal{L}}\right) = \sum_{\{T \subseteq N \,:\, \mathrm{ex}(S) \subseteq T \subseteq S\}} d_T(v).$$

Next, this formula is extended to the case when the graph is a connected block graph. Indeed, the formula holds in every partition convex geometry and can be shown by means of the Minkowski-Krein-Milman property.

Proposition 5.4.2 *Let $(N, \mathcal{L})$ be a partition convex geometry and let (N, v) be a game. The dividends of $S \in \mathcal{L}$ in the restricted game $v^{\mathcal{L}}$ are*

$$d_S\left(v^{\mathcal{L}}\right) = \sum_{\{T \subseteq N : \overline{T} = S\}} d_T(v) = \sum_{\{T \subseteq N \,:\, \mathrm{ex}(S) \subseteq T \subseteq S\}} d_T(v).$$

Proof. By formula (5.1), $v^{\mathcal{L}} = \sum_{T \subseteq N} d_T(v) u_T^{\mathcal{L}}$ and Theorem 5.2.1 implies that $v^{\mathcal{L}} = \sum_{S \in \mathcal{L}} d_S\left(v^{\mathcal{L}}\right) u_S$. We claim that $u_T^{\mathcal{L}} = u_{\overline{T}}$ for every $\emptyset \neq T \subseteq N$. To verify this claim, consider the following equivalent conditions for all $S \subseteq N$:

$$u_T^{\mathcal{L}}(S) = 1 \Longleftrightarrow \exists C \in \mathcal{L} \text{ such that } T \subseteq C \subseteq S \Longleftrightarrow \overline{T} \subseteq S \Longleftrightarrow u_{\overline{T}}(S) = 1.$$

Then, the coefficients satisfy

$$d_S\left(v^{\mathcal{L}}\right) = \sum_{\{T \subseteq N : \overline{T} = S\}} d_T(v).$$

Next, we show that for all $S \in \mathcal{L}$,

$$\{T \subseteq N : \overline{T} = S\} = \{T \subseteq N : \text{ex}(S) \subseteq T \subseteq S\}.$$

Let $T \subseteq N$ be such that $\overline{T} = S$. Then, $T \subseteq \overline{T} = S$, and $S \in \mathcal{L}$. But since every convex set is the closure of its extreme points, the set $\text{ex}(S)$ is a minimal subset of S such that $\overline{\text{ex}(S)} = S$, hence $\text{ex}(S) \subseteq T \subseteq S$. Conversely, if $\text{ex}(S) \subseteq T \subseteq S$, we obtain $S = \overline{\text{ex}(S)} \subseteq \overline{T} \subseteq \overline{S} = S$. □

The relation between the dividends of Harsanyi in the restricted game $v^{\mathcal{L}}$ and the worths in the original game v is given by the next result.

Proposition 5.4.3 *Let $(N, \mathcal{L})$ be a partition convex geometry and let $v \in \Gamma^N$. If $v^{\mathcal{L}}$ is the restricted game associated to v then*

$$d_S\left(v^{\mathcal{L}}\right) = \sum_{T \in [S^-, S]} (-1)^{|S|-|T|} v(T), \quad \textit{where } S^- = S \setminus \text{ex}(S).$$

Proof. It follows from Theorem 5.2.1 that for all $S \in \mathcal{L}$,

$$v(S) = v^{\mathcal{L}}(S) = \sum_{T \in \mathcal{L}} d_T\left(v^{\mathcal{L}}\right) u_T(S) = \sum_{\{T \in \mathcal{L} : T \subseteq S\}} d_T\left(v^{\mathcal{L}}\right).$$

The Möbius function of a convex geometry is given by [65, Theorem 4.3]:

$$\mu(T, S) = \begin{cases} (-1)^{|S|-|T|}, & \text{if } S \setminus T \subseteq \text{ex}(S) \\ 0, & \text{otherwise.} \end{cases}$$

Then, the Möbius inversion formula of $\mathcal{L}$ implies (see [224, p. 116]) that

$$\begin{aligned} d_S\left(v^{\mathcal{L}}\right) &= \sum_{\{T \in \mathcal{L} : T \subseteq S\}} v(T)\mu(T, S) \\ &= \sum_{\{T \in \mathcal{L} : S \setminus T \subseteq \text{ex}(S)\}} (-1)^{|S|-|T|} v(T). \end{aligned}$$

We know that $\{T \in \mathcal{L} : S \setminus T \subseteq \text{ex}(S)\} = [S^-, S]$ and so, we obtain the formula for the dividends. □

5.5 The Shapley and Banzhaf values

Let (N, v) be a game and let $(N, \mathcal{L})$ be a partition system. The *Shapley value* for player i in the restricted game $v^{\mathcal{L}}$ is given by $\Phi_i\left(N, v^{\mathcal{L}}\right)$, for all $i \in N$. The *Banzhaf value* for player i in game $v^{\mathcal{L}}$ is given by $\beta_i'\left(N, v^{\mathcal{L}}\right)$, for all $i \in N$. In terms of dividends [195, p. 212], we have that

$$\Phi_i\left(N, v^{\mathcal{L}}\right) = \sum_{\{S \subseteq N : i \in S\}} \frac{d_S\left(v^{\mathcal{L}}\right)}{|S|}, \tag{5.3}$$

$$\beta_i'\left(N, v^{\mathcal{L}}\right) = \sum_{\{S \subseteq N : i \in S\}} \frac{d_S\left(v^{\mathcal{L}}\right)}{2^{|S \setminus i|}}.$$

Edelman and Jamison [65, Theorem 4.2] proved that if $(N, \mathcal{L})$ is a convex geometry and $S \in \mathcal{L}$, then the interval $[S^-, S] = \{C \in \mathcal{L} : S^- \subseteq C \subseteq S\}$ is a Boolean algebra, where $S^- = S \setminus \mathrm{ex}(S)$. Thus $[S^-, S]$ is isomorphic to $2^{\mathrm{ex}(S)}$. We consider the interval $[T, T^+]$, where $T \in \mathcal{L}$ and $T^+ = \{i \in N : T \cup i \in \mathcal{L}\}$.

Proposition 5.5.1 *Let $(N, \mathcal{L})$ be a partition convex geometry. Then we have:*

(a) *If $T \in \mathcal{L}$ and $T \neq \emptyset$, then $[T, T^+]$ is a Boolean algebra isomorphic to $2^{T^+ \setminus T}$.*

(b) *If $T = \emptyset$, then $[T, T^+] = \mathcal{L}$.*

Proof. (a) Since the interval $[T, T^+] = \{S \in \mathcal{L} : T \subseteq S \subseteq T^+\}$ is isomorphic to subsets of

$$T^+ \setminus T = \{j \in N \setminus T : T \cup j \in \mathcal{L}\},$$

the result is obtained.
(b) Property (P1) implies $\{i\} \in \mathcal{L}$ for all $i \in N$. If $T = \emptyset$, then $T^+ = N$. □

In the next theorem two *explicit formulas,* in terms of v, for the Shapley and Banzhaf values of the players in the restricted game $v^{\mathcal{L}}$, are proved. For this, we need the following lemma.

Lemma 5.5.2 *Let $(N, \mathcal{L})$ be a partition convex geometry. If $T \in \mathcal{L}$, $T \neq \emptyset$ then $\{S \in \mathcal{L} : T \in [S^-, S]\} = [T, T^+]$.*

Proof. We first show that if $S \in \mathcal{L}$ and $S^- \subseteq T \subseteq S$, then $S \in [T, T^+]$. Since $T \subseteq S$ it is sufficient to prove that $S \setminus T \subseteq T^+ \setminus T = \{j \in N \setminus T : T \cup j \in \mathcal{L}\}$. For any $j \in S \setminus T$ we have that $T \cup j \in [S^-, S]$ and we know that the interval is a Boolean algebra. This implies that $T \cup j \in \mathcal{L}$.

Conversely, suppose $S \in [T, T^+]$. Then by Proposition 5.5.1(a) we have $S \in \mathcal{L}$. We shall show that $S \setminus T \subseteq \mathrm{ex}(S)$, i.e., $S^- \subseteq T \subseteq S$. Since $[T, T^+]$ is a Boolean algebra, we have that $j \in S \setminus T$ implies $S \setminus j \in [T, T^+]$ and hence $S \setminus j \in \mathcal{L}$. Therefore $j \in \mathrm{ex}(S)$. $\square$

Theorem 5.5.3 *Let $(N, \mathcal{L})$ be a partition convex geometry and let (N, v) be a game. If we consider the following collections,*

$$\begin{aligned}\mathcal{L}_i &= \{T \in \mathcal{L} : i \in T\},\\ \mathcal{L}_i^+ &= \{T \in \mathcal{L} : i \in \mathrm{ex}(T),\ (T \setminus i)^+ = T^+\},\\ \mathcal{L}_i^\star &= \{T \in \mathcal{L} : i \notin T,\ T \cup i \in \mathcal{L},\ T^+ \neq (T \cup i)^+\},\end{aligned}$$

for all $i \in N$, then:

(a) *The Shapley value for player i in the restricted game $v^{\mathcal{L}}$ satisfies*

$$\begin{aligned}\Phi_i\left(N, v^{\mathcal{L}}\right) &= \sum_{T \in \mathcal{L}_i^+} \frac{(t-1)!(t^+ - t)!}{t^+!}\left(v(T) - v(T \setminus i)\right)\\ &+ \sum_{T \in \mathcal{L}_i \setminus \mathcal{L}_i^+} \frac{(t-1)!(t^+ - t)!}{t^+!} v(T)\\ &- \sum_{T \in \mathcal{L}_i^\star} \frac{(t)!(t^+ - t - 1)!}{t^+!} v(T).\end{aligned}$$

(b) *The Banzhaf value for player i in the restricted game $v^{\mathcal{L}}$ satisfies*

$$\begin{aligned}\beta_i'\left(N, v^{\mathcal{L}}\right) &= \sum_{T \in \mathcal{L}_i^+} \frac{1}{2^{t^+ - 1}}\left(v(T) - v(T \setminus i)\right)\\ &+ \sum_{T \in \mathcal{L}_i \setminus \mathcal{L}_i^+} \frac{1}{2^{t^+ - 1}} v(T) - \sum_{T \in \mathcal{L}_i^\star} \frac{1}{2^{t^+ - 1}} v(T),\end{aligned}$$

where $t = |T|$, and $t^+ = |T^+|$.

Proof. (a) By Theorem 5.2.1, we know that $d_S\left(v^{\mathcal{L}}\right) = 0$ unless $S \in \mathcal{L}$. We use the formula (5.3) and Proposition 5.4.3 for computing

$$\Phi_i\left(N, v^{\mathcal{L}}\right) = \sum_{\{S \in \mathcal{L} : i \in S\}} \frac{d_S\left(v^{\mathcal{L}}\right)}{|S|} = \sum_{\{S \in \mathcal{L} : i \in S\}} \frac{1}{|S|}\left[\sum_{T \in [S^-, S]} (-1)^{|S| - |T|} v(T)\right].$$

Reversing the order of summation, we obtain

$$\Phi_i\left(N, v^{\mathcal{L}}\right) = \sum_{T \in \mathcal{L}}\left[\sum_{\{S \in \mathcal{L} : i \in S,\ T \in [S^-, S]\}} \frac{(-1)^{|S| - |T|}}{|S|}\right] v(T) = \sum_{T \in \mathcal{L}} c_i(T) v(T).$$

We apply Lemma 5.5.2 to the term in brackets, and denote $s = |S|$ and $t = |T|$. Thus,

$$c_i(T) = \sum_{\{S\in[T,T^+]\,:\,i\in S\}} \frac{(-1)^{s-t}}{s} = \sum_{\{S\in\mathcal{L}\,:\,T\cup i\subseteq S\subseteq T^+\}} \frac{(-1)^{s-t}}{s}.$$

First, the case $i \in T$ is considered. The interval $[T, T^+]$ is a Boolean algebra, hence the summation index is $\{S \in 2^N : T \subseteq S \subseteq T^+\}$. Now, we consider $S = T \cup R$, where $R = S \setminus T$ and $r = |R|$. Then,

$$\begin{aligned} c_i(T) &= \sum_{R\subseteq T^+\setminus T} \frac{(-1)^r}{t+r} = \sum_{r=0}^{t^+-t} \binom{t^+-t}{r} \frac{(-1)^r}{t+r} \\ &= \sum_{r=0}^{t^+-t} \binom{t^+-t}{r} (-1)^r \int_0^1 x^{t+r-1}\,dx \\ &= \int_0^1 x^{t-1} \sum_{r=0}^{t^+-t} \binom{t^+-t}{r} (-x)^r\,dx \\ &= \int_0^1 x^{t-1}(1-x)^{t^+-t}\,dx \\ &= \frac{(t-1)!(t^+-t)!}{t^+!}. \end{aligned}$$

Next, assume that $i \notin T$, hence the index is $\{S \in \mathcal{L} : T \cup i \subseteq S \subseteq T^+\}$. Then $i \in T^+$ and $T^+ = T \cup \{j \notin T : T \cup j \in \mathcal{L}\}$ implies that $T \cup i \in \mathcal{L}$. Now, the previous result yields (note that $[T \cup i, T^+]$ is a Boolean algebra):

$$c_i(T) = -\sum_{\{S\in 2^N\,:\,T\cup i\subseteq S\subseteq T^+\}} \frac{(-1)^{s-(t+1)}}{s} = -\frac{t!(t^+-t-1)!}{t^+!}.$$

Inserting the coefficients, we have

$$\begin{aligned} \Phi_i\left(N, v^{\mathcal{L}}\right) &= \sum_{T\in\mathcal{L}_i} \frac{(t-1)!(t^+-t)!}{t^+!} v(T) \\ &\quad - \sum_{\{T\in\mathcal{L}\,:\,i\notin T,\ T\cup i\in\mathcal{L}\}} \frac{(t)!(t^+-t-1)!}{t^+!} v(T). \end{aligned} \tag{5.4}$$

For any $T \in \mathcal{L}_i$, if $i \in \mathrm{ex}(T)$ and $(T \setminus i)^+ = T^+$, then $T \in \mathcal{L}_i^+$, hence $T \setminus i \in \mathcal{L}$ and $c_i(T \setminus i) = -c_i(T)$. Consequently, its contribution to the sum is $c_i(T)[v(T) - v(T \setminus i)]$. If $T \in \mathcal{L}_i \setminus \mathcal{L}_i^+$, then its term of the sum is $c_i(T)v(T)$.

Finally, for any $T \in \mathcal{L}$ with $i \notin T$ and $T \cup i \in \mathcal{L}$, such that $T^+ \neq (T \cup i)^+$, i.e., $T \in \mathcal{L}_i^\star$, the coefficients $c_i(T)$ and $-c_i(T \cup i)$ are different. Therefore, its contribution is $c_i(T)v(T)$ (where $i \notin T$ implies $c_i(T) < 0$).

(b) The proof of the formula of the Banzhaf value is similar to the proof of part (a). The only difference is that the coefficients are:

$$\begin{aligned} c_i(T) &= \sum_{r=0}^{t^+-t} \binom{t^+ - t}{r} (-1)^r \left(\frac{1}{2}\right)^{t+r-1} = \left(\frac{1}{2}\right)^{t^+-1}, \quad \text{if } i \in T, \\ c_i(T) &= -\left(\frac{1}{2}\right)^{t^+-1}, \quad \text{if } i \notin T \text{ and } T \cup i \in \mathcal{L}. \end{aligned}$$

□

Notice that if $\mathcal{L} = 2^N$, then $\text{ex}(T) = T$, and $T^+ = N$ for every $T \in \mathcal{L}$. Thus, the formulas obtained in the above theorem are equal to the Shapley and Banzhaf values. Moreover, the equation (5.4) is equal to the equation of Shapley (see reprint in [212, p. 35]). The formulas for computing these values can be further simplified when the player is an extreme point of N. Before doing so, we will need a lemma.

Lemma 5.5.4 *Let $(N, \mathcal{L})$ be a partition convex geometry. If $i \in \text{ex}(N)$, then we obtain*

$$\{T \in \mathcal{L}_i : |T| \geq 2\} = \{T \in \mathcal{L}_i^+ : |T| \geq 2\}.$$

Proof. If $T \in \mathcal{L}_i$, then $T \setminus i = T \cap (N \setminus i) \in \mathcal{L}$ and so $i \in \text{ex}(T)$. For all $T \in \mathcal{L}_i$ with $|T| \geq 2$, note that $T \in \mathcal{L}_i^+ \iff (T \setminus i)^+ = T^+$. We prove that these sets are equal.

First, if $j \in T^+$, with $j \notin T$ then $(T \setminus i) \cup j = (N \setminus i) \cap (T \cup j) \in \mathcal{L}$. Hence, we stated that $j \in (T \setminus i)^+$.

Conversely, take $j \in (T \setminus i)^+$. If $j = i$ we obtain the result and if $j \neq i$, observe that $((T \setminus i) \cup j) \cap T = T \setminus i \neq \emptyset$ because $|T| \geq 2$. Then, Proposition 5.1.1 implies that $(T \setminus i) \cup j \cup T = T \cup j \in \mathcal{L}$, and so $j \in T^+$. □

Theorem 5.5.5 *Let $(N, \mathcal{L})$ be a partition convex geometry and let (N, v) be a game such that $v(i) = 0$ for all $i \in N$. If $i \in \text{ex}(N)$, then the values for player i in $v^{\mathcal{L}}$ satisfy*

$$\begin{aligned} \Phi_i\left(N, v^{\mathcal{L}}\right) &= \sum_{T \in \mathcal{L}} \frac{(t-1)!(t^+ - t)!}{t^+!} \left(v(T) - v(T \setminus i)\right), \\ \beta_i'\left(N, v^{\mathcal{L}}\right) &= \sum_{T \in \mathcal{L}} \frac{1}{2^{t^+-1}} \left(v(T) - v(T \setminus i)\right), \end{aligned}$$

where $t = |T|$ and $t^+ = |T^+|$.

Proof. By Lemma 5.5.4, if $T \neq \emptyset$, $i \notin T$, and $T \cup i \in \mathcal{L}$ then $(T \cup i)^+ = T^+$. Therefore, $\mathcal{L}_i^\star = \{T \in \mathcal{L} \ : \ i \notin T, \ \ T \cup i \in \mathcal{L}, \ (T \cup i)^+ \neq T^+\} = \{\emptyset\}$. The index of $c_i(T)\,(v(T) - v(T \setminus i))$ in Theorem 5.5.3 is $\{T \in \mathcal{L}_i \ : \ |T| \geq 2\}$, and $v(T) - v(T \setminus i) = 0$, if $i \notin T$ or $|T| \leq 1$. □

The explicit formula for the potential of $v^{\mathcal{L}}$ can be obtained by a similar method to the one that is used in Theorem 5.5.3.

Theorem 5.5.6 *Let $(N, \mathcal{L})$ be a partition convex geometry and let (N, v) be a game. Then the potential of $v^{\mathcal{L}}$ satisfies*

$$P\left(N, v^{\mathcal{L}}\right) = \sum_{T \in \mathcal{L}} \frac{(t-1)!(t^+ - t)!}{t^+!} v(T), \ \textit{where} \ t = |\,T\,|, \ \ t^+ = |\,T^+\,|.$$

Example: Let (N, v) be the *four-person apex* game, that is, the simple game with the winning coalitions,

$$\mathcal{W} = \{\{1,2\}, \{1,3\}, \{1,4\}, \{1,2,3\}, \{1,2,4\}, \{1,3,4\}, \{2,3,4\}, N\}.$$

The Shapley and Banzhaf values are [196, p. 143]:

$$\begin{aligned} \Phi(N, v) &= \left(\frac{1}{2}, \frac{1}{6}, \frac{1}{6}, \frac{1}{6}\right), \\ \beta'(N, v) &= \left(\frac{3}{4}, \frac{1}{4}, \frac{1}{4}, \frac{1}{4}\right). \end{aligned}$$

Next, we consider the chain $N = (1 < 2 < 3 < 4)$ and the partition convex geometry $\mathcal{L} = Co\,(N)$ given by

$$\mathcal{L} = \{\emptyset, \{1\}, \{2\}, \{3\}, \{4\}, \{1,2\}, \{2,3\}, \{3,4\}, \{1,2,3\}, \{2,3,4\}, N\}\,.$$

The collection of coalitions which are both winning and convex is

$$\mathcal{W} \cap \mathcal{L} = \{\{1,2\}, \{1,2,3\}, \{2,3,4\}, N\}\,.$$

Theorem 5.5.5 is used to compute the values in $v^{\mathcal{L}}$ of players 1 and 4,

$$\begin{aligned} \Phi_1\left(N, v^{\mathcal{L}}\right) &= \frac{1}{3!}[v(12) - v(2)] + \frac{2}{4!}[v(123) - v(23)] = \frac{1}{4}, \\ \Phi_4\left(N, v^{\mathcal{L}}\right) &= \frac{2}{4!}[v(234) - v(23)] = \frac{1}{12}, \\ \beta'_1\left(N, v^{\mathcal{L}}\right) &= \left(\frac{1}{2}\right)^2 + \left(\frac{1}{2}\right)^3 = \frac{3}{8}, \quad \beta'_4\left(N, v^{\mathcal{L}}\right) = \left(\frac{1}{2}\right)^3 = \frac{1}{8}. \end{aligned}$$

To calculate the values in $v^{\mathcal{L}}$ of players 2 and 3, we use Theorem 5.5.3. If $i = 2$, then $\mathcal{W} \cap \mathcal{L}_2^+ = \emptyset$, $\mathcal{W} \cap \left(\mathcal{L}_2 \setminus \mathcal{L}_2^+\right) = \mathcal{W} \cap \mathcal{L}$ and $\mathcal{W} \cap \mathcal{L}_2^\star = \emptyset$. Now, it follows

$$\begin{aligned}\Phi_2\left(N, v^{\mathcal{L}}\right) &= \frac{1}{3!}[v(12)] + \frac{2}{4!}[v(123) + v(234)] + \frac{3!}{4!}v(N) = \frac{7}{12}, \\ \beta_2'\left(N, v^{\mathcal{L}}\right) &= \left(\frac{1}{2}\right)^2 + 2\left(\frac{1}{2}\right)^3 + \left(\frac{1}{2}\right)^3 = \frac{5}{8}.\end{aligned}$$

For $i = 3$, we obtain $\mathcal{W} \cap \mathcal{L}_3^+ = \emptyset$, $\mathcal{W} \cap \left(\mathcal{L}_3 \setminus \mathcal{L}_3^+\right) = \{\{1,2,3\},\{2,3,4\},N\}$, and $\mathcal{W} \cap \mathcal{L}_3^\star = \{\{1,2\}\}$. Hence,

$$\begin{aligned}\Phi_3\left(N, v^{\mathcal{L}}\right) &= \frac{2}{4!}[v(123) + v(234)] + \frac{3!}{4!}v(N) - \frac{2}{3!}v(12) = \frac{1}{12}, \\ \beta_3'\left(N, v^{\mathcal{L}}\right) &= \frac{1}{8}.\end{aligned}$$

Then, the Shapley and Banzhaf values in the restricted game $v^{\mathcal{L}}$ are

$$\begin{aligned}\Phi\left(N, v^{\mathcal{L}}\right) &= \left(\frac{1}{4}, \frac{7}{12}, \frac{1}{12}, \frac{1}{12}\right), \\ \beta'\left(N, v^{\mathcal{L}}\right) &= \left(\frac{3}{8}, \frac{5}{8}, \frac{1}{8}, \frac{1}{8}\right).\end{aligned}$$

The *multilinear extension* (MLE) of the game (N, v) is the function of n real variables (see Owen [196]),

$$f(v)(q_1, \ldots, q_n) = \sum_{S \subseteq N} \prod_{j \in S} q_j \; d_S(v),$$

where $d_S(v)$ is the dividend of S in the game (N, v). Owen showed that

$$\begin{aligned}\Phi_i(N, v) &= \int_0^1 \frac{\partial f(v)}{\partial q_i}(t, \ldots, t)\; dt, \\ \beta_i'(N, v) &= \frac{\partial f(v)}{\partial q_i}\left(\frac{1}{2}, \ldots, \frac{1}{2}\right).\end{aligned}$$

Proposition 5.5.7 *Let $(N, \mathcal{L})$ be a partition convex geometry and let (N, v) be a game. Then, the MLE of $v^{\mathcal{L}}$ is given by*

$$f\left(v^{\mathcal{L}}\right)(q_1, \ldots, q_n) = \sum_{S \in \mathcal{L}} \prod_{j \in S} q_j \left(\sum_{T \in [S^-, S]} (-1)^{|S|-|T|} v(T) \right).$$

Chapter 6

Restricted games by union stable systems

Several models of partial cooperation have been proposed, among which are those derived from *communication situations* as introduced by Myerson [182]. In Myerson's model, the bilateral relations among the players are represented by means of an undirected graph and the feasible coalitions are those that induce connected subgraphs. However, partial cooperation can not always be modelled by a graph, so the communication model has been generalized in several directions, for instance towards *conference structures* by Myerson [183] and *hypergraph communication situations* by van den Nouweland, Borm and Tijs [188].

In this chapter, we first introduce the formal definition of a *union stable system*, and the notions of *basis* and *support*. Next, the *Myerson value* and the *position value* for restricted games by union stable systems are defined. Moreover, an axiomatic characterization of the position value is provided for a specific class of union stable systems.

In the last section, the conditions under which the convexity is inherited from the underlying game to the conference game are investigated. For that, first a characterization of convexity in cooperative games due to Shapley [217] is extended. Furthermore, we show that under these same conditions the position value belongs to the core of the restricted game.

The results of this chapter are obtained by Algaba in her Ph. D. Thesis [1] and Algaba, Bilbao, Borm and López [2] [3].

6.1 Union stable systems

We study partial cooperation structures, which satisfy the following property: Given any two feasible coalitions with a nonempty intersection, the union is another feasible coalition. This type of feasible coalitions systems will be called union stable systems. In the partial cooperation context, this condition means that if two feasible coalitions have common elements, these ones will act as intermediaries between the two coalitions in order to establish meaningful cooperation in the whole group. A particular case of union stable systems are the communication situations. Throughout this chapter N denotes a finite set, and we use $\mathcal{F}$ to denote the set system $(N, \mathcal{F})$.

Definition 6.1.1 *A set system $\mathcal{F}$ is called union stable if for all $A, B \in \mathcal{F}$ with $A \cap B \neq \emptyset$ it is satisfied that $A \cup B \in \mathcal{F}$.*

A communication situation is a triple (N, v, E), where (N, v) is a game and $G = (N, E)$ is a graph. It is easy to see that the collection $\mathcal{F}$, defined by

$$\mathcal{F} = \{S \subseteq N : (S, E(S)) \text{ is a connected subgraph of } G\},$$

is a union stable system. Notice that a union stable system can not always be modelled by a communication situation. For example, let $N = \{1, 2, 3, 4\}$ and

$$\mathcal{F} = \{\emptyset, \{1\}, \{2\}, \{3\}, \{4\}, \{1, 2, 3\}, \{2, 3, 4\}, N\}.$$

This set system is union stable, but does not coincide with the connected subgraph family of any graph.

Let $\mathcal{F}$ be a union stable system and $\mathcal{G} \subseteq \mathcal{F}$. We define inductively the families

$$\mathcal{G}^{(0)} := \mathcal{G}, \quad \mathcal{G}^{(n)} := \left\{S \cup T : S, T \in \mathcal{G}^{(n-1)},\ S \cap T \neq \emptyset\right\} \quad (n = 1, 2, \ldots)$$

Notice that $\mathcal{G}^{(0)} \subseteq \mathcal{G}^{(n-1)} \subseteq \mathcal{G}^{(n)} \subseteq \mathcal{F}$, since $\mathcal{G} \subseteq \mathcal{F}$ and $\mathcal{F}$ is union stable. The inductive process is finite because $\mathcal{F}$ is finite.

Definition 6.1.2 *Let $\mathcal{F}$ be a union stable system and let $\mathcal{G} \subseteq \mathcal{F}$. We define the closure $\overline{\mathcal{G}}$ by $\overline{\mathcal{G}} = \mathcal{G}^{(k)}$, where k is the smallest integer such that $\mathcal{G}^{(k+1)} = \mathcal{G}^{(k)}$.*

For any $S \in \mathcal{F}$, $\overline{\{S\}} = \{S\}$, the collection $\overline{\mathcal{G}}$ is union stable, and $\overline{\mathcal{F}} = \mathcal{F}$.

Example: Let $N = \{1, 2, 3, 4\}$ and consider the union stable system given by:

$$\mathcal{F} = \{\{1\}, \{3\}, \{1, 2\}, \{2, 3\}, \{2, 4\}, \{1, 2, 3\}, \{1, 2, 4\}, \{2, 3, 4\}, N\}.$$

For the subsystem $\mathcal{G} = \{\{1, 2\}, \{2, 3\}, \{2, 4\}\} \subseteq \mathcal{F}$, note that

$$\begin{aligned} \mathcal{G}^{(1)} &= \{\{1,2\}, \{2,3\}, \{2,4\}, \{1,2,3\}, \{1,2,4\}, \{2,3,4\}\}, \\ \mathcal{G}^{(2)} &= \overline{\mathcal{G}} = \{\{1,2\}, \{2,3\}, \{2,4\}, \{1,2,3\}, \{1,2,4\}, \{2,3,4\}, N\}. \end{aligned}$$

Let $\mathcal{F}$ be a union stable system and $\mathcal{G} \subseteq \mathcal{F}$. If $\mathcal{G}$ is union stable, there can be feasible coalitions which can be written as the union of two feasible coalitions with a nonempty intersection. So, we can consider the following set:

$$D(\mathcal{G}) := \{G \in \mathcal{G} \, : \, G = A \cup B, \, A \neq G, \, B \neq G, \, A, B \in \mathcal{G}, \, A \cap B \neq \emptyset\}.$$

Note that $D(\mathcal{G})$ is composed of those feasible coalitions which can be written as the union of two distinct feasible coalitions with a nonempty intersection.

Definition 6.1.3 *Let $\mathcal{F}$ be a union stable system. The set $B(\mathcal{F}) = \mathcal{F} \setminus D(\mathcal{F})$ is called the basis of $\mathcal{F}$, and the elements of $B(\mathcal{F})$ are called supports of $\mathcal{F}$.*

Example: Let $N = \{1, 2, 3, 4\}$ and consider the union stable system

$$\mathcal{F} = \{\{1\}, \{2\}, \{3\}, \{4\}, \{1, 2\}, \{1, 3\}, \{3, 4\}, \{1, 2, 3\}, \{1, 3, 4\}, \{2, 3, 4\}, N\}.$$

Clearly, the set $D(\mathcal{F}) = \{\{1, 2, 3\}, \{1, 3, 4\}, N\}$, and hence the basis of $\mathcal{F}$ is

$$B(\mathcal{F}) = \{\{1\}, \{2\}, \{3\}, \{4\}, \{1, 2\}, \{1, 3\}, \{3, 4\}, \{2, 3, 4\}\}.$$

By construction, $B(\mathcal{F})$ is unique, nonempty if $\mathcal{F}$ is nonempty, and satisfies the following properties:

(a) If $\emptyset \in \mathcal{F}$, then $\emptyset \in B(\mathcal{F})$.

(b) If $\{j\} \in \mathcal{F}$ for some $j \in N$, then $\{j\} \in B(\mathcal{F})$.

(c) If $S \in \mathcal{F}$ is a minimal element in $(\mathcal{F}, \subseteq)$, then $S \in B(\mathcal{F})$.

(d) Let $S \in \mathcal{F}$, $|S| > 1$. If $|T| \in \{0, 1, |S|\}$ for all $T \in 2^S \cap \mathcal{F}$, then $S \in B(\mathcal{F})$.

In particular, if $S \in \mathcal{F}$ and $|S| \leq 2$, then $S \in B(\mathcal{F})$.

Proposition 6.1.1 *Let $\mathcal{F}$ be a union stable system. The map $\varphi : 2^{\mathcal{F}} \to 2^{\mathcal{F}}$, defined by $\varphi(\mathcal{G}) = \overline{\mathcal{G}}$ is a closure operator, i.e.,*

(a) *For all $\mathcal{G} \in 2^{\mathcal{F}}$, $\mathcal{G} \subseteq \varphi(\mathcal{G})$.*

(b) *The relation $\mathcal{G} \subseteq \mathcal{R} \subseteq \mathcal{F}$ implies $\varphi(\mathcal{G}) \subseteq \varphi(\mathcal{R})$.*

(c) *For all $\mathcal{G} \in 2^{\mathcal{F}}$, $\varphi(\varphi(\mathcal{G})) = \varphi(\mathcal{G})$.*

This result assures that $\left(\mathcal{F}, \varphi\left(2^{\mathcal{F}}\right)\right)$, where $\varphi\left(2^{\mathcal{F}}\right) = \left\{\varphi(\mathcal{G}) : \mathcal{G} \in 2^{\mathcal{F}}\right\}$, is a *closure space*. From now on the closure space $\left(\mathcal{F}, \varphi\left(2^{\mathcal{F}}\right)\right)$ will be denoted by $(\mathcal{F}, -)$ and the elements of $\varphi\left(2^{\mathcal{F}}\right)$ will be called *closed*. We can obtain:

Proposition 6.1.2 *Let $(\mathcal{F}, -)$ be the above closure space, and $\mathcal{G} \in 2^{\mathcal{F}}$. Then, $\mathcal{G}$ is closed if and only if $\mathcal{G}$ is union stable.*

Next, we provide two characterizations of the basis of a union stable system.

Proposition 6.1.3 *Let $\mathcal{F}$ be a union stable system. Then, $B(\mathcal{F})$ is the minimal subset of $\mathcal{F}$ such that $\overline{B(\mathcal{F})} = \mathcal{F}$.*

Proof. We first prove that $\overline{B(\mathcal{F})} = \mathcal{F}$. We have that $\overline{B(\mathcal{F})} \subseteq \mathcal{F}$, since $B(\mathcal{F}) \subseteq \mathcal{F}$ and $\mathcal{F}$ is union stable. In order to prove the reverse inclusion, we use induction on the number of elements of feasible coalitions in $\mathcal{F}$. Clearly, the minimal elements in $(\mathcal{F}, \subseteq)$ belong to the basis and hence to $\overline{B(\mathcal{F})}$. Now, suppose $F \in \overline{B(\mathcal{F})}$ for all $F \in \mathcal{F}$ with $|F| < p$. Then, given $F \in \mathcal{F}$ with $|F| = p$, we have either $F \in B(\mathcal{F})$ or $F \notin B(\mathcal{F})$. In the first case $F \in \overline{B(\mathcal{F})}$. Otherwise, $F \in D(\mathcal{F})$ and hence, there are two feasible coalitions $S, T \in \mathcal{F}$, $S \neq F$, $T \neq F$, $S \cap T \neq \emptyset$ such that $S \cup T = F$. By using the induction hypothesis, since $|S| < p$ and $|T| < p$, we have that $S, T \in \overline{B(\mathcal{F})}$, and the union stability implies that $F = S \cup T \in \overline{B(\mathcal{F})}$. Finally, we note that $B(\mathcal{F})$ is a minimal subset of $\mathcal{F}$ such that $\overline{B(\mathcal{F})} = \mathcal{F}$ by construction. □

Definition 6.1.4 *Let $\mathcal{F}$ be a union stable system and let $\mathcal{G} \subseteq \mathcal{F}$ be a union stable subsystem. The collection of the extremal sets is defined by*

$$\mathrm{ex}(\mathcal{G}) := \{G \in \mathcal{G} : \mathcal{G} \setminus \{G\} \text{ is union stable}\}.$$

Proposition 6.1.4 *Let $\mathcal{F}$ be a union stable system and let $\mathcal{G} \subseteq \mathcal{F}$ be a union stable subsystem. Then $\mathrm{ex}(\mathcal{G}) = B(\mathcal{G})$.*

Proof. We first prove $\mathrm{ex}(\mathcal{G}) \subseteq B(\mathcal{G})$. As $\overline{B(\mathcal{G})} = \mathcal{G}$, $\mathcal{G}$ is the smallest union stable system that contains $B(\mathcal{G})$. Let $G \in \mathrm{ex}(\mathcal{G})$. Then $\mathcal{G} \setminus \{G\}$ is union stable. If $G \notin B(\mathcal{G})$, then $B(\mathcal{G}) \subseteq \mathcal{G} \setminus \{G\} \subset \mathcal{G}$ and hence $\mathcal{G}$ would not be the smallest union stable system that contains $B(\mathcal{G})$.

It remains to prove that $B(\mathcal{G}) \subseteq \mathrm{ex}(\mathcal{G})$. For this, let $B \in B(\mathcal{G})$. We show that $\mathcal{G} \setminus \{B\}$ is union stable. Indeed, let $S, T \in \mathcal{G} \setminus \{B\}$, with $S \cap T \neq \emptyset$. Since $\mathcal{G}$ is union stable, $S \cup T \in \mathcal{G}$. On the other hand, $S \cup T \neq B$ since $S \cup T = B$ would imply $B \notin B(\mathcal{G})$. Hence, $S \cup T \in \mathcal{G} \setminus \{B\}$. □

Remark 6.1.1 *It has been shown that, if $\mathcal{G} \subseteq \mathcal{F}$ is union stable system, then $\mathrm{ex}(\mathcal{G}) = B(\mathcal{G}) = \mathcal{G}$. Therefore, the closure space $(\mathcal{F}, -)$ is a convex geometry (see Section 1.4).*

6.2 Components of sets

In this section we will study the properties of the maximal feasibles subsets of a set and the relationship between Myerson's conference structures and union stable systems.

Definition 6.2.1 *Consider $\mathcal{F} \subseteq 2^N$ and let $S \subseteq N$. A set $T \subseteq S$ is called a $\mathcal{F}$-component of S if it is satisfied that $T \in \mathcal{F}$ and there exists no $T' \in \mathcal{F}$ such that $T \subset T' \subseteq S$.*

The $\mathcal{F}$-components of S are the maximal coalitions that belong to $\mathcal{F}$ and are contained in S. We denote by $C_{\mathcal{F}}(S)$ the set of the $\mathcal{F}$-components of S. Observe that the set $C_{\mathcal{F}}(S)$ may be the empty set.

Remark 6.2.1 *Let (N, v, E) be a communication situation and*

$$\mathcal{F} = \{S \subseteq N : (S, E(S)) \text{ is a connected subgraph of } G = (N, E)\}.$$

Then, the $\mathcal{F}$-components of any coalition $S \subseteq N$ are the connected components of the subgraph $(S, E(S))$ and form a partition of S. Moreover, the collection $\{B \in B(\mathcal{F}) : |B| \geq 2\}$ consists of the edges of graph $G = (N, E)$.

Proposition 6.2.1 *The set system $\mathcal{F} \subseteq 2^N$ is union stable if and only if for any $S \subseteq N$ such that $C_{\mathcal{F}}(S) \neq \emptyset$, the $\mathcal{F}$-components of S form a partition of a subset of S.*

Proof. Let $\mathcal{F}$ be a union stable system. Let S^1, S^2, $S^1 \neq S^2$, be maximal feasible coalitions of S. If $S^1 \cap S^2 \neq \emptyset$, then $S^1 \cup S^2 \in \mathcal{F}$ since $\mathcal{F}$ is union stable and $S^1 \cup S^2 \subseteq S$. This contradicts the fact that S^1 and S^2 are $\mathcal{F}$-components of S.

Conversely, assume for any S such that $C_{\mathcal{F}}(S) \neq \emptyset$, that its $\mathcal{F}$-components form a partition of a subset of S. Suppose that $\mathcal{F}$ is not union stable, then there are $A, B \in \mathcal{F}$, with $A \cap B \neq \emptyset$ and $A \cup B \notin \mathcal{F}$. Hence, there must be an $\mathcal{F}$-component $C_1 \in C_{\mathcal{F}}(A \cup B)$, with $A \subseteq C_1$ and an $\mathcal{F}$-component $C_2 \in C_{\mathcal{F}}(A \cup B)$, with $B \subseteq C_2$ such that $C_1 \neq C_2$. This contradicts the fact that the $\mathcal{F}$-components of $A \cup B$ are disjoint. □

Notice that in general the $\mathcal{F}$-components of S do not form a partition of S, but if $\mathcal{F}$ is a union stable system such that $\{i\} \in \mathcal{F}$ for all $i \in N$, then they do.

Remark 6.2.2 *Proposition 5.1.1 shows that a union stable system $\mathcal{F} \subseteq 2^N$ is a partition system if and only if $\emptyset \in \mathcal{F}$ and $\{i\} \in \mathcal{F}$ for all $i \in N$.*

Now, some relations between $\mathcal{F}$-components, feasible coalitions, and supports of a union stable system are studied, which are used in the next sections.

Proposition 6.2.2 *Let $\mathcal{F}$ be a union stable system and $B(\mathcal{F})$ its basis. Then*

(a) *If $N \notin \mathcal{F}$, we define the partition $\{\mathcal{B}_1, \mathcal{B}_2, \ldots, \mathcal{B}_p\}$ of the basis $B(\mathcal{F})$ by $\mathcal{B}_i = \{B \in B(\mathcal{F}) : B \subseteq N_i,\ N_i \in C_{\mathcal{F}}(N)\}$. Then, for all $B \in \mathcal{B}_i$ and $B' \in \mathcal{B}_j$, with $i \neq j$, $1 \leq i, j \leq p$, we have that $B \cap B' = \emptyset$.*

(b) *Let $\mathcal{I} \subseteq B(\mathcal{F})$, $\mathcal{J} \subseteq B(\mathcal{F})$ such that for all $B \in \mathcal{I}$ and for all $B' \in \mathcal{J}$, we have $B \cap B' = \emptyset$. Then*

- **(i)** *For all $S \in \overline{\mathcal{I}}$ and for all $S' \in \overline{\mathcal{J}}$, $S \cap S' = \emptyset$.*
- **(ii)** *$\overline{\mathcal{I} \cup \mathcal{J}} = \overline{\mathcal{I}} \cup \overline{\mathcal{J}}$.*
- **(iii)** *$C_{\overline{\mathcal{I} \cup \mathcal{J}}}(N) = C_{\overline{\mathcal{I}}}(N) \cup C_{\overline{\mathcal{J}}}(N)$.*

Proposition 6.2.3 *Let $\mathcal{F}$ be a union stable system and $B(\mathcal{F})$ its basis. Let $F \in \mathcal{F}$ with $|F| \geq 2$. Then, F can be written as a union of supports of size at least two.*

Proof. It is clear that F is a union of supports B_i, $i \in I$, with nonempty intersections. If $|B_j| = 1$, there is B_k, $k \in I$, $k \neq j$ such that $B_k \cap B_j \neq \emptyset$, and hence $B_j \subseteq B_k$ and $|B_k| \geq 2$. Therefore F can be written as a union of supports of size at least two with a nonempty intersection. □

We have also the following consequence of the definitions.

Proposition 6.2.4 *Let $\mathcal{F}$ be a union stable system. Let $S \subseteq N$ and consider the collection $\mathcal{F}_S = \{F \in \mathcal{F} : F \subseteq S\}$. Then, we have*

(a) *$\mathcal{F}_S$ is union stable.*

(b) *$C_{\mathcal{F}}(S) = C_{\mathcal{F}_S}(N)$.*

(c) *$\mathcal{B}(\mathcal{F}_S) = \{B \in \mathcal{B}(\mathcal{F}) : B \subseteq S\}$.*

In order to establish a relation between Myerson's conference structures and union stable systems, we will give the following results. Moreover, the next theorem will be essential in order to prove the uniqueness in the axiomatization of Myerson value in union stable systems.

Definition 6.2.2 *Let $\mathcal{F}$ be a union stable system. The players $i, j \in N$, are called connected by $\mathcal{B}(\mathcal{F})$ if there exists a sequence of supports $(B_1, \ldots, B_k)$, such that $i \in B_1$, $j \in B_k$ and if $k \geq 2$, $B_p \cap B_{p+1} \neq \emptyset$, for all $p = 1, \ldots, k-1$.*

Theorem 6.2.5 *Let $\mathcal{F}$ be a union stable system. Let $S \in \mathcal{F}$ and $i, j \in N$, $i \neq j$. Then $\{i, j\} \subseteq S$ if and only if i and j are connected by supports in $\mathcal{C}(\mathcal{F})$ contained in S, where $\mathcal{C}(\mathcal{F}) = \{B \in \mathcal{B}(\mathcal{F}) : |B| \geq 2\}$.*

Proof. Let $\{i, j\} \subseteq S$. If $S \in \mathcal{C}(\mathcal{F})$, it suffices to take $k = 1$ and $B_1 = S$. If $S \notin \mathcal{C}(\mathcal{F})$, then $S = A \cup B$, with $A, B \in \mathcal{F}$, and $A \cap B \neq \emptyset$. If $A, B \in \mathcal{C}(\mathcal{F})$ then we obtain the result. Otherwise, we repeat this decomposition and proceeding in this manner, we obtain the sequence of supports. The converse is obvious.□

Corollary 6.2.6 *Let $\mathcal{F}$ be a union stable system. Let $i, j \in N$, $i \neq j$. Then i and j are in the same $\mathcal{F}$-component of N if and only if i and j are connected by $\mathcal{C}(\mathcal{F})$.*

Example: Myerson [183] introduced the term *conference*, to refer to any set of two or more players who might meet together to discuss their cooperative plans. A *conference structure* CS is any collection $\mathcal{Q} \subseteq \{S \subseteq N : |S| \geq 2\}$. Given a conference structure $\mathcal{Q} \in CS$, two players i and j are connected by $\mathcal{Q}$ if $i = j$ or there exists some sequence of conferences $(S_1, \ldots, S_k)$ such that $i \in S_1$, $j \in S_k$, $\{S_1, \ldots, S_k\} \subseteq \mathcal{Q}$, and $S_p \cap S_{p+1} \neq \emptyset$ for all $p = 1, \ldots, k-1$.

If $\mathcal{F}$ is a union stable system then the set formed by the non-unitary supports is a Myerson's conference structure. Conversely, given a Myerson's conference structure, the set system

$$\mathcal{F} = \{S \subseteq N : \text{each pair of players } i, j \in S \text{ are connected by conferences in } S\}$$

is union stable.

6.3 The Myerson value

This section deals with a solution concept for games restricted by union stable structures: the *Myerson value.*

Definition 6.3.1 *Let (N, v) be a game and let $\mathcal{F} \subseteq 2^N$ be a union stable system. The $\mathcal{F}$-restricted game $v^{\mathcal{F}} : 2^N \to \mathbb{R}$, is defined by*

$$v^{\mathcal{F}}(S) := \sum_{T \in C_{\mathcal{F}}(S)} v(T).$$

A *union stable structure* is a triple $(N, v, \mathcal{F})$ where (N, v) is a game and $\mathcal{F} \subseteq 2^N$ is a union stable system.

Definition 6.3.2 *The Myerson value of a union stable structure $(N, v, \mathcal{F})$ is given by the vector $\mu(N, v, \mathcal{F}) := \Phi\left(N, v^{\mathcal{F}}\right)$, where Φ is the Shapley value.*

The following example illustrates the concepts introduced above.

Example: Consider the set $N = \{1,2,3,4\}$ and the union stable system given by $\mathcal{F} = \{\{1\},\{1,2,3\},\{2,3,4\},N\}$. Let $v : 2^N \to \mathbb{R}$ be the game defined by $v(S) = |S| - 1$, $S \neq \emptyset$, and $v(\emptyset) = 0$. Then, $\mathcal{B}(\mathcal{F}) = \{\{1\},\{1,2,3\},\{2,3,4\}\}$ and $\mathcal{C}(\mathcal{F}) = \{\{1,2,3\},\{2,3,4\}\}$. In this case, it is clear that

$$v^{\mathcal{F}}(S) = \begin{cases} |S| - 1, & \text{if } S \in \mathcal{F} \\ 0, & \text{otherwise,} \end{cases}$$

and the Myerson value is $\mu(N,v,\mathcal{F}) = \frac{1}{12}(5,13,13,5)$.

We now consider some properties that would be desirable for *an allocation rule*, and we focus on the study of these properties for the Myerson value. The set of all union stable structures with player set N will be denoted by US^N.

Definition 6.3.3 *An allocation rule on US^N is a map $\gamma : US^N \to \mathbb{R}^n$, such that it is component-efficient and component-dummy, that is, respectively,*

1. *For all $(N,v,\mathcal{F}) \in US^N$ and $M \in C_{\mathcal{F}}(N)$, we have*

$$\sum_{i \in M} \gamma_i(N,v,\mathcal{F}) = v(M).$$

2. *For all $i \notin \bigcup_{M \in C_{\mathcal{F}}(N)} M$, we have $\gamma_i(N,v,\mathcal{F}) = 0$.*

Proposition 6.3.1 *The Myerson value $\mu : US^N \to \mathbb{R}^n$ is an allocation rule.*

Proof. Let $(N,v,\mathcal{F}) \in US^N$. If $N \in \mathcal{F}$ then N is its unique $\mathcal{F}$-component, and hence $\sum_{i \in N} \mu(N,v,\mathcal{F}) = \sum_{i \in N} \Phi_i(N,v^{\mathcal{F}}) = v^{\mathcal{F}}(N) = v(N)$. Suppose, that $N \notin \mathcal{F}$ and, therefore, consider the set $C_{\mathcal{F}}(N)$. To each $\mathcal{F}$-component M of N is associated the game $u^M : 2^N \to \mathbb{R}$, which is defined with M fixed, for all $T \subseteq N$,

$$u^M(T) = v^{\mathcal{F}}(T \cap M) = \sum_{H \in C_{\mathcal{F}}(T \cap M)} v(H).$$

Moreover, for any coalition $T \subseteq N$, $C_{\mathcal{F}}(T) = \bigcup_{R \in C_{\mathcal{F}}(N)} C_{\mathcal{F}}(T \cap R)$, and hence, it is immediate that $v^{\mathcal{F}} = \sum_{R \in C_{\mathcal{F}}(N)} u^R$. Taking into account the above considerations for the game $(N, v^{\mathcal{F}})$, we find

$$\sum_{i \in M} \mu(N,v,\mathcal{F}) = \sum_{i \in M} \Phi_i(N,u^M) + \sum_{\{R \in C_{\mathcal{F}}(N) : R \neq M\}} \left[\sum_{i \in M} \Phi_i(N,u^R) \right].$$

Since $\sum_{i \in M} \Phi_i(N, u^M) = v^{\mathcal{F}}(M)$, and $\Phi_i(N, u^R) = 0,\ R \neq M,\ i \in M$, the above expression implies that $\sum_{i \in M} \mu(N, v, \mathcal{F}) = v^{\mathcal{F}}(M) = v(M)$.

Component-dummy is immediate since if $i \notin \bigcup_{M \in C_{\mathcal{F}}(N)} M$ then we have $C_{\mathcal{F}}(S) = C_{\mathcal{F}}(S \setminus i)$, for all $S \in \mathcal{F}$. Hence, the marginal contributions are $v^{\mathcal{F}}(S) - v^{\mathcal{F}}(S \setminus i) = 0$, and $\mu_i(N, v, \mathcal{F}) = 0$. □

Definition 6.3.4 *An allocation rule γ is fair if for all $(N, v, \mathcal{F}) \in US^N$ and $B \in \mathcal{B}(\mathcal{F})$, there exists $c \in \mathbb{R}$ such that $\gamma_j(N, v, \mathcal{F}) - \gamma_j(N, v, \mathcal{F}') = c$, for all $j \in B$, where $\mathcal{F}' = \overline{\mathcal{B}(\mathcal{F}) \setminus \{B\}}$.*

So, according to a fair allocation rule all players in a support B lose or gain the same amount if the support B is deleted. We now extend the axiomatization of the Myerson value to union stable structures.

Theorem 6.3.2 *The Myerson value is the unique fair allocation rule on US^N.*

Proof. Uniqueness: Let $(N, v, \mathcal{F}) \in US^N$. Suppose γ^1 and γ^2 are two fair allocation rules on US^N. We will prove by induction to the number $|\mathcal{C}(\mathcal{F})|$ of non-unitary supports in the basis of $\mathcal{F}$, that $\gamma^1(N, v, \mathcal{F}) = \gamma^2(N, v, \mathcal{F})$.

If $|\mathcal{C}(\mathcal{F})| = 0$ then $C_{\mathcal{F}}(N) = \{\{i\} : \{i\} \in \mathcal{F}\}$. Applying component-efficiency and component-dummy we obtain that $\gamma^1(N, v, \mathcal{F}) = \gamma^2(N, v, \mathcal{F})$.

Now, assume that $\gamma^1(N, v, \mathcal{G}) = \gamma^2(N, v, \mathcal{G})$ for all $\mathcal{G}$ with $|\mathcal{C}(\mathcal{G})| \leq k - 1$, and let $|\mathcal{C}(\mathcal{F})| = k$. Consider $C \in \mathcal{C}(\mathcal{F})$. Fairness implies that there exist numbers $c \in \mathbb{R}$ and $d \in \mathbb{R}$ such that

$$\gamma_j^1(N, v, \mathcal{F}) - \gamma_j^1\left(N, v, \overline{\mathcal{B}(\mathcal{F}) \setminus \{C\}}\right) = c,$$

$$\gamma_j^2(N, v, \mathcal{F}) - \gamma_j^2\left(N, v, \overline{\mathcal{B}(\mathcal{F}) \setminus \{C\}}\right) = d,$$

for all $j \in C$. Note that by the induction hypothesis

$$\gamma_j^1\left(N, v, \overline{\mathcal{B}(\mathcal{F}) \setminus \{C\}}\right) = \gamma_j^2\left(N, v, \overline{\mathcal{B}(\mathcal{F}) \setminus \{C\}}\right).$$

So there is a constant $\alpha = c - d$ such that

$$\gamma_j^1(N, v, \mathcal{F}) - \gamma_j^2(N, v, \mathcal{F}) = \alpha, \quad \text{for all } j \in C. \tag{6.1}$$

Given $M \in C_{\mathcal{F}}(N)$, by component-efficiency for γ^1 and γ^2, we obtain

$$\sum_{i \in M} \left[\gamma_i^1(N, v, \mathcal{F}) - \gamma_i^2(N, v, \mathcal{F})\right] = 0.$$

Applying Theorem 6.2.5 and equality (6.1) recursively, we get

$$\gamma_i^1(N, v, \mathcal{F}) - \gamma_i^2(N, v, \mathcal{F}) = \alpha,$$

for all $i \in M$, with $M \in C_{\mathcal{F}}(N)$, and this implies

$$\sum_{i \in M} \left[\gamma_i^1(N, v, \mathcal{F}) - \gamma_i^2(N, v, \mathcal{F})\right] = |M| \, \alpha.$$

Therefore $|M| \, \alpha = 0$, and hence $\gamma^1(N, v, \mathcal{F}) = \gamma^2(N, v, \mathcal{F})$.

Next, we show that the Myerson value is fair. Consider the game (N, w) given by $w(S) = v^{\mathcal{F}}(S) - v^{\mathcal{F}'}(S)$, for all $S \subseteq N$, where $\mathcal{F}' = \overline{\mathcal{B}(\mathcal{F}) \setminus \{B\}}$. Let $k \in B$. We may deduce that $w(S) = 0$, for all $S \subseteq N$, $B \nsubseteq S$, and, since for all coalitions $S \subseteq N$ with $B \subseteq S$ we have $B \nsubseteq S \setminus k$ and so $w(S \setminus k) = 0$. Thus, we can write for $k \in B$,

$$\Phi_k(N, w) = \sum_{\{S : B \subseteq S\}} \frac{(s-1)!\,(n-s)!}{n!} w(S),$$

where $s = |S|$, $n = |N|$. It follows that $\Phi_k(N, w) = \Phi_p(N, w)$, for all $p \in B$, and we obtain that the Myerson value is fair. □

Definition 6.3.5 *An allocation rule γ is basis monotonic if for all $(N, v, \mathcal{F})$, for all $B \in \mathcal{B}(\mathcal{F})$, and for all $j \in B$ it holds $\gamma_j(N, v, \mathcal{F}) \geq \gamma_j(N, v, \mathcal{F}')$, where $\mathcal{F}' = \overline{\mathcal{B}(\mathcal{F}) \setminus \{B\}}$.*

This condition asserts that all the players always benefit from reaching an agreement and cooperate.

Proposition 6.3.3 *Let $(N, v, \mathcal{F}) \in US^N$. If v is superadditive and zero-normalized, i.e. $v(i) = 0$ for all $i \in N$, then $\mu(N, v, \mathcal{F})$ is basis monotonic.*

Proof. It suffices to prove that $w(S) \geq 0$ for any $S \subseteq N$ such that $B \subseteq S$, where for all $S \subseteq N$, $w(S) = v^{\mathcal{F}}(S) - v^{\mathcal{F}'}(S)$, with $\mathcal{F}' = \overline{\mathcal{B}(\mathcal{F}) \setminus \{B\}}$. Any maximal feasible coalition of S in $\mathcal{F}'$ is either a maximal feasible coalition of S in $\mathcal{F}$ or it is contained in an $\mathcal{F}$-component of S. Then, taking the $\mathcal{F}'$-components of S and taking into consideration that the game (N, v) is superadditive and zero-normalized, we obtain

$$\begin{aligned} v^{\mathcal{F}'}(S) &= \sum_{T' \in C_{\mathcal{F}'}(S)} v(T') \\ &\leq \sum_{T \in C_{\mathcal{F}}(S)} \left[v \left(\bigcup_{\{T' \in C_{\mathcal{F}'}(S) : T' \subseteq T\}} T' \right) \right] \\ &\leq v^{\mathcal{F}}(S). \end{aligned}$$

□

To provide other axiomatic characterizations for the Myerson value, the next definitions are introduced (see van den Nouweland [189]). We use $\mathcal{C}_i(\mathcal{F})$ to denote the collection $\{C \in \mathcal{C}(\mathcal{F}) : i \in C\}$.

Definition 6.3.6 *A union stable structure $(N, v, \mathcal{F})$ is called point anonymous if there exists a function $f : \{0, 1, \dots, |D|\} \to \mathbb{R}$ such that $v^{\mathcal{F}}(S) = f(|S \cap D|)$ for all $S \subseteq N$, where $D = \{i \in N : \mathcal{C}_i(\mathcal{F}) \neq \emptyset\}$.*

Definition 6.3.7 *An allocation rule γ satisfies point anonymity if for all point anonymous structure $(N, v, \mathcal{F})$, there exists $\alpha \in \mathbb{R}$ such that*

$$\gamma_i(N, v, \mathcal{F}) = \begin{cases} \alpha, & \textit{if } i \in D \\ 0, & \textit{otherwise.} \end{cases}$$

Proposition 6.3.4 *The Myerson value $\mu(N, v, \mathcal{F})$ satisfies point anonymity.*

Proof. Let $(N, v, \mathcal{F}) \in US^N$ be point anonymous. If $D = \emptyset$, then the restricted game $v^{\mathcal{F}}(S) = f(|S \cap \emptyset|) = f(0) = 0$, for all $S \subseteq N$. Hence, $\mu_i(N, v, \mathcal{F}) = 0$ for all $i \in N$. Let $D \neq \emptyset$. If $i \notin D$, obviously $S \cap D = (S \setminus i) \cap D$ and $\mu_i(N, v, \mathcal{F}) = 0$. On the other hand, if $i, j \in D$ applying the symmetry property of the Shapley value we have $\mu_i(N, v, \mathcal{F}) = \mu_j(N, v, \mathcal{F})$, and hence

$$f(|D|) = \sum_{i \in D} \mu_i(N, v, \mathcal{F}) = |D| \, \mu_i(N, v, \mathcal{F}).$$

This obviously implies that $\mu_i(N, v, \mathcal{F}) = f(|D|)/|D| = \alpha$, for all $i \in D$ and $\mu_i(N, v, \mathcal{F}) = 0$, otherwise. □

Definition 6.3.8 *Let $(N, v, \mathcal{F}) \in US^N$. Then player $i \in N$ is called superfluous for $(N, v, \mathcal{F})$ if $v^{\mathcal{F}}(S) = v^{\mathcal{F}}(S \setminus i)$, for all $S \subseteq N$. An allocation rule γ satisfies the superfluous player property if for all $(N, v, \mathcal{F})$ and every player $i \in N$ that is superfluous for $(N, v, \mathcal{F})$ it holds $\gamma(N, v, \mathcal{F}) = \gamma(N, v, \mathcal{F}_{N \setminus i})$, where $\mathcal{F}_{N \setminus i} = \{F \in \mathcal{F} : F \subseteq N \setminus i\}$.*

Proposition 6.3.5 *The Myerson value $\mu(N, v, \mathcal{F})$ satisfies the superfluous player property.*

Proof. Let $i \in N$ be a superfluous player for $(N, v, \mathcal{F})$. We have to prove $\mu(N, v, \mathcal{F}) = \mu(N, v, \mathcal{F}_{N \setminus i})$. We observe that $v^{\mathcal{F}}(S) - v^{\mathcal{F}}(S \setminus i) = 0$, for all $S \subseteq N$ and hence $\mu_i(N, v, \mathcal{F}) = 0$. Since $i \notin \bigcup_{M \in C_{\mathcal{F}_{N \setminus i}}(N)} M$ and Proposition 6.3.1 implies that μ satisfies component-dummy, we have $\mu_i(N, v, \mathcal{F}_{N \setminus i}) = 0$.

For the other players, it suffices to show that $v^{\mathcal{F}}(S) = v^{\mathcal{F}_{N\setminus i}}(S)$, or equivalently, as i is a superfluous player for $(N, v, \mathcal{F})$, that $v^{\mathcal{F}}(S \setminus i) = v^{\mathcal{F}_{N\setminus i}}(S)$, for all $S \subseteq N$. The components satisfy $C_{\mathcal{F}}(S \setminus i) = C_{\mathcal{F}_{N\setminus i}}(S)$, and hence

$$\begin{aligned} v^{\mathcal{F}}(S \setminus i) &= \sum_{T \in C_{\mathcal{F}}(S\setminus i)} v(T) \\ &= \sum_{T \in C_{\mathcal{F}_{N\setminus i}}(S)} v(T) \\ &= v^{\mathcal{F}_{N\setminus i}}(S), \end{aligned}$$

for all $S \subseteq N$. □

Definition 6.3.9 *An allocation rule $\gamma : US^N \to \mathbb{R}^n$ is additive if*

$$\gamma(N, v + w, \mathcal{F}) = \gamma(N, v, \mathcal{F}) + \gamma(N, w, \mathcal{F})$$

for all $(N, v, \mathcal{F}), (N, w, \mathcal{F}) \in US^N$.

We obtain immediately that the Myerson value is additive.

Lemma 6.3.6 *If γ is an additive allocation rule that satisfies the superfluous player property, then $\gamma(N, v, \mathcal{F}) = \gamma\left(N, v^{\mathcal{F}}, \mathcal{F}\right)$, for all $(N, v, \mathcal{F}) \in US^N$.*

Proof. By additivity of γ, it suffices to show that $\gamma\left(N, v - v^{\mathcal{F}}, \mathcal{F}\right) = 0$, for all $(N, v, \mathcal{F}) \in US^N$. Indeed, for any $S \subseteq N$,

$$\begin{aligned} \left(v - v^{\mathcal{F}}\right)^{\mathcal{F}}(S) &= \sum_{T \in C_{\mathcal{F}}(S)} \left(v - v^{\mathcal{F}}\right)(T) \\ &= \sum_{T \in C_{\mathcal{F}}(S)} \left[v(T) - v^{\mathcal{F}}(T)\right] = 0. \end{aligned}$$

Therefore, all players are superfluous for any $\left(N, v - v^{\mathcal{F}}, \mathcal{F}\right) \in US^N$. Hence, taking recursively all players in the same maximal component $M \in C_{\mathcal{F}}(N)$ we have

$$\gamma\left(N, v - v^{\mathcal{F}}, \mathcal{F}\right) = \gamma\left(N, v - v^{\mathcal{F}}, \mathcal{F}_{N\setminus M}\right).$$

For all $i \in M$, we obtain

$$\gamma_i\left(N, v - v^{\mathcal{F}}, \mathcal{F}\right) = \gamma_i\left(N, v - v^{\mathcal{F}}, \mathcal{F}_{N\setminus M}\right) - 0,$$

since $i \notin \bigcup_{H \in C_{\mathcal{F}_{N\setminus M}}(N)} H$. It follows that $\gamma_i\left(N, v - v^{\mathcal{F}}, \mathcal{F}\right) = 0$, for all $i \in N$. □

Theorem 6.3.7 *The Myerson value is the unique allocation rule on US^N that satisfies additivity, the superfluous player property and point anonymity.*

Proof. Let γ be an allocation rule on US^N that also satisfies additivity, the superfluous player property and point anonymity. From Lemma 6.3.6 we deduce $\gamma(N, v, \mathcal{F}) = \gamma\left(N, v^{\mathcal{F}}, \mathcal{F}\right)$. The unanimity games $\{u_S : S \in \mathcal{F},\ S \neq \emptyset\}$ form a basis for the vector space of the $\mathcal{F}$-restricted games (see Bilbao [10]), that is,

$$v^{\mathcal{F}} = \sum_{\{S \in \mathcal{F} : S \neq \emptyset\}} \alpha_S u_S$$

for some coefficients α_S. Applying additivity, it suffices to show that the vector $\gamma(N, \alpha u_S, \mathcal{F})$, is uniquely determined for all $S \in \mathcal{F},\ S \neq \emptyset$ and $\alpha \in \mathbb{R}$. Fix S and α. If $i \in N \setminus S$ then for all coalitions $T \subseteq N$,

$$\alpha u_S(T) = \alpha \Longleftrightarrow S \subseteq T \Longleftrightarrow S \subseteq T \setminus i \Longleftrightarrow \alpha u_S(T \setminus i) = \alpha.$$

We deduce that any player that is not in S is superfluous and hence by the superfluous player property:

$$\gamma(N, \alpha u_S, \mathcal{F}) = \gamma\left(N, \alpha u_S, \mathcal{F}_{N \setminus (N \setminus S)}\right) = \gamma(N, \alpha u_S, \mathcal{F}_S).$$

Since $C_{\mathcal{F}_S}(N) = C_{\mathcal{F}}(S) = \{S\}$, component-dummy property implies that for all $i \in N \setminus S$, we have $\gamma_i(N, \alpha u_S, \mathcal{F}_S) = 0$. It only remains to compute $\gamma_i(N, \alpha u_S, \mathcal{F}_S)$ for all $i \in S$. First, for all $T \subseteq N$, we have

$$(\alpha u_S)^{\mathcal{F}_S}(T) = \sum_{H \in C_{\mathcal{F}_S}(T)} \alpha u_S(H) = \alpha \Longleftrightarrow \exists H \in \mathcal{F}_S,\ S \subseteq H \subseteq T.$$

If $H \in \mathcal{F}_S$ then $H \subseteq S$, and hence $(\alpha u_S)^{\mathcal{F}_S}(T) = \alpha$ if and only if $S \subseteq T$. Therefore, $(\alpha u_S)^{\mathcal{F}_S} = \alpha u_S$ implies

$$(\alpha u_S)^{\mathcal{F}_S}(T) = \alpha u_S(T) = \alpha \Longleftrightarrow S \subseteq T \Longleftrightarrow S \cap T = S.$$

It follows that there exists a function $f : \{0, 1, \ldots, |S|\} \rightarrow \mathbb{R}$, such that $(\alpha u_S)^{\mathcal{F}_S}(T) = f(|S \cap T|)$, for all $T \subseteq N$, where $f(0) = \cdots = f(|S| - 1) = 0$, and $f(|S|) = \alpha$. Hence $(N, \alpha u_S, \mathcal{F}_S)$ is point anonymous and applying point anonymity to the rule γ, there exists $\beta \in \mathbb{R}$ such that

$$\gamma_i(N, \alpha u_S, \mathcal{F}_S) = \begin{cases} \beta, & \text{if } i \in S \\ 0, & \text{otherwise.} \end{cases}$$

Further, $C_{\mathcal{F}_S}(N) = \{S\}$, and using component-efficiency we get

$$\sum_{i \in S} \gamma_i(N, \alpha u_S, \mathcal{F}_S) = \alpha = |S|\ \beta.$$

Then $\beta = \alpha / |S|$ and we deduce that $\gamma(N, v, \mathcal{F})$ is the Myerson value. □

6.4 The position value

We now introduce the *conference game*, derived from a cooperative game (N, v) and a union stable system $\mathcal{F} \subseteq 2^N$. The $\mathcal{F}$-restricted game $(N, v^{\mathcal{F}})$ measures the economic values of the coalitions assuming that the entire cooperation structure given by $\mathcal{F}$ can be considered. Thus, the $\mathcal{F}$-restricted game focuses on the role of a player in creating economic possibilities and establishing meaningful communication among the players.

Recall that $\mathcal{B}(\mathcal{F})$ is the basis of $\mathcal{F}$ and $\mathcal{C}(\mathcal{F}) = \{B \in \mathcal{B}(\mathcal{F}) : |B| \geq 2\}$. We use $\mathcal{C}$ to denote $\mathcal{C}(\mathcal{F})$.

Definition 6.4.1 *Let (N, v) be a game and $\mathcal{F} \subseteq 2^N$ a union stable system. The conference game is the game $(\mathcal{C}, v^{\mathcal{C}})$ where $v^{\mathcal{C}} : 2^{\mathcal{C}} \to \mathbb{R}$, is defined by $v^{\mathcal{C}}(\mathcal{A}) := v^{\overline{\mathcal{A}}}(N)$ for all $\mathcal{A} \subseteq \mathcal{C}$.*

The game $(\mathcal{C}, v^{\mathcal{C}})$ is well defined since for each $\mathcal{A} \subseteq \mathcal{C}$, its closure $\overline{\mathcal{A}}$ is a union stable system. The conference game measures the economic value of the grand coalition when specific parts of the cooperation structure are considered. Note that if the game (N, v) is zero-normalized, then $v^{\mathcal{C}}(\mathcal{C}) = v^{\overline{\mathcal{C}}}(N) = v^{\mathcal{F}}(N)$.

Remark 6.4.1 *Let (N, v, E) be a communication situation. The family*

$$\mathcal{F} = \{S \subseteq N : (S, E(S)) \text{ is a connected subgraph of } G = (N, E)\}$$

is a union stable system. The two above definitions extend the point game and the arc game respectively. These games were introduced by Borm, Owen, and Tijs [26], and for this situation we have that $\mathcal{C} = \{\{i, j\} : \{i, j\} \in E\}$.

Let $(N, v, \mathcal{F})$ be a union stable structure. We assume from now on that the underlying game (N, v) is zero-normalized. The *position value* for graph communication situations was first introduced in Meesen [172] and studied in Borm, Owen and Tijs [26]. This value was extended to hypergraphs communication situations in van den Nouweland, Borm and Tijs [188]. Now, this value will be defined in a union stable structure, by assigning to each support its Shapley value in the conference game and then dividing the value of each support equally among its participants.

Let $(N, v, \mathcal{F})$ be a union stable structure. On $\mathcal{C} \subseteq \mathcal{B}(\mathcal{F})$, we can consider the Shapley value $\Phi(\mathcal{C}, v^{\mathcal{C}}) \in \mathbb{R}^{\mathcal{C}}$ associated to the conference game $(\mathcal{C}, v^{\mathcal{C}})$.

Definition 6.4.2 *Let $(N, v, \mathcal{F})$ be a union stable structure. For $i \in N$ the position value $\pi_i(N, v, \mathcal{F})$ is given by*

$$\pi_i(N, v, \mathcal{F}) := \sum_{C \in \mathcal{C}_i} \frac{1}{|C|} \Phi_C(\mathcal{C}, v^{\mathcal{C}}),$$

where $\mathcal{C}_i = \{C \in \mathcal{C} : i \in C\}$ *denotes the set of all supports with at least two players in which* i *participates.*

Example: Let $N = \{1,2,3,4\}$ and $\mathcal{F} = \{\{1\},\{1,2,3\},\{2,3,4\},N\}$. Then $\mathcal{B}(\mathcal{F}) = \{\{1\},\{1,2,3\},\{2,3,4\}\}$ and $\mathcal{C} = \{\{1,2,3\},\{2,3,4\}\}$. Let $v : 2^N \to \mathbb{R}$ be the game defined by $v(S) = |S| - 1$, for all nonempty $S \subseteq N$, and $v(\emptyset) = 0$. Thus, $v^{\mathcal{F}}(S) = |S| - 1$, if $S \in \mathcal{F}$, and $v^{\mathcal{F}}(S) = 0$, otherwise. The conference game is given by

$\mathcal{A} \subseteq \mathcal{C}$	$\overline{\mathcal{A}}$	$C_{\overline{\mathcal{A}}}(N)$	$v^{\mathcal{C}}(\mathcal{A})$
$\{\{1,2,3\}\}$	$\{\{1,2,3\}\}$	$\{\{1,2,3\}\}$	2
$\{\{2,3,4\}\}$	$\{\{2,3,4\}\}$	$\{\{2,3,4\}\}$	2
$\{\{1,2,3\},\{2,3,4\}\}$	$\{\{1,2,3\},\{2,3,4\},N\}$	$\{N\}$	3

In this game, $\Phi_{\{1,2,3\}}(\mathcal{C}, v^{\mathcal{C}}) = \Phi_{\{2,3,4\}}(\mathcal{C}, v^{\mathcal{C}}) = 3/2$, and hence the position value is $\pi(N, v, \mathcal{F}) = (1/2, 1, 1, 1/2)$.

In order to prove that the position value is an allocation rule, we need the following result.

Lemma 6.4.1 *Let* $(N, v, \mathcal{F})$ *be a union stable structure with* $N \notin \mathcal{F}$ *and* $(\mathcal{C}, v^{\mathcal{C}})$ *the associated conference game. Define the partition* $\{\mathcal{D}_1, \mathcal{D}_2, \ldots, \mathcal{D}_p\}$ *of* $\mathcal{C}$ *by* $\mathcal{D}_i = \{D \in \mathcal{C} : D \subseteq N_i,\ N_i \in C_{\mathcal{F}}(N)\}$. *Then* $v^{\mathcal{C}} = \sum_{i=1}^{p} v^{\mathcal{D}_i}$, *where the games* $(\mathcal{C}, v^{\mathcal{D}_i})$ *are defined by* $v^{\mathcal{D}_i}(\mathcal{A}) = v^{\mathcal{C}}(\mathcal{A} \cap \mathcal{D}_i)$, *for all* $\mathcal{A} \subseteq \mathcal{C}$, $1 \le i \le p$.

Proof. If $\mathcal{A} \subseteq \mathcal{C}$, then

$$\begin{aligned} v^{\mathcal{C}}(\mathcal{A}) &= v^{\mathcal{C}}\left(\bigcup_{i=1}^{p} (\mathcal{A} \cap \mathcal{D}_i)\right) = v^{\overline{\bigcup_{i=1}^{p}(\mathcal{A}\cap\mathcal{D}_i)}}(N) \\ &= \sum_{M \in C_{\overline{\bigcup_{i=1}^{p}(\mathcal{A}\cap\mathcal{D}_i)}}(N)} v(M) = \sum_{M \in \bigcup_{i=1}^{p} C_{\overline{\mathcal{A}\cap\mathcal{D}_i}}(N)} v(M) \\ &= \sum_{i=1}^{p} \left[\sum_{M \in C_{\overline{\mathcal{A}\cap\mathcal{D}_i}}(N)} v(M) \right] = \sum_{i=1}^{p} v^{\mathcal{C}}(\mathcal{A} \cap \mathcal{D}_i) \\ &= \sum_{i=1}^{p} v^{\mathcal{D}_i}(\mathcal{A}). \end{aligned}$$

The equality in the second line follows from Proposition 6.2.2 (b) (iii). □

Theorem 6.4.2 *The position value $\pi : US^N \longrightarrow \mathbb{R}^n$ is an allocation rule.*

Proof. We first prove that the position value satisfies component-dummy. If $i \notin \bigcup_{M \in C_{\mathcal{F}}(N)} M$, this player is not in any feasible coalition. Therefore $\mathcal{C}_i = \emptyset$ and $\pi_i (N, v, \mathcal{F}) = 0$, by definition.

To prove component-efficiency, let $M \in C_{\mathcal{F}}(N)$. Thus, either $|M| = 1$ or $|M| > 1$. If $|M| = 1$, then $M = \{i\}$ and $\mathcal{C}_i = \emptyset$. Hence,

$$\sum_{i \in M} \pi_i (N, v, \mathcal{F}) = 0 = v(i),$$

since v is zero-normalized. If $|M| > 1$, then $\mathcal{C}_i \neq \emptyset$, for all $i \in M$. Moreover as $M \in C_{\mathcal{F}}(N)$, denoting all basis elements which are contained in M by B_k, $k \in K$, we have $M = \bigcup_{k \in K} B_k$. Hence, if $i \in M$, then $\mathcal{C}_i \subseteq \{B_k\}_{k \in K}$. Suppose on the contrary that $C \in \mathcal{C}_i$ and $C \neq B_k$ for all $k \in K$. This implies that $C \cap (N \setminus M) \neq \emptyset$. Moreover, since $i \in M \cap C$, it follows that $M \cup C \in \mathcal{F}$, but this contradicts that $M \in C_{\mathcal{F}}(N)$. Therefore, applying Lemma 6.4.1 and using the properties of the Shapley value:

$$\begin{aligned}
\sum_{i \in M} \pi_i (N, v, \mathcal{F}) &= \sum_{i \in M} \left[\sum_{C \in \mathcal{C}_i} \frac{1}{|C|} \Phi_C \left(\mathcal{C}, v^{\mathcal{C}}\right) \right] \\
&= \sum_{\{\{B_k\}_{k \in K} : |B_k| \geq 2\}} \left[|B_k| \frac{1}{|B_k|} \right] \Phi_{B_k} \left(\mathcal{C}, v^{\mathcal{C}}\right) \\
&= \sum_{\{\{B_k\}_{k \in K} : |B_k| \geq 2\}} \Phi_{B_k} \left(\mathcal{C}, v^{\mathcal{C}}\right) \\
&= \sum_{\{\{B_k\}_{k \in K} : |B_k| \geq 2\}} \Phi_{B_k} \left(\mathcal{C}, v^{\mathcal{D}_k}\right) = v^{\mathcal{D}_k} \left(\{B_k\}_{k \in K}\right) \\
&= v^{\mathcal{C}} \left(\mathcal{D}_k \bigcap \{B_k\}_{k \in K}\right) = v^{\mathcal{C}} \left(\{B_k\}_{k \in K}\right) = v(M).
\end{aligned}$$

□

Definition 6.4.3 *The support $H \in \mathcal{C}$ is called superfluous for $(N, v, \mathcal{F}) \in US^N$ if $v^{\mathcal{C}} (\mathcal{A}) = v^{\mathcal{C}} (\mathcal{A} \setminus \{H\})$, for all $\mathcal{A} \subseteq \mathcal{C}$, i.e., if H is a null player in the conference game.*

Definition 6.4.4 *An allocation rule $\gamma : US^N \longrightarrow \mathbb{R}^n$ has the superfluous support property if $\gamma (N, v, \mathcal{F}) = \gamma \left(N, v, \overline{\mathcal{B}(\mathcal{F}) \setminus \{H\}}\right)$, for all $(N, v, \mathcal{F}) \in US^N$ and for all superfluous supports $H \in \mathcal{C}$.*

Theorem 6.4.3 *The position value π satisfies additivity and the superfluous support property.*

Proof. Additivity follows directly from the additivity of the Shapley value. Let $(N, v, \mathcal{F}) \in US^N$. Let $H \in \mathcal{C}$ be a superfluous support. We have to prove that $\pi(N, v, \mathcal{F}) = \pi\left(N, v, \overline{\mathcal{B}(\mathcal{F}) \setminus \{H\}}\right)$. For each $i \in N$, by definition

$$\begin{aligned} \pi_i(N, v, \mathcal{F}) &= \sum_{C \in \mathcal{C}_i} \frac{1}{|C|} \Phi_C\left(\mathcal{C}, v^{\mathcal{C}}\right), \\ \pi_i\left(N, v, \overline{\mathcal{B}(\mathcal{F}) \setminus \{H\}}\right) &= \sum_{C \in \mathcal{C}_i \setminus \{H\}} \frac{1}{|C|} \Phi_C\left(\mathcal{C} \setminus \{H\}, v^{\mathcal{C} \setminus \{H\}}\right). \end{aligned}$$

If $i \in H$, we can deduce

$$\pi_i(N, v, \mathcal{F}) = \sum_{C \in \mathcal{C}_i \setminus \{H\}} \frac{1}{|C|} \Phi_C\left(\mathcal{C}, v^{\mathcal{C}}\right) + \frac{1}{|H|} \Phi_H\left(\mathcal{C}, v^{\mathcal{C}}\right),$$

where $\Phi_H\left(\mathcal{C}, v^{\mathcal{C}}\right) = 0$, since H is a superfluous support and hence a null player in the conference game $\left(\mathcal{C}, v^{\mathcal{C}}\right)$. Therefore, for all $i \in N$,

$$\pi_i(N, v, \mathcal{F}) = \sum_{C \in \mathcal{C}_i \setminus \{H\}} \frac{1}{|C|} \Phi_C\left(\mathcal{C}, v^{\mathcal{C}}\right).$$

This implies that it suffices to prove that

$$\Phi_C\left(\mathcal{C}, v^{\mathcal{C}}\right) = \Phi_C\left(\mathcal{C} \setminus \{H\}, v^{\mathcal{C} \setminus \{H\}}\right),$$

for all $C \in \mathcal{C} \setminus \{H\}$. Applying the Shapley formula,

$$\begin{aligned} \Phi_C\left(\mathcal{C}, v^{\mathcal{C}}\right) &= \sum_{\{\mathcal{S} \subseteq \mathcal{C} : C \in \mathcal{S}\}} \gamma(\mathcal{S}) \left[v^{\mathcal{C}}(\mathcal{S}) - v^{\mathcal{C}}(\mathcal{S} \setminus \{C\})\right] \\ &= \sum_{\{\mathcal{S} \subseteq \mathcal{C} : C \in \mathcal{S},\, H \in \mathcal{S}\}} \gamma(\mathcal{S}) \left[v^{\mathcal{C}}(\mathcal{S}) - v^{\mathcal{C}}(\mathcal{S} \setminus \{C\})\right] \\ &\quad + \sum_{\{\mathcal{S} \subseteq \mathcal{C} : C \in \mathcal{S},\, H \notin \mathcal{S}\}} \gamma(\mathcal{S}) \left[v^{\mathcal{C}}(\mathcal{S}) - v^{\mathcal{C}}(\mathcal{S} \setminus \{C\})\right], \end{aligned}$$

where

$$\gamma(\mathcal{S}) = \frac{(s-1)!\,(c-s)!}{c!}, \quad s = |\mathcal{S}|, \quad c = |\mathcal{C}|.$$

To each coalition $\mathcal{S} \subseteq \mathcal{C}$ which contains the superfluous support H corresponds, when H is deleted, a coalition which does not contain it, and this relation is bijective.

Therefore, $\Phi_C\left(\mathcal{C}, v^{\mathcal{C}}\right)$

$$\begin{aligned}
&= \sum_{\{\mathcal{S}\subseteq\mathcal{C}:\,C\in\mathcal{S},\,H\in\mathcal{S}\}} \gamma(\mathcal{S})\left[v^{\mathcal{C}}(\mathcal{S}\setminus\{H\}) - v^{\mathcal{C}}((\mathcal{S}\setminus\{C\})\setminus\{H\})\right] \\
&\quad + \sum_{\{\mathcal{S}\subseteq\mathcal{C}:\,C\in\mathcal{S},\,H\notin\mathcal{S}\}} \gamma(\mathcal{S})\left[v^{\mathcal{C}}(\mathcal{S}) - v^{\mathcal{C}}(\mathcal{S}\setminus\{C\})\right] \\
&= \sum_{\{\mathcal{S}\subseteq\mathcal{C}:\,C\in\mathcal{S},\,H\notin\mathcal{S}\}} \left(\gamma(\mathcal{S}\cup\{H\})+\gamma(\mathcal{S})\right)\left[v^{\mathcal{C}}(\mathcal{S}) - v^{\mathcal{C}}(\mathcal{S}\setminus\{C\})\right] \\
&= \sum_{\{\mathcal{S}\subseteq\mathcal{C}\setminus\{H\}:\,C\in\mathcal{S}\}} \frac{(s-1)!\,(c-s-1)!}{(c-1)!}\left[v^{\mathcal{C}}(\mathcal{S}) - v^{\mathcal{C}}(\mathcal{S}\setminus\{C\})\right] \\
&= \sum_{\{\mathcal{S}\subseteq\mathcal{C}\setminus\{H\}:\,C\in\mathcal{S}\}} \frac{(s-1)!\,(c-s-1)!}{(c-1)!}\left[v^{\mathcal{C}\setminus\{H\}}(\mathcal{S}) - v^{\mathcal{C}\setminus\{H\}}(\mathcal{S}\setminus\{C\})\right] \\
&= \Phi_C\left(\mathcal{C}\setminus\{H\}, v^{\mathcal{C}\setminus\{H\}}\right).
\end{aligned}$$

Note that we use that $v^{\mathcal{C}}(\mathcal{S}) = v^{\mathcal{C}\setminus\{H\}}(\mathcal{S})$ whenever $\mathcal{S}\subseteq\mathcal{C}\setminus\{H\}$. □

Let $(N, v, \mathcal{F})$ be a union stable structure. The *influence* of a player i is given by $I_i(N, v, \mathcal{F}) = \sum_{C\in\mathcal{C}_i}(1/|C|)$.

The union stable structure $(N, v, \mathcal{F})$ is called *support anonymous* if there exists a function $f : \{0, 1, \ldots, |\mathcal{C}|\} \to \mathbb{R}$ such that $v^{\mathcal{C}}(\mathcal{A}) = f(|\mathcal{A}|)$, for all $\mathcal{A}\subseteq\mathcal{C}$, i.e. if the conference game $\left(\mathcal{C}, v^{\mathcal{C}}\right)$ is anonymous.

An allocation rule $\gamma : US^N \to \mathbb{R}^n$ has the *influence property* if for each $(N, v, \mathcal{F}) \in US^N$ that is support anonymous, there exists an $\alpha\in\mathbb{R}$ such that $\gamma_i(N, v, \mathcal{F}) = \alpha I_i(N, v, \mathcal{F})$, for all $i\in N$.

So, if the value of the grand coalition only depends on the number of supports that are present, the payoffs to the players are proportional to its influence.

Theorem 6.4.4 *The position value π satisfies the influence property.*

Proof. Consider $(N, v, \mathcal{F}) \in US^N$ that is support anonymous. This means that the game $\left(\mathcal{C}, v^{\mathcal{C}}\right)$ is anonymous, hence $\Phi_C\left(\mathcal{C}, v^{\mathcal{C}}\right) = v^{\mathcal{C}}(\mathcal{C})/|\mathcal{C}|$, for all $C\in\mathcal{C}$. Thus, for all $i\in N$, we have

$$\begin{aligned}
\pi_i(N, v, \mathcal{F}) &= \sum_{C\in\mathcal{C}_i} \frac{1}{|C|}\Phi_C\left(\mathcal{C}, v^{\mathcal{C}}\right) \\
&= \sum_{C\in\mathcal{C}_i} \frac{1}{|C|}\left(\frac{1}{|\mathcal{C}|}v^{\mathcal{C}}(\mathcal{C})\right)
\end{aligned}$$

$$= \frac{v^{\mathcal{C}}(\mathcal{C})}{|\mathcal{C}|} \sum_{C \in \mathcal{C}_i} \frac{1}{|\mathcal{C}|}$$
$$= \frac{v^{\mathcal{C}}(\mathcal{C})}{|\mathcal{C}|} I_i(N, v, \mathcal{F}).$$

We may conclude with the choice of $\alpha = v^{\mathcal{C}}(\mathcal{C}) / |\mathcal{C}|$. □

The following example illustrates the influence property of the position value in a union stable structure.

Example: Let $(N, v, \mathcal{F})$ be the union stable structure defined in example on page 143. Since $\mathcal{C}_1 = \{\{1,2,3\}\}$ the influence of player 1 is $I_1 = 1/3$, and since $\mathcal{C}_2 = \mathcal{C}_3 = \{\{1,2,3\}, \{2,3,4\}\}$, we obtain $I_2 = I_3 = 2/3$. Further, $\mathcal{C}_4 = \{\{2,3,4\}\}$ implies $I_4 = 1/3$. Then $I(N, v, \mathcal{F}) = (1/3, 2/3, 2/3, 1/3)$. In this example, $(N, v, \mathcal{F})$ is a support anonymous system. Indeed, let $f : \{0,1,2\} \to \mathbb{R}$ be such that $f(0) = 0,\ f(1) = 2$, and $f(2) = 3$, then $v^{\mathcal{C}}(\mathcal{A}) = f(|\mathcal{A}|),\ \mathcal{A} \subseteq \mathcal{C}$. The position value is

$$\pi(N, v, \mathcal{F}) = \frac{3}{2}\left(\frac{1}{3}, \frac{2}{3}, \frac{2}{3}, \frac{1}{3}\right),$$

where the number $\alpha = 3/2$, as it was shown in the proof of Theorem 6.4.4.

Lemma 6.4.5 *Let $\mathcal{F}$ be a union stable system, such that the expression of each non-unitary feasible coalition as a union of non-unitary supports is unique. If $S = \bigcup_{i \in I} S_i$, and $T = \bigcup_{j \in J} T_j$ with $S_i, T_j \in \mathcal{C}$, for all i, j, then $S \subseteq T$ if and only if $\{S_i\}_{i \in I} \subseteq \{T_j\}_{j \in J}$.*

Proof. Let $S \subseteq T$. If $\{S_i\}_{i \in I}$ is not contained in $\{T_j\}_{j \in J}$, then there is S_k, $k \in I$, such that $S_k \neq T_j$, for all $j \in J$. Hence,

$$T = \bigcup_{j \in J} T_j = \left(\bigcup_{j \in J} T_j\right) \cup S_k,$$

since $S_k \subseteq S \subseteq T$, and consequently, the expression of T as a union of supports is not unique in contradiction with the hypothesis. The converse is obvious. □

Next, we obtain an axiomatic characterization of the position value on the class of the union stable structures $(N, v, \mathcal{F})$ such that:

1. For all $S, T \in \mathcal{F}$, with $|S \cap T| \geq 2$ we have $S \cap T \in \mathcal{F}$.

2. All non-unitary feasible coalitions can be written in a unique way as a union of non-unitary supports.

We will denote by USI^N the subclass of US^N where the above two conditions are true. The following result is a generalization of the characterization for the position value given by Borm, Owen and Tijs [26], since in a communication situation the non-unitary supports are the edges of the graph and the feasible coalitions are the connected ones and the subclass of communication situations for which the communication graphs do not contain cycles satisfy the two above conditions.

Theorem 6.4.6 *The position value is the unique allocation rule on USI^N that satisfies additivity, the superfluous support property and the influence property.*

Proof. To prove uniqueness, let $\pi, \gamma : USI^N \to \mathbb{R}^n$ be two allocation rules that satisfy the three properties. We prove that $\pi(N, v, \mathcal{F}) = \gamma(N, v, \mathcal{F})$. The game (N, v) is zero-normalized and $v = \sum_{\{T \subseteq N : |T| \geq 2\}} \alpha_T u_T$, where u_T are the unanimity games. Moreover, since π and γ are additive allocation rules, it suffices to show $\pi(N, \alpha u_T, \mathcal{F}) = \gamma(N, \alpha u_T, \mathcal{F})$ for all $T \subseteq N$, $|T| \geq 2$, $\alpha \in \mathbb{R}$. Two cases will be distinguished:
(a) There exists a coalition $S \in \mathcal{F}$ such that $T \subseteq S$.
(b) There is no coalition $S \in \mathcal{F}$ such that $T \subseteq S$.
We first consider (b). As there is no coalition $S \in \mathcal{F}$ such that $T \subseteq S$,

$$(\alpha u_T)^{\mathcal{C}}(\mathcal{A}) = \alpha \sum_{M \in C_{\overline{\mathcal{A}}}(N)} u_T(M) = 0, \quad \text{for all } \mathcal{A} \subseteq \mathcal{C},$$

which implies that the conference game associated to αu_T is the null game, and consequently, $\Phi_C\left(\mathcal{C}, (\alpha u_T)^{\mathcal{C}}\right) = 0$, for all $C \in \mathcal{C}$. So $\pi_i(N, \alpha u_T, \mathcal{F}) = 0$, for all $i \in N$. On the other hand, if $(\alpha u_T)^{\mathcal{C}} = 0$, it means that each support of $\mathcal{C}$ is superfluous and hence $\gamma(N, \alpha u_T, \mathcal{F}) = \gamma\left(N, \alpha u_T, \overline{\mathcal{B}(\mathcal{F}) \setminus \mathcal{C}}\right)$. Moreover, the triple $\left(N, \alpha u_T, \overline{\mathcal{B}(\mathcal{F}) \setminus \mathcal{C}}\right)$ is support anonymous and then the influence property implies that there is a $\beta \in \mathbb{R}$ such that

$$\gamma\left(N, \alpha u_T, \overline{\mathcal{B}(\mathcal{F}) \setminus \mathcal{C}}\right) = \beta I\left(N, \alpha u_T, \overline{\mathcal{B}(\mathcal{F}) \setminus \mathcal{C}}\right) = \mathbf{0}.$$

Therefore, $\pi(N, \alpha u_T, \mathcal{F}) = \gamma(N, \alpha u_T, \mathcal{F}) = \mathbf{0}$.

Now consider (a). By assumption, the set $\{F \in \mathcal{F} : T \subseteq F\}$ is nonempty and define $\overline{T} = \bigcap \{F \in \mathcal{F} : T \subseteq F\}$. We have that $\overline{T} \neq \emptyset$ and by condition 1 of USI^N it follows that $\overline{T} \in \mathcal{F}$. It is also immediate that $\overline{T}$ is the minimal feasible set that contains the set T. Proposition 6.2.3 implies there are $B_i \in \mathcal{C}$ such that $\overline{T} = \bigcup_{i \in I} B_i$. Thus, the conference game $(\alpha u_T)^{\mathcal{C}} : 2^{\mathcal{C}} \to \mathbb{R}$, associated to αu_T is given by

$$(\alpha u_T)^{\mathcal{C}}(\mathcal{A}) = \begin{cases} \alpha, & \text{if } \overline{T} \in \overline{\mathcal{A}} \\ 0, & \text{otherwise.} \end{cases}$$

Moreover, $\overline{T} \in \overline{\mathcal{A}} \Leftrightarrow \{B_i\}_{i\in I} \subseteq \mathcal{A}$, since each non-unitary feasible coalition can be written in a unique way as a union of non-unitary supports. Hence

$$(\alpha u_T)^{\mathcal{C}}(\mathcal{A}) = \begin{cases} \alpha, & \text{if } \{B_i\}_{i\in I} \subseteq \mathcal{A} \\ 0, & \text{otherwise.} \end{cases}$$

Notice that all supports $B \in \mathcal{C}$ such that $B \notin \{B_i\}_{i\in I}$, are superfluous in the conference game, and so the superfluous support property implies that

$$\begin{aligned} \pi(N, \alpha u_T, \mathcal{F}) &= \pi(N, \alpha u_T, \mathcal{F}'), \\ \gamma(N, \alpha u_T, \mathcal{F}) &= \gamma(N, \alpha u_T, \mathcal{F}'), \end{aligned}$$

where $\mathcal{F}' = \left(\bigcup\{\{j\} : \{j\} \in \mathcal{F}\}\right) \cup \left(\overline{\{B_i\}}_{i\in I}\right)$.

The conference game associated to αu_T in $\mathcal{F}'$ is support anonymous since,

$$(\alpha u_T)^{\{B_i\}_{i\in I}}(\mathcal{A}) = \begin{cases} \alpha, & \text{if } \mathcal{A} = \{B_i\}_{i\in I} \\ 0, & \text{otherwise.} \end{cases}$$

Hence, applying the influence property for the allocation rules π and γ, there exist $\beta, \delta \in \mathbb{R}$ such that

$$\begin{aligned} \pi(N, \alpha u_T, \mathcal{F}') &= \delta\, I(N, \alpha u_T, \mathcal{F}'), \\ \gamma(N, \alpha u_T, \mathcal{F}') &= \beta\, I(N, \alpha u_T, \mathcal{F}'). \end{aligned}$$

So, if $i \in N \setminus \overline{T}$ then $I_i(N, \alpha u_T, \mathcal{F}') = 0$, and therefore

$$\pi_i(N, \alpha u_T, \mathcal{F}') = \gamma_i(N, \alpha u_T, \mathcal{F}') = 0.$$

Let $i \in \overline{T}$. Since $\overline{T} \in C_{\mathcal{F}'}(N)$, component-efficiency implies

$$\sum_{i\in\overline{T}} \gamma_i(N, \alpha u_T, \mathcal{F}') = \alpha u_T\left(\overline{T}\right) = \sum_{i\in\overline{T}} \pi_i(N, \alpha u_T, \mathcal{F}'),$$

and thus,

$$\sum_{i\in\overline{T}} \beta\, I_i(N, \alpha u_T, \mathcal{F}') = \sum_{i\in\overline{T}} \delta\, I_i(N, \alpha u_T, \mathcal{F}').$$

Subtracting

$$(\beta - \delta) \sum_{i\in\overline{T}} I_i(N, \alpha u_T, \mathcal{F}') = 0,$$

and as for $\overline{T} \in C_{\mathcal{F}'}(N)$ it is satisfied that $\sum_{i\in\overline{T}} I_i(N, \alpha u_T, \mathcal{F}') \neq 0$, it is deduced that $\beta = \delta$. □

6.5 Convexity of the conference game

We study now the conditions under which the convexity is inherited from the underlying game to the conference game and we show that under the same conditions the position value is in the core of the restricted game. Shapley [217] showed that a game (N, v) is convex if and only if

$$v(T_1 \cup T_2) - v(T_1) - v(T_2) \geq v(S_1 \cup S_2) - v(S_1) - v(S_2),$$

where $T_1 \cap T_2 = \emptyset$, $S_1 \subseteq T_1$, $S_2 \subseteq T_2$. In the following result we will extend this condition. For this we will use that a game (N, v) is convex if and only if

$$v(T \cup R) - v(T) \geq v(S \cup R) - v(S),$$

for all $S \subseteq T \subseteq N \setminus R$.

Lemma 6.5.1 *If the game (N, v) is convex, then*

$$v\left(\bigcup_{i=1}^{k} T_i\right) - \sum_{i=1}^{k} v(T_i) \geq v\left(\bigcup_{i=1}^{k} S_i\right) - \sum_{i=1}^{k} v(S_i),$$

for all $k = 2, 3, \ldots$, where $T_i \cap T_j = \emptyset$, $i \neq j$, and $S_i \subseteq T_i$, for all $i \in N$.

Proof. Let $S_i \subseteq T_i$ for all i and $T_i \cap T_j = \emptyset$. It is satisfied that

$$v\left(T_1 \cup T_2 \cup T_3 \cup \cdots \cup T_k\right) - v(T_1) \geq v\left(S_1 \cup T_2 \cup T_3 \cup \cdots \cup T_k\right) - v(S_1),$$

since $S_1 \subseteq T_1 \subseteq N \setminus \{T_2 \cup \cdots \cup T_k\}$. Similarly,

$$v\left(S_1 \cup T_2 \cup T_3 \cup \cdots \cup T_k\right) - v(T_2) \geq v\left(S_1 \cup S_2 \cup T_3 \cup \cdots \cup T_k\right) - v(S_2),$$

since $S_2 \subseteq T_2 \subseteq N \setminus \{S_1 \cup T_3 \cup \cdots \cup T_k\}$. Following in the same way:

$$v\left(S_1 \cup S_2 \cup \cdots \cup S_{k-1} \cup T_k\right) - v(T_k) \geq v\left(S_1 \cup S_2 \cup \cdots \cup S_{k-1} \cup S_k\right) - v(S_k),$$

since $S_k \subseteq T_k \subseteq N \setminus \{S_1 \cup S_2 \cup \cdots \cup S_{k-1}\}$. Now, by adding all k inequalities we find

$$v\left(T_1 \cup T_2 \cup \cdots \cup T_k\right) - \sum_{i=1}^{k} v(T_i) \geq v\left(S_1 \cup S_2 \cup \cdots \cup S_k\right) - \sum_{i=1}^{k} v(S_i).$$

□

Theorem 6.5.2 *Let $(N, v, \mathcal{F}) \in USI^N$. If the game (N, v) is convex, then the conference game $\left(\mathcal{C}, v^{\mathcal{C}}\right)$ is convex.*

Proof. We have to prove that given $C \in \mathcal{C}$ and $\mathcal{S} \subseteq \mathcal{T} \subseteq \mathcal{C} \setminus \{C\}$ it holds that $v^{\mathcal{C}}(\mathcal{T} \cup \{C\}) - v^{\mathcal{C}}(\mathcal{T}) \geq v^{\mathcal{C}}(\mathcal{S} \cup \{C\}) - v^{\mathcal{C}}(\mathcal{S})$. We see from the definition of the conference game that the inequality is equivalent to

$$\sum_{T \in C_{\overline{\mathcal{T} \cup \{C\}}}(N)} v(T) - \sum_{T \in C_{\overline{\mathcal{T}}}(N)} v(T) \geq \sum_{S \in C_{\overline{\mathcal{S} \cup \{C\}}}(N)} v(S) - \sum_{S \in C_{\overline{\mathcal{S}}}(N)} v(S).$$

Let $C_{\overline{\mathcal{T}}}(N) = \{T_1, \ldots, T_k\} \cup \{T_{k+1}, \ldots, T_h\}$ such that $T_i \cap C \neq \emptyset$ for all $i = 1, \ldots, k$ and $T_i \cap C = \emptyset$ for all $i = k+1, \ldots, h$. Let $i = 1, \ldots, k$, we first show that $|T_i \cap C| = 1$. Suppose $|T_i \cap C| \geq 2$, then $T_i \cap C \in \mathcal{F}$, since $(N, v, \mathcal{F}) \in USI^N$ and $T_i \cap C \neq C$ because otherwise $C \subseteq T_i$ and then $C \in \overline{\mathcal{T}}$ but this contradicts the fact that $\mathcal{T} \subseteq \mathcal{C} \setminus \{C\}$. As $T_i \cap C \subseteq T_i \in \overline{\mathcal{T}}$ and $T_i \cap C \in \mathcal{F}$, then $T_i \cap C$ is either a support or a union of non-unitary supports of $\mathcal{T}$. Hence, since $C = C \cup (T_i \cap C)$ and $T_i \cap C \neq C$, we deduce that C could be expressed in two different ways as a union of supports contradicting that $(N, v, \mathcal{F}) \in USI^N$. Clearly, $T_1 \cap C \neq \emptyset, \ldots, T_k \cap C \neq \emptyset$, and

$$T_1 \cup C \in \mathcal{F}, \ldots, T_k \cup C \in \mathcal{F},$$

hence $T_1 \cup \cdots \cup T_k \cup C \in \mathcal{F}$. Moreover, as we showed above, every $T_1, \ldots, T_k$ contains a unique element of C. For this reason, the maximal feasible coalitions of N in $\overline{\mathcal{T} \cup \{C\}}$ will be $\{T_1 \cup \cdots \cup T_k \cup C, T_{k+1}, \ldots, T_h\}$. Therefore, we obtain

$$\sum_{T \in C_{\overline{\mathcal{T} \cup \{C\}}}(N)} v(T) - \sum_{T \in C_{\overline{\mathcal{T}}}(N)} v(T) = v(T_1 \cup \cdots \cup T_k \cup C) - \sum_{i=1}^{k} v(T_i).$$

An analogous reasoning leads to

$$\sum_{S \in C_{\overline{\mathcal{S} \cup \{C\}}}(N)} v(S) - \sum_{S \in C_{\overline{\mathcal{S}}}(N)} v(S) = v(S_1 \cup \cdots \cup S_p \cup C) - \sum_{i=1}^{p} v(S_i),$$

with $S_i \in C_{\overline{\mathcal{S}}}(N)$, $S_i \cap C \neq \emptyset$, $i = 1, \ldots, p$ and each S_i contains a unique element of C. Consequently, the following inequality remains to be proved:

$$v(T_1 \cup \cdots \cup T_k \cup C) - \sum_{i=1}^{k} v(T_i) \geq v(S_1 \cup \cdots \cup S_p \cup C) - \sum_{i=1}^{p} v(S_i). \quad (6.2)$$

First, $T_1 \cup \cdots \cup T_k \cup C = T_1 \cup \cdots \cup T_k \cup C'$, where $C' \subseteq C$ is defined by $C' = \{i \in C : i \notin T_1 \cup \cdots \cup T_k\}$. Moreover, $S_1 \cup \cdots \cup S_p \cup C = S_1 \cup \cdots \cup S_p \cup C''$, where $C'' \subseteq C$ is defined by $C'' = \{i \in C : i \notin S_1 \cup \cdots \cup S_p\}$.

On the other hand, since $\mathcal{S} \subseteq \mathcal{T}$ we have $\overline{\mathcal{S}} \subseteq \overline{\mathcal{T}}$ and if S_i is a maximal feasible coalition of N in $\overline{\mathcal{S}}$, then $S_i \subseteq T_i$ for a unique maximal feasible coalition T_i of N in $\overline{\mathcal{T}}$. So, $p \leq k$ and without loss of generality $S_1 \subseteq T_1, \ldots, S_p \subseteq T_p$.

Since $S_i \subseteq T_i$ they must contain the same elements of C, and $S_i \cap C = T_i \cap C$. We observe that $C' \subseteq C'' \subseteq C$ implies $C'' = C' \cup (C'' \setminus C') \subseteq C$, and for each T_j, where $j = p+1, \ldots, k$, there is an element $s(j) \in C'' \setminus C'$ such that $s(j) \in T_j$. Consequently $s(j) \neq s(i)$ for all $i, j \in \{p+1, \ldots, k\}$, $i \neq j$. By Lemma 6.5.1, convexity of (N, v) implies

$$\begin{aligned} v & \left(T_1 \cup \cdots \cup T_p \cup T_{p+1} \cup \cdots \cup T_k \cup C'\right) - v(T_1) - \cdots - v(T_k) - v\left(C'\right) \\ \geq & \ v\left(S_1 \cup \cdots \cup S_p \cup \{s(p+1)\} \cup \ldots \cup \{s(k)\} \cup C'\right) - v(S_1) \\ - & \ \cdots - v(S_p) - v\left(\{s(p+1)\}\right) - \cdots - v\left(\{s(k)\}\right) - v\left(C'\right). \end{aligned}$$

By simplifying the above expression and taking into account that the game (N, v) is zero-normalized, we find

$$\begin{aligned} v & \left(T_1 \cup \cdots \cup T_p \cup T_{p+1} \cup \cdots \cup T_k \cup C'\right) - \sum_{i=1}^{k} v(T_i) \\ \geq & \ v\left(S_1 \cup \cdots \cup S_p \cup (C'' \setminus C') \cup C'\right) - \sum_{i=1}^{p} v(S_i). \end{aligned}$$

Since $T_1 \cup \cdots \cup T_k \cup C' = T_1 \cup \cdots \cup T_k \cup C$, and also

$$S_1 \cup \cdots \cup S_p \cup (C'' \setminus C') \cup C' = S_1 \cup \cdots \cup S_p \cup C,$$

we may conclude that the inequality (6.2) is satisfied. □

Theorem 6.5.3 *Let $(N, v, \mathcal{F}) \in USI^N$ such that (N, v) is a convex game. Then, $\pi(N, v, \mathcal{F})$ belongs to the core $C\left(v^{\mathcal{F}}\right)$.*

Proof. By Theorem 6.5.2, $\left(\mathcal{C}, v^{\mathcal{C}}\right)$ is convex and, therefore $\Phi\left(\mathcal{C}, v^{\mathcal{C}}\right) \in C\left(v^{\mathcal{C}}\right)$. We first show that $\Phi\left(\mathcal{C}, v^{\mathcal{C}}\right) \geq \mathbf{0}$. By definition

$$\Phi_C\left(\mathcal{C}, v^{\mathcal{C}}\right) = \sum_{\{\mathcal{S} \subseteq \mathcal{C} : C \notin \mathcal{S}\}} \gamma(\mathcal{S}) \left[v^{\mathcal{C}}\left(\mathcal{S} \cup \{C\}\right) - v^{\mathcal{C}}\left(\mathcal{S}\right)\right],$$

for all $C \in \mathcal{C}$. So, we have to prove $v^{\mathcal{C}}\left(\mathcal{S} \cup \{C\}\right) - v^{\mathcal{C}}\left(\mathcal{S}\right) \geq 0$ for all $C \in \mathcal{C}$. By convexity of $v^{\mathcal{C}}$ we have $v^{\mathcal{C}}\left(\mathcal{S} \cup \{C\}\right) - v^{\mathcal{C}}\left(\mathcal{S}\right) \geq v^{\mathcal{C}}\left(\{C\}\right)$, therefore it suffices to show that $v^{\mathcal{C}}\left(\{C\}\right) \geq 0$. Indeed

$$v^{\mathcal{C}}\left(\{C\}\right) = \sum_{T \in C_{\overline{\{C\}}}(N)} v(T) = v(C) \geq 0,$$

since v is superadditive and zero-normalized. From $\Phi_C\left(\mathcal{C}, v^{\mathcal{C}}\right) \geq 0$ for all $C \in \mathcal{C}$, we deduce that $\pi(N, v, \mathcal{F}) \in \mathbb{R}_+^N$. We prove that $\sum_{i \in N} \pi_i(N, v, \mathcal{F}) = v^{\mathcal{F}}(N)$.

Putting $C_{\mathcal{F}}(N) = \{N_1, \ldots, N_k\}$, we have

$$\begin{aligned}
\sum_{i \in N} \pi_i(N, v, \mathcal{F}) &= \sum_{i \notin \bigcup_{j=1}^k N_j} \pi_i(N, v, \mathcal{F}) + \sum_{i \in \bigcup_{j=1}^k N_j} \pi_i(N, v, \mathcal{F}) \\
&= \sum_{i \in \bigcup_{j=1}^k N_j} \pi_i(N, v, \mathcal{F}) \\
&= \sum_{j=1}^{k} \left[\sum_{i \in N_j} \pi_i(N, v, \mathcal{F}) \right] \\
&= \sum_{j=1}^{k} v(N_j) = v^{\mathcal{F}}(N),
\end{aligned}$$

since the position value is component-efficient and satisfies component-dummy because it is an allocation rule.

Finally, we prove that $\sum_{i \in S} \pi_i(N, v, \mathcal{F}) \geq v(S)$, for all $S \in \mathcal{F}$, $S \neq \emptyset$. If S is a unitary coalition $\{i\}$, then $\pi_i(N, v, \mathcal{F}) \geq v(i) = 0$. If S is not a unitary coalition, as S is feasible, it is a union of non-unitary supports. Moreover, this expression is unique. Thus, let $S = \bigcup_{k \in K} S_k$, $S_k \in \mathcal{C}$. Then

$$\sum_{i \in S} \pi_i(N, v, \mathcal{F}) = \sum_{i \in S} \left[\sum_{C \in \mathcal{C}_i} \frac{1}{|C|} \Phi_C\left(\mathcal{C}, v^{\mathcal{C}}\right) \right].$$

As $\mathcal{C}_i \supseteq \{S_k : i \in S_k\}$ and $\Phi_C\left(\mathcal{C}, v^{\mathcal{C}}\right) \geq 0$ for all $C \in \mathcal{C}$, we obtain that

$$\begin{aligned}
\sum_{i \in S} \pi_i(N, v, \mathcal{F}) &\geq \sum_{i \in S} \left[\sum_{\{S_k : i \in S_k\}} \frac{1}{|S_k|} \Phi_{S_k}\left(\mathcal{C}, v^{\mathcal{C}}\right) \right] \\
&= \sum_{k \in K} \left[\frac{1}{|S_k|} |S_k| \right] \Phi_{S_k}\left(\mathcal{C}, v^{\mathcal{C}}\right) \\
&\geq v^{\mathcal{C}}\left(\{S_k\}_{k \in K}\right) \\
&= \sum_{M \in C_{\overline{\{S_k\}_{k \in K}}}(N)} v(M) \\
&= v(S),
\end{aligned}$$

since $S \in \overline{\{S_k\}_{k \in K}}$ and $S = \bigcup_{k \in K} S_k$. The last inequality is due to the fact that $\Phi\left(\mathcal{C}, v^{\mathcal{C}}\right) \in C\left(v^{\mathcal{C}}\right)$. □

6.6 Project: Solution continuation games

Hamiache [120] [121] proposed a new generalization of the Shapley value for graph-restricted games. This extension for limited communication between players modeled by a simple graph, is called *value with incomplete communication* and it is different from the Myerson value and from the position value.

Recall that if (N, v) is a game and $G = (N, E)$ is a graph, we define its associated graph-restricted game $v^G : 2^N \to \mathbb{R}$ by:

$$v^G(S) := \sum_{T \in \Pi(S)} v(T),$$

where $\Pi(S)$ is the collection of the connected components of the subgraph $G[S]$ induced by the vertex set S. Note that the components form a partition of S and $|\Pi(S)| = 1$ if and only if S is connected. The Myerson value is the Shapley value of thegraph-restricted game (N, v^G) and Hamiache used a new associated game for obtaining his value. First, we introduce his framework.

Definition 6.6.1 *Let* $\Upsilon_v^G = \{(S, v|_{2^S}, G[S]) : S \subseteq N\}$ *be the set of all subgames of* (N, v) *with communication structure induced by the graph* $G = (N, E)$. *A value on* Υ_v^G *is a function* σ *which associates with each subgame with communication structure* $(S, v|_{2^S}, G[S])$ *exactly one vector* $\sigma(S, v|_{2^S}, G[S]) \in \mathbb{R}^{|S|}$.

Definition 6.6.2 *Given a value* σ *on* Υ_v^G, *its associated game, denoted by* (N, v_σ^*, G), *is defined by*

$$\begin{aligned} v_\sigma^*(S) &= v(S) + \sum_{j \in S^* \setminus S} [\sigma_j(S \cup j, v|_{2^{S \cup j}}, G[S \cup j]) - v(j)] \quad \text{if } |\Pi(S)| = 1, \\ v_\sigma^*(S) &= (v_\sigma^*)^G(S) = \sum_{T \in \Pi(S)} v_\sigma^*(T) \quad \text{otherwise,} \end{aligned}$$

where S^* *is the set of all the vertices of the graph* G *which are adjacent with at least one vertex of* S.

The definition means that players in a connected coalition S can only communicate directly with players in $S^* \setminus S$ and $\sigma_j(S \cup j, v|_{2^{S \cup j}}, G[S \cup j]) - v(j)$ is the surplus obtained by the player j. Moreover, the following system of axioms for values are introduced:

1. *Component efficiency*: For all games (N, v, G) and all $S \in C_G(N)$, we have $\sum_{i \in S} \sigma_i(N, v, G) = v(S)$.
2. *Linearity:* For all numbers $a, b \in \mathbb{R}$ and games (N, v, G), (N, w, G), we have $\sigma(N, av + bw, G) = a\,\sigma(N, v, G) + b\,\sigma(N, w, G)$.

3. *Independence of irrelevant players:* For all connected coalitions $T \subseteq S$ and for all players $i \in S$, we have $\sigma_i(N, u_T, G) = \sigma_i(S, u_T, G[S])$.

4. *Positivity:* For all connected coalitions T and for all players $i \in T$, the values $\sigma_i(T, u_T, G[T])$ are positive.

5. *Associated consistency:* For all games (N, v, G) we have

$$\sigma(N, v, G) = \sigma(N, v_\sigma^*, G).$$

Hamiache proved that these axioms are sufficient to ensure the existence and uniqueness of values on Υ_v^G and that this value coincides with the Shapley value when G is the complete graph of order $|N|$.

Let $(N, v, \mathcal{F})$ be a union stable structure and let Φ be a value on the collection

$$\Upsilon_v^{\mathcal{F}} = \{(S, v|_{2^S}, \mathcal{F}_S) : S \subseteq N\},$$

where $\mathcal{F}_S = \{F \in \mathcal{F} : F \subseteq S\}$. For a feasible set $S \in \mathcal{F}$, we consider the set of feasible continuations of S, given by

$$\Gamma(S) = \{j \in N \setminus S : S \cup j \in \mathcal{F}\}.$$

As an example consider a graph G and let

$$\mathcal{F} = \{S \subseteq N : G[S] \text{ is a connected subgraph of } G\}.$$

In this case for all connected coalitions $S \in \mathcal{F}$ we have $\Gamma(S) = S^* \setminus S$.

Now we are ready to give a definition of the associated game in the framework of union stable structures.

Definition 6.6.3 *Given a value Φ on $\Upsilon_v^{\mathcal{F}}$ we say that the game $(N, v_\Phi^*, \mathcal{F})$ is a Φ-continuation game if it is defined by*

$$\begin{aligned} v_\Phi^*(S) &= v(S) + \sum_{j \in \Gamma(S)} \left[\Phi_j(S \cup j, v|_{2^{S \cup j}}, \mathcal{F}_{S \cup j}) - v(j)\right] \quad \textit{if } S \in \mathcal{F}, \\ v_\Phi^*(S) &= (v_\Phi^*)^{\mathcal{F}}(S) = \sum_{T \in C_{\mathcal{F}}(S)} v_\Phi^*(T) \quad \textit{otherwise.} \end{aligned}$$

Thereby, different notions of limited communication between players have natural representations using union stable structures $\mathcal{F} \subseteq 2^N$.

Chapter 7

Values for games on convex geometries

In the classical model of cooperative games it is generally assumed that there are no restrictions on cooperation and hence, every subset of players is a feasible coalition. However, in many social and economic situations, this model does not apply. Examples are provided by local public goods which are supplied by local communities, social and sports clubs, labor unions, political parties, and other institutions. Situations like these were studied by Myerson [182] [183] and this line of research was continued by Owen [195], van den Nouweland and Borm [187], and Borm, Owen and Tijs [26].

Faigle and Kern [78] proposed a model in which cooperation among players is restricted to some family of subsets of players, the feasible coalitions of the game. Their idea is to restrict the allowable coalitions by using underlying partially ordered sets. In our model, we will define the feasible coalitions by using the combinatorial theory of convexity called *convex geometries*, which includes the model of Faigle and Kern. The idea in this model of abstract convexity is that if the players form coalitions because they agree on certain propositions, then intersections of feasible coalitions should also be feasible, since they agree on a larger set of propositions. On the other hand, the big coalition N is reached by the following processes: a given feasible coalition can be enlarged to a feasible coalition by using an augmentation property.

In this chapter, we extend the work of Weber [244] on probabilistic values to games on convex geometries. We introduce *values* for these games and observe the axioms which characterize such values in detail. In particular, compatible-order values are considered. Finally, we show how to derive the Shapley value and the Banzhaf index from a set of axioms.

7.1 Games on convex geometries

Recall that a *cooperative game* is a pair (N, v) where N is a finite set of n elements and $v : 2^N \to \mathbb{R}$, is a function such that $v(\emptyset) = 0$. The *players* are the elements of N and the *coalitions* are the elements $S \subseteq N$ of the Boolean algebra 2^N. Taking into account that there exist situations in which there are restrictions in the cooperation among the players, we introduce games defined on families of feasible coalitions.

In this chapter, we consider the following combinatorial structure. A family $\mathcal{L}$ of subsets of N is a *convex geometry* if it satisfies the following properties:

(C1) $\emptyset \in \mathcal{L}$,

(C2) $\mathcal{L}$ is closed under intersection,

(C3) If $S \in \mathcal{L}$ and $S \neq N$, then there exists $j \in N \setminus S$ such that $S \cup j \in \mathcal{L}$.

Property (C2) implies that intersections of feasible coalitions should also be feasible, since the players agree on a profile of cooperation and (C3) is the augmentation property.

In the model of *conference structure* by Myerson [183], two players are connected if they can be coordinated by meeting in separate conferences which have some members in common to serve as intermediaries. In our model, the sets of intermediaries are in the cooperation structure.

We call the sets in a convex geometry *convex sets*. A *maximal chain* of $\mathcal{L} \subseteq 2^N$ is an ordered collection of convex sets that is not contained in any larger chain. Edelman and Jamison [65] showed that every maximal chain contains $n + 1$ convex sets

$$\emptyset = S_0 \subset S_1 \subset \cdots \subset S_{n-1} \subset S_n = N,$$

and the cardinal $|S_i| = i$, for all $i = 0, 1, \ldots, n$. Moreover, the *hierarchical situations* by Moulin [178], when users pay their incremental costs according to an ordering of N, can be modeled by convex geometries.

Definition 7.1.1 *A game on a convex geometry $\mathcal{L}$ is a function $v : \mathcal{L} \to \mathbb{R}$, with $v(\emptyset) = 0$.*

The coalitions are the convex sets of $\mathcal{L}$ and the players are the elements $i \in N$. We denote by $\Gamma(\mathcal{L})$ the vector space of all games defined on the convex geometry $\mathcal{L} \subseteq 2^N$.

Example: A *communication structure* is a triple (N, G, v), where (N, v) is a game and $G = (N, E)$ is a graph. If $G = (N, E)$ is a connected block graph and

$$\mathcal{L} = \{S \subseteq N : (S, E(S)) \text{ is a connected subgraph of } G\},$$

then we have that $\mathcal{L}$ is a convex geometry.

Example: A subset S of a poset $(P, \leq)$ is *convex* whenever $a \in S$, $b \in S$ and $a \leq b$ imply that the interval

$$[a, b] = \{x \in P : a \leq x \leq b\} \subseteq S.$$

If $(P, \leq)$ is finite, then ex(C) is the union of the maximal and minimal elements of C. Birkhoff and Bennett [20] proved that the family of the convex sets $Co(P)$ is a convex geometry. Edelman [67] studied *voting games* defined on the convex geometry $Co\left(\mathbf{n}\right)$, where $\mathbf{n} = (1 < 2 < \cdots < n)$ is the linear order given by the policy order left-right.

Example: Let $N = \{1, 2, \ldots, n\}$ be voters and $\mathbb{R}^d$ be an d-dimensional issue space. We will denote by x_i the ideal point of the voter i in this space. Let $u : \mathbb{R}^d \to \mathbb{R}$ be a convex function such that $u(x) = u(-x)$, for all $x \in \mathbb{R}^d$. For each $y \in \mathbb{R}^d$ we will let $u_i(y) = u(y - x_i)$ be the utility of the outcome y to voter i.

Assume that each voter i will vote in favor of an outcome y if $u_i(y) > \epsilon_y$. Suppose that every voter $i \in A \subset N$ will vote in favor of y and that for some $j \notin A$ we have that x_j is in the convex hull of $\{x_i : i \in A\}$. It follows from the convexity of the function u that $u_j(y) > \epsilon_y$ and hence j will vote in favor of y as well. Thus coalitions $A \subseteq N$ that form in this model have the property that x_j in the convex hull of $\{x_i : i \in A\}$ implies that $j \in A$.

Since we do not know very much about the function u or the threshold values ϵ_y, it would be reasonable to assume that all coalitions with this closure property might form. This collection of subsets is a convex geometry (Edelman and Jamison [65]).

Example: Let $(P, \leq)$ be a poset. For any $X \subseteq P$,

$$X \longmapsto \overline{X} := \{y \in P : y \leq x \text{ for some } x \in X\},$$

defines a closure operator on $(P, \leq)$. Its closed sets are the *order ideals* (down sets) of $(P, \leq)$, and we use $J(P)$ to denote this convex geometry closed under set-union. Note that ex(S) is the set of all maximal points $Max(S)$ of the subposet $S \in J(P)$.

The games $(\mathcal{C}, v)$ and $(\mathcal{A}, c)$ of Faigle and Kern [78] [77], where $\mathcal{C}$ is the family of down sets of $(P, \leq)$ and $\mathcal{A}$ is the set of antichains of a *rooted tree*, are games on convex geometries.

7.2 Probabilistic values for games

We will follow the work of Weber [244] to obtain an axiomatic development of the *probabilistic values* for games on convex geometries. We consider the following games on the convex geometry $\mathcal{L}$. For any $T \in \mathcal{L}, T \neq \emptyset$, the *unanimity game* $u_T : \mathcal{L} \to \mathbb{R}$, is defined by

$$u_T(S) := \begin{cases} 1, & \text{if } T \subseteq S \\ 0, & \text{otherwise.} \end{cases}$$

The *identity game* $\delta_T : \mathcal{L} \to \mathbb{R}$, is defined by

$$\delta_T(S) := \begin{cases} 1, & \text{if } S = T \\ 0, & \text{if } S \neq T. \end{cases}$$

It is a simple matter to check that the collections of these games are two different bases of the vector space $\Gamma(\mathcal{L})$. Faigle and Kern [78] observed that

$$u_T = \sum_{\{S \in \mathcal{L} : S \supseteq T\}} \delta_S.$$

Definition 7.2.1 *A probabilistic value for $i \in N$ is a function $\Phi_i : \Gamma(\mathcal{L}) \to \mathbb{R}$, which satisfies*

$$\Phi_i(v) = \sum_{\{S \in \mathcal{L} : i \in \text{ex}(S)\}} p_S^i \left(v(S) - v(S \setminus i)\right), \quad p_S^i \geq 0, \qquad \sum_{\{S \in \mathcal{L} : i \in \text{ex}(S)\}} p_S^i = 1.$$

Remark 7.2.1 *If we take $T = S \setminus i$, then we obtain the equivalent expression*

$$\Phi_i(v) = \sum_{\{T \in \mathcal{L} : i \notin T, \, T \cup i \in \mathcal{L}\}} p_T^i \left(v(T \cup i) - v(T)\right).$$

In this definition we only consider the coalitions $S \in \mathcal{L}$ such that $i \in S$ and $S \setminus i \in \mathcal{L}$. The Shapley value for games on $J(P)$ defined by Faigle and Kern [78], and its generalization to convex geometries by Bilbao and Edelman [16] are probabilistic values.

We will introduce a number of properties for values for games on convex geometries. First, we consider the linearity property.

Linearity axiom. For all $\alpha, \beta \in \mathbb{R}$, and $v, w \in \Gamma(\mathcal{L})$, we have for all $i \in N$,

$$\Phi_i(\alpha v + \beta w) = \alpha \Phi_i(v) + \beta \Phi_i(w).$$

Theorem 7.2.1 *Let $\Phi_i : \Gamma(\mathcal{L}) \to \mathbb{R}$ be a value for i which satisfies the linearity axiom. Then there exists a unique set of coefficients $\{a^i_S : S \in \mathcal{L},\ S \neq \emptyset\}$ such that for each game $v \in \Gamma(\mathcal{L})$,*

$$\Phi_i(v) = \sum_{S \in \mathcal{L}} a^i_S v(S).$$

Proof. The collection $\{\delta_S : S \in \mathcal{L},\ S \neq \emptyset\}$ is a basis of $\Gamma(\mathcal{L})$. Let $a^i_S = \Phi_i(\delta_S)$, for all $S \in \mathcal{L}$. Then $v = \sum_{S \in \mathcal{L}} v(S)\delta_S$, and the linearity axiom implies

$$\Phi_i(v) = \sum_{S \in \mathcal{L}} \Phi_i(\delta_S) v(S) = \sum_{S \in \mathcal{L}} a^i_S v(S).$$

□

We introduce the concept of *dummy player.*

Definition 7.2.2 *The player $i \in N$ is a dummy in the game $v \in \Gamma(\mathcal{L})$ if, for every $T \in \mathcal{L}$ such that $i \notin T$ and $T \cup i \in \mathcal{L}$, we have*

$$v(T \cup i) - v(T) = \begin{cases} v(i), & \text{if } \{i\} \in \mathcal{L} \\ 0, & \text{otherwise.} \end{cases}$$

Remark 7.2.2 *The following definition is equivalent: For all $S \in \mathcal{L}$ such that $i \in \text{ex}(S)$, we have $v(S) - v(S \setminus i) = v(i)$, if $\{i\} \in \mathcal{L}$, and 0, otherwise.*

We need some properties of dummy players in the unanimity and identity games.

Proposition 7.2.2 *Let $\mathcal{L}$ be a convex geometry and let $S \in \mathcal{L}$ be a nonempty convex set. Then*

1. *If $i \notin \text{ex}(S)$ then i is a dummy player in the unanimity game u_S.*
2. *If $i \in S \setminus \text{ex}(S)$ then i is a dummy player in the identity game δ_S.*
3. *If $i \notin S$ and $S \cup i \notin \mathcal{L}$ then i is a dummy player in the identity game δ_S.*

Proof. Let $S \in \mathcal{L}$, $S \neq \emptyset$ and let $i \in N$ be such that $i \notin \text{ex}(S)$. Note that if $\{i\} \in \mathcal{L}$ then $S \neq \{i\}$, hence $u_S(i) = 0$ and $\delta_S(i) = 0$.

1. Let $i \notin \text{ex}(S)$. We suppose that there exists $T \in \mathcal{L}$, and $i \notin T$ such that $T \cup i \in \mathcal{L}$ and satisfies $u_S(T \cup i) \neq u_S(T)$. Then $u_S(T \cup i) = 1$ and $u_S(T) = 0$, hence $S \subseteq T \cup i$ and $S \nsubseteq T$. Thus $S \setminus i = S \cap T \in \mathcal{L}$, and we obtain $i \in \text{ex}(S)$, which is a contradiction. Hence, $u_S(T \cup i) - u_S(T) = 0$.

2. Let $i \in S \setminus \text{ex}(S)$. If $T \in \mathcal{L}$, $i \notin T$ and $T \cup i \in \mathcal{L}$, we have $T \cup i \neq S$ since $i \notin \text{ex}(S)$ and $T \neq S$ since $i \in S$. Then, it is obvious that $\delta_S(T \cup i) = \delta_S(T) = 0$.

3. Let $i \notin S$ and $S \cup i \notin \mathcal{L}$. If $T \in \mathcal{L}$, $i \notin T$ and $T \cup i \in \mathcal{L}$, it is clear that $\delta_S(T \cup i) = \delta_S(T) = 0$. □

Dummy axiom. If the player $i \in N$ is a dummy in $v \in \Gamma(\mathcal{L})$, then

$$\Phi_i(v) = \begin{cases} v(i), & \text{if } \{i\} \in \mathcal{L} \\ 0, & \text{otherwise.} \end{cases}$$

Theorem 7.2.3 *Let $\Phi_i : \Gamma(\mathcal{L}) \to \mathbb{R}$ be a value for the player i defined by $\Phi_i(v) = \sum_{S \in \mathcal{L}} a_S^i v(S)$ for every game $v \in \Gamma(\mathcal{L})$, and suppose that Φ_i satisfies the dummy axiom. Then,*

$$\Phi_i(v) = \sum_{\{S \in \mathcal{L} : i \in \text{ex}(S)\}} a_S^i \left(v(S) - v(S \setminus i) \right).$$

Moreover, if $\{i\} \in \mathcal{L}$ then

$$\sum_{\{S \in \mathcal{L} : i \in \text{ex}(S)\}} a_S^i = 1.$$

Proof. Let $E_i : \Gamma(\mathcal{L}) \to \mathbb{R}$ be defined for $i \in N$ by

$$E_i(v) := \sum_{\{S \in \mathcal{L} : i \in \text{ex}(S)\}} a_S^i \left(v(S) - v(S \setminus i) \right).$$

The operators E_i and Φ_i are linear and the unanimity games form a basis of $\Gamma(\mathcal{L})$. Then it will be enough to show that $\Phi_i(u_T) = E_i(u_T)$, for every convex set $T \in \mathcal{L}$, $T \neq \emptyset$. Let us fix $i \in N$ and $T \in \mathcal{L}$, and we consider two cases.

First, if $i \notin \text{ex}(T)$ then Proposition 7.2.2 implies that i is a dummy player in u_T. Then, for all $S \in \mathcal{L}$ such that $i \in \text{ex}(S)$, we have $u_T(S) - u_T(S \setminus i) = 0$, since $u_T(i) = 0$, whenever $\{i\} \in \mathcal{L}$. Therefore, the definition of E_i yields $E_i(u_T) = 0$. Furthermore, the dummy axiom implies that $\Phi_i(u_T) = 0$.

Now, if we suppose that $i \in \text{ex}(T)$ then $i \in T$ and hence $u_T(S \setminus i) = 0$, for every $S \in \mathcal{L}$ such that $i \in \text{ex}(S)$. Therefore, we obtain the equivalence $u_T(S) - u_T(S \setminus i) = 1$ if and only if $S \in \mathcal{L}$ and $S \supseteq T$. Observe that

$$\begin{aligned} E_i(u_T) &= \sum_{\{S \in \mathcal{L} : i \in \text{ex}(S),\, S \supseteq T\}} a_S^i \\ &= \sum_{\{S \in \mathcal{L} : i \in \text{ex}(S),\, S \supseteq T\}} \Phi_i(\delta_S) \end{aligned}$$

$$= \Phi_i\left(\sum_{\{S\in\mathcal{L}:\, S\supseteq T\}} \delta_S\right)$$
$$= \Phi_i(u_T),$$

where the last but one equation follows from $\Phi_i(\delta_S)=0$, when $i \in S\setminus \mathrm{ex}(S)$.

Finally, if $\{i\} \in \mathcal{L}$ then i is a dummy in the game $u_{\{i\}}$. Let T be a convex set such that $i \notin T$ and $T\cup i \in \mathcal{L}$. Note that $u_{\{i\}}(T\cup i)=1$, $u_{\{i\}}(T)=0$, and $u_{\{i\}}(i)=1$. Because Φ_i satisfies the dummy axiom, it follows from the above equation that

$$\begin{aligned}\sum_{\{S\in\mathcal{L}:i\in\mathrm{ex}(S)\}} a_S^i &= \sum_{\{S\in\mathcal{L}:i\in\mathrm{ex}(S)\}} \Phi_i(\delta_S)\\ &= \sum_{\{S\in\mathcal{L}:i\in S\}} \Phi_i(\delta_S)\\ &= \Phi_i\left(\sum_{\{S\in\mathcal{L}:\, i\in S\}} \delta_S\right)\\ &= \Phi_i(u_{\{i\}}) = u_{\{i\}}(i) = 1.\end{aligned}$$

□

We consider a convex geometry with $\{i\} \notin \mathcal{L}$ for some $i \in N$. In this case, note that
$$\sum_{\{S\in\mathcal{L}:i\in\mathrm{ex}(S)\}} a_S^i = \sum_{\{S\in\mathcal{L}:i\in S\}} a_S^i = \Phi_i\left(u_{\overline{\{i\}}}\right),$$
where the unanimity game on the closure of $\{i\}$ satisfies

$$u_{\overline{\{i\}}}(T) = \begin{cases} 1, & \text{if } \overline{\{i\}} \subseteq T \\ 0, & \text{otherwise.}\end{cases} = \begin{cases} 1, & \text{if } i \in T \\ 0, & \text{otherwise.}\end{cases}$$

We obtain the following result by the application of Theorems 7.2.1 and 7.2.3.

Theorem 7.2.4 *Let $\Phi_i : \Gamma(\mathcal{L}) \to \mathbb{R}$ be a value for the player i which satisfies the linearity and dummy axioms. Then for every game v, there is a a unique set of coefficients $\left\{a_S^i : S \in \mathcal{L},\ i \in \mathrm{ex}(S)\right\}$ such that*

$$\Phi_i(v) = \sum_{\{S\in\mathcal{L}:i\in\mathrm{ex}(S)\}} a_S^i\left(v(S) - v(S\setminus i)\right).$$

Moreover, if $\Phi_i\left(u_{\overline{\{i\}}}\right) = 1$ then $\sum_{\{S\in\mathcal{L}:i\in\mathrm{ex}(S)\}} a_S^i = 1$.

Remark 7.2.3 *For a convex geometry which satisfies $\{i\} \in \mathcal{L}$, for all $i \in N$, condition $\Phi_i\left(u_{\overline{\{i\}}}\right) = 1$ is not necessary, because it follows from the dummy axiom.*

Definition 7.2.3 *A game $v \in \Gamma(\mathcal{L})$ is monotone if $v(S) \leq v(T)$ for every $S, T \in \mathcal{L}$ such that $S \subseteq T$.*

Monotonicity axiom. If $v \in \Gamma(\mathcal{L})$ is monotone then $\Phi_i(v) \geq 0$.

Theorem 7.2.5 *Let $\Phi_i : \Gamma(\mathcal{L}) \to \mathbb{R}$ be a value for i defined, for every game v, by*

$$\Phi_i(v) = \sum_{\{S \in \mathcal{L}\,:\, i \in \mathrm{ex}(S)\}} a_S^i \left(v(S) - v(S \setminus i)\right).$$

If the value Φ_i satisfies the monotonicity axiom, then $a_S^i \geq 0$, for all $S \in \mathcal{L}$ such that $i \in \mathrm{ex}(S)$.

Proof. For any $T \in \mathcal{L}$, we consider the game $\widehat{u}_T : \mathcal{L} \to \mathbb{R}$, defined by

$$\widehat{u}_T(S) := \begin{cases} 1, & \text{if } T \subset S \\ 0, & \text{otherwise.} \end{cases}$$

The game $\widehat{u}_T$ is monotone, hence $\Phi_i\left(\widehat{u}_T\right) \geq 0$. On the other hand, for every $T \in \mathcal{L}$ such that $i \notin T$ and $T \cup i \in \mathcal{L}$, we obtain the identity

$$\Phi_i\left(\widehat{u}_T\right) = \sum_{\{S \in \mathcal{L}\,:\, i \in \mathrm{ex}(S)\}} a_S^i \left(\widehat{u}_T(S) - \widehat{u}_T(S \setminus i)\right) = a_{T \cup i}^i.$$

□

It is easy to check that every probabilistic value satisfies the previous axioms. Thus, we have the following theorems.

Theorem 7.2.6 *Let $\Phi_i : \Gamma(\mathcal{L}) \to \mathbb{R}$ be a value for the player i such that $\{i\} \in \mathcal{L}$. Then Φ_i satisfies the linearity, dummy, and monotonicity axioms if and only if Φ_i is a probabilistic value.*

Theorem 7.2.7 *Let $\Phi_i : \Gamma(\mathcal{L}) \to \mathbb{R}$ be a value for the player i. Then Φ_i satisfies the linearity, dummy, monotonicity axioms and $\Phi_i\left(u_{\overline{\{i\}}}\right) = 1$ if and only if Φ_i is a probabilistic value.*

7.3 Efficiency and compatible-order values

Let Φ be a group value, that is, $\Phi : \Gamma(\mathcal{L}) \to \mathbb{R}^n,\ v \mapsto (\Phi_1(v), \ldots, \Phi_n(v))$. If the payoff vector $\Phi(v)$ is a distribution of the available resources of the coalition N, then Φ satisfies the following axiom:

Efficiency axiom. For each game $v \in \Gamma(\mathcal{L})$ we have that

$$\sum_{i \in N} \Phi_i(v) = v(N).$$

Theorem 7.3.1 *Let $\Phi : \Gamma(\mathcal{L}) \to \mathbb{R}^n$ be a group value that satisfies efficiency axiom. If Φ_i satisfies linearity, dummy, and monotonicity axioms for every $i \in N$, then Φ_i is a probabilistic value.*

Proof. Theorem 7.2.7 implies that it is enough to show that $\Phi_i\left(u_{\overline{\{i\}}}\right) = 1$, for all $i \in N$. We claim that $\mathrm{ex}\left(\overline{\{i\}}\right) = \{i\}$, for each $i \in N$. To verify this claim note that in a convex geometry $\mathrm{ex}\left(\overline{\{i\}}\right) \neq \emptyset$. Suppose $j \in \mathrm{ex}\left(\overline{\{i\}}\right)$ for some $j \neq i$, then $\overline{\{i\}} \setminus j \in \mathcal{L}$. Thus, we obtain $i \in \overline{\{i\}} \setminus j$, hence by definition of the closure operator $\overline{\{i\}} \subseteq \overline{\{i\}} \setminus j$, and this is a contradiction. It follows from Proposition 7.2.2 that every $j \neq i$ satisfies $\Phi_j\left(u_{\overline{\{i\}}}\right) = 0$, because $j \notin \mathrm{ex}\left(\overline{\{i\}}\right)$. Therefore, the efficiency axiom implies that

$$\sum_{j \in N} \Phi_j\left(u_{\overline{\{i\}}}\right) = \Phi_i\left(u_{\overline{\{i\}}}\right) = 1.$$

□

The efficiency axiom implies the following properties for the coefficients of the values that satisfy the linearity and dummy axioms.

Theorem 7.3.2 *Let $\Phi : \Gamma(\mathcal{L}) \to \mathbb{R}^n$ be a group value defined, for every game v and for all $i \in N$, by*

$$\Phi_i(v) = \sum_{\{S \in \mathcal{L} \,:\, i \in \mathrm{ex}(S)\}} a_S^i \left(v(S) - v(S \setminus i)\right).$$

Then the group value Φ satisfies the efficiency axiom if and only if

$$\sum_{i \in \mathrm{ex}(N)} a_N^i = 1,$$

$$\sum_{i \in \mathrm{ex}(S)} a_S^i = \sum_{\{j \notin S \,:\, S \cup j \in \mathcal{L}\}} a_{S \cup j}^j,$$

for all $S \in \mathcal{L}$, $\emptyset \neq S \neq N$.

Proof. For any $v \in \Gamma(\mathcal{L})$ we have

$$\begin{aligned}\sum_{i \in N} \Phi_i(v) &= \sum_{i \in N} \sum_{\{S \in \mathcal{L} : i \in \mathrm{ex}(S)\}} a_S^i \left(v(S) - v(S \setminus i)\right) \\ &= \sum_{S \in \mathcal{L}} v(S) \left(\sum_{i \in \mathrm{ex}(S)} a_S^i - \sum_{\{j \notin S : S \cup j \in \mathcal{L}\}} a_{S \cup j}^j \right).\end{aligned}$$

If the coefficients satisfy the relations of the hypothesis of theorem, then $\sum_{i \in N} \Phi_i(v) = v(N)$ and the group value Φ satisfies the efficiency axiom.

Conversely, let us fix a nonempty set $T \in \mathcal{L}$, and consider the identity game δ_T. If we apply the previous equality to the game δ_T, then we have

$$\sum_{i \in N} \Phi_i(\delta_T) = \begin{cases} \sum_{i \in \mathrm{ex}(N)} a_N^i, & \text{if } T = N \\ \sum_{i \in \mathrm{ex}(S)} a_S^i - \sum_{\{j \notin S : S \cup j \in \mathcal{L}\}} a_{S \cup j}^j, & \text{if } T = S \neq N. \end{cases}$$

Thus, if Φ satisfies the efficiency axiom then the relations are true. □

Edelman and Jamison [65] defined a *compatible ordering* of a convex geometry $\mathcal{L} \subseteq 2^N$ as a total ordering of the elements of N, $i_1 < i_2 < \cdots < i_n$ such that the set $\{i_1, i_2, \ldots, i_j\} \in \mathcal{L}$, for all $1 \leq j \leq n$. A compatible ordering of $\mathcal{L}$ corresponds exactly to a maximal chain in $\mathcal{L}$ [66, Proposition 2.2]. We use $c(\mathcal{L})$ to denote the total number of maximal chains and $\mathcal{C}(\mathcal{L})$ to denote the set of all compatible orderings. Given $i \in N$ and a compatible ordering $C = (i_1, i_2, \ldots, i_n)$ of $\mathcal{L}$, the set

$$C(i) := \{j \in N : j \leq i \text{ in } C\}$$

is the convex set formed by the player i and her/his predecessors in the chain C. Since $C(i) \setminus i \in \mathcal{L}$, the player $i \in \mathrm{ex}\left(C(i)\right)$.

Definition 7.3.1 *Let $\{p_C : C \in \mathcal{C}(\mathcal{L})\}$ be a probability distribution on the set of all compatible orderings. A compatible-order value on $\Gamma(\mathcal{L})$ is a function*

$$\Omega : \Gamma(\mathcal{L}) \to \mathbb{R}^n, \quad v \mapsto \left(\Omega_1(v), \ldots, \Omega_n(v)\right),$$

with components defined, for all $i \in N$, by

$$\Omega_i(v) = \sum_{C \in \mathcal{C}(\mathcal{L})} p_C \left[v\left(C(i)\right) - v\left(C(i) \setminus i\right)\right].$$

The relationship between compatible-order values and probabilistic values that satisfy the efficiency axiom is a matter of the following two theorems.

Theorem 7.3.3 *Let $\Omega = (\Omega_1, \ldots, \Omega_n)$ be a compatible-order value on $\Gamma(\mathcal{L})$. Then Ω satisfies the efficiency axiom and each component Ω_i is a probabilistic value for every $i \in N$.*

Proof. For every $i \in N$ and $v \in \Gamma(\mathcal{L})$, we have

$$\begin{aligned}\Omega_i(v) &= \sum_{C \in \mathcal{C}(\mathcal{L})} p_C \left[v\left(C(i)\right) - v\left(C(i) \setminus i\right)\right] \\ &= \sum_{\{S \in \mathcal{L} : i \in \mathrm{ex}(S)\}} \left(\sum_{\{C \in \mathcal{C}(\mathcal{L}) : C(i) = S\}} p_C \right) \left(v(S) - v(S \setminus i)\right).\end{aligned}$$

The coefficients satisfy

$$\sum_{\{S \in \mathcal{L} : i \in \mathrm{ex}(S)\}} \left(\sum_{\{C \in \mathcal{C}(\mathcal{L}) : C(i) = S\}} p_C \right) = \sum_{C \in \mathcal{C}(\mathcal{L})} p_C = 1,$$

hence Ω_i is a probabilistic value. Furthermore, for every game $v \in \Gamma(\mathcal{L})$ we have

$$\begin{aligned}\sum_{i \in N} \Omega_i(v) &= \sum_{i \in N} \sum_{C \in \mathcal{C}(\mathcal{L})} p_C \left[v\left(C(i)\right) - v\left(C(i) \setminus i\right)\right] \\ &= \sum_{C \in \mathcal{C}(\mathcal{L})} p_C \left(\sum_{i \in N} \left[v\left(C(i)\right) - v\left(C(i) \setminus i\right)\right] \right) \\ &= \sum_{C \in \mathcal{C}(\mathcal{L})} p_C \left(v(N) - v(\emptyset)\right) \\ &= v(N).\end{aligned}$$

□

Theorem 7.3.4 *Let $\Phi : \Gamma(\mathcal{L}) \to \mathbb{R}^n$ be a group value that satisfies the efficiency axiom such that each component Φ_i is a probabilistic value for every $i \in N$. Then Φ is a compatible-order value.*

Proof. The hypothesis implies that for all $i \in N$,

$$\Phi_i(v) = \sum_{\{S \in \mathcal{L} : i \in \mathrm{ex}(S)\}} p_S^i \left(v(S) - v(S \setminus i)\right).$$

For any $T \in \mathcal{L}$ and $i \notin T$ such that $T \cup i \in \mathcal{L}$, we define

$$A(i,T)=\begin{cases}\dfrac{p^i_{T\cup i}}{\sum\limits_{\{j\notin T\,:\,T\cup j\in\mathcal{L}\}} p^j_{T\cup j}}, & \text{if the denominator is not equal to 0}\\ 0, & \text{otherwise.}\end{cases}$$

For any maximal chain $C\in\mathcal{C}(\mathcal{L})$ such that the associated compatible order is $i_1<i_2<\cdots<i_n$, we define

$$p_C=p^{i_1}_{\{i_1\}}\,A\left(i_2,\{i_1\}\right)\,A\left(i_3,\{i_1,i_2\}\right)\cdots A\left(i_n,\{i_1,i_2,\ldots,i_{n-1}\}\right),$$

where $C=\left(i_1,i_2,\ldots,i_n\right)$. Note that the last factor is 1. The collection of nonnegative numbers $\{p_C\,:\,C\in\mathcal{C}(\mathcal{L})\}$ satisfies

$$\begin{aligned}\sum_{C\in\mathcal{C}(\mathcal{L})} p_C &= \sum_{\{i_1\,:\,\{i_1\}\in\mathcal{L}\}}\ \sum_{\{i_2\notin\{i_1\}\,:\,\{i_1,i_2\}\in\mathcal{L}\}}\cdots\sum_{i_n\notin\{i_1,\ldots,i_{n-1}\}} p_{(i_1,\ldots,i_n)}\\ &= \sum_{\{i_1\,:\,\{i_1\}\in\mathcal{L}\}} p^{i_1}_{\{i_1\}}\\ &= \sum_{i\in N}\Phi_i\left(\widehat{u}_\emptyset\right)\\ &= 1,\end{aligned}$$

where the last equation follows from the efficiency axiom applied to the game $\widehat{u}_\emptyset$, defined by $\widehat{u}_\emptyset(S)=1$, for every nonempty $S\in\mathcal{L}$. Hence $\{p_C:C\in\mathcal{C}(\mathcal{L})\}$ is a probability distribution, and the compatible-order value associated is

$$\Omega_i(v)=\sum_{\{S\in\mathcal{L}\,:\,i\in\mathrm{ex}(S)\}}\left(\sum_{\{C\in\mathcal{C}(\mathcal{L})\,:\,C(i)=S\}} p_C\right)\left(v(S)-v(S\setminus i)\right),$$

for all $i\in N$. Thus, $\Phi_i=\Omega_i$ if for every $S\in\mathcal{L}$ with $i\in\mathrm{ex}(S)$, the coefficients satisfy

$$p^i_S=\sum_{\{C\in\mathcal{C}(\mathcal{L})\,:\,C(i)=S\}} p_C.$$

Let $S\in\mathcal{L}$ be such that $i\in\mathrm{ex}(S)$, let $C_1=(i_1,\ldots,i_{s-1})$ be a maximal chain from $\emptyset$ to $S\setminus i$ and let $C_2=(i_{s+1},\ldots,i_n)$ be a maximal chain from S to N, where $s=|S|$. These chains can be concatenated with i to make the chain $C_1\cup\{i\}\cup C_2\in\mathcal{C}(\mathcal{L})$. Then we obtain the following equations,

$$\sum_{\{C\in\mathcal{C}(\mathcal{L})\,:\,C(i)=S\}} p_C$$

$$= \sum_{i_{s-1}\in \text{ex}(S\setminus i)} \sum_{i_{s-2}\in \text{ex}(S\setminus\{i,i_{s-1}\})} \cdots \sum_{i_1\in \text{ex}(S\setminus\{i,i_{s-1},\ldots,i_2\})} \sum_{\{i_{s+1}\notin S:\, S\cup i_{s+1}\in\mathcal{L}\}}$$

$$\cdots \sum_{\{i_n\notin S\cup\{i_{s+1},\ldots,i_{n-1}\}:\, S\cup\{i_{s+1},\ldots,i_n\}\in\mathcal{L}\}} p_{(i_1,\ldots,i_{s-1},i,i_{s+1},\ldots,i_n)}$$

$$= A(i, S\setminus i) \sum_{i_{s-1}\in \text{ex}(S\setminus i)} A\left(i_{s-1}, S\setminus\{i,i_{s-1}\}\right)\cdots$$

$$\sum_{i_1\in \text{ex}(S\setminus\{i,i_{s-1}\ldots,i_2\})} p^{i_1}_{\{i_1\}} \sum_{\{i_{s+1}\notin S:\, S\cup i_{s+1}\in\mathcal{L}\}} A(i_{s+1}, S)\cdots$$

$$\sum_{\{i_n\notin S\cup\{i_{s+1},\ldots,i_{n-1}\}:\, S\cup\{i_{s+1},\ldots,i_n\}\in\mathcal{L}\}} A\left(i_n,\, S\cup\{i_{s+1},\ldots,i_{n-1}\}\right).$$

The product of the first s factors are equal to

$$\frac{p^i_S}{\sum_{\{j\notin S\setminus i:\, (S\setminus i)\cup j\in\mathcal{L}\}} p^j_{(S\setminus i)\cup j}}$$

$$\cdot \sum_{i_{s-1}\in \text{ex}(S\setminus i)} \left(\frac{p^{i_{s-1}}_{S\setminus i}}{\sum_{\{j\notin S\setminus\{i,i_{s-1}\}:\, (S\setminus\{i,i_{s-1}\})\cup j\in\mathcal{L}\}} p^j_{(S\setminus\{i,i_{s-1}\})\cup j}}\right)$$

$$\cdots \sum_{i_1\in \text{ex}(S\setminus\{i,i_{s-1},\ldots,i_2\})} p^{i_1}_{\{i_1\}}$$

$$= p^i_S,$$

because the equations of Theorem 7.3.2 imply that the denominator of a factor is equal to the next numerator. Furthermore, the following $n-s$ factors for $s+1 \le k \le n$, where $T_k = S\cup\{i_{s+1},\ldots,i_k\}$, satisfy

$$\sum_{\{i_k\notin T_{k-1}:\, T_{k-1}\cup i_k\in\mathcal{L}\}} A(i_k, T_{k-1}) =$$

$$\sum_{\{i_k\notin T_{k-1}:\, T_{k-1}\cup i_k\in\mathcal{L}\}} \left(\frac{p^{i_k}_{T_{k-1}\cup i_k}}{\sum_{\{j\notin T_{k-1}:\, T_{k-1}\cup j\in\mathcal{L}\}} p^j_{T_{k-1}\cup j}}\right) = 1.$$

Therefore, the result holds. □

7.4 Axioms for the Shapley value

A set $U \in 2^N$ is called a *carrier* for a game v if $v(S) = v(S \cap U)$ for all $S \in 2^N$. The carrier axiom states: If U is a carrier of v then $\sum_{i \in U} \Phi_i(v) = v(U)$. In the classical characterization, the *Shapley value* is the only value that satisfies the carrier, symmetry and additivity axioms on the class of superadditive games [212]. For the class of all games, Weber [244, Theorem 15] considered the linearity, dummy, symmetry, and efficiency axioms, and proved the uniqueness of the Shapley value. This value for player $i \in N$ is

$$\Phi_i(N, v) := \sum_{\{S \in 2^N \,:\, i \in S\}} \frac{(s-1)!(n-s)!}{n!} \left(v(S) - v(S \setminus i)\right),$$

where $n = |N|$ and $s = |S|$.

Let $S \in \mathcal{L}$ and $i \in S$. We define the *hierarchical strength* $h_S(i)$ as

$$h_S(i) := \frac{|\{C \in \mathcal{C}(\mathcal{L}) : C(i) \cap S = S\}|}{|\mathcal{C}(\mathcal{L})|},$$

i.e., $h_S(i)$ is the average number of compatible orderings of $\mathcal{L}$ in which i is the last member of S in the ordering. Note that $h_S(i)$ satisfies $h_S(i) \neq 0$ if and only if $i \in \mathrm{ex}(S)$. For games on the distributive lattice $J(P)$ of the order ideals of a poset $(P, \leq)$, Faigle and Kern [78] proposed the following axiom.

Hierarchical strength axiom. For any $S \in J(P)$ and $i, j \in S$, it holds that $h_S(i)\Phi_j(u_S) = h_S(j)\Phi_i(u_S)$.

These authors provided [78, Theorems 1 and 2] the following characterization of the Shapley value for games $v : J(P) \to \mathbb{R}$.

> "There is a unique function $\Phi : v \mapsto (\Phi_1(v), \ldots, \Phi_n(v))$ satisfying the axioms of linearity, carrier and hierarchical strength. Moreover, for every $i \in P$,
>
> $$\Phi_i(v) = \sum_{\{T \in J(P) \,:\, i \in Max(T)\}} \frac{e(T \setminus i)\, e(P \setminus T)}{e(P)} \left(v(T) - v(T \setminus i)\right),$$
>
> where $e(\cdot)$ is the number of linear extensions of the corresponding subposets of $(P, \leq)$."

Bilbao and Edelman [16] extended the formula for the Shapley value to games on a convex geometry $\mathcal{L}$. In this context, the number of linear extensions $e(P)$ is the number of maximal chains of $\mathcal{L}$. We denote by $c([T, S])$ the number of maximal chains from T to S, where $T \subset S$ and $c(S) := c([\emptyset, S])$ is the number of maximal chains from $\emptyset$ to $S \neq \emptyset$.

Definition 7.4.1 *Let $v \in \Gamma(\mathcal{L})$ be a game on a convex geometry. The Shapley value of the game v is given by, for all $i \in N$,*

$$\Phi_i(v) := \sum_{\{S \in \mathcal{L}:\, i \in \mathrm{ex}(S)\}} \frac{c(S \setminus i)\, c([S,N])}{c(N)} \left(v(S) - v(S \setminus i)\right),$$

where $c([N,N]) = c(\emptyset) = 1$.

Now, we introduce a new axiom, in which the value of the player depends on the number of maximal chains (see Bilbao [11]).

Chain axiom. For any $S \in \mathcal{L}$ and $i, j \in \mathrm{ex}(S)$, $c(S \setminus i)\Phi_j(\delta_S) = c(S \setminus j)\, \Phi_i(\delta_S)$.

By using our previous results, we prove the following characterization for the Shapley value on convex geometries.

Theorem 7.4.1 *The Shapley value is the unique function $\Phi : \Gamma(\mathcal{L}) \to \mathbb{R}^n$ that satisfies the linearity, dummy, efficiency, and chain axioms.*

Proof. It is obvious that the Shapley value satisfies the properties. Conversely, it follows from Theorems 7.2.4 and 7.3.2 that, for every $i \in N$,

$$\left\{a_S^i : S \in \mathcal{L},\ i \in \mathrm{ex}(S)\right\}$$

is a collection such that

$$\begin{aligned}
\Phi_i(v) &= \sum_{\{S \in \mathcal{L}:\, i \in \mathrm{ex}(S)\}} a_S^i \left(v(S) - v(S \setminus i)\right), \\
\sum_{i \in \mathrm{ex}(N)} a_N^i &= 1, \\
\sum_{i \in \mathrm{ex}(S)} a_S^i &= \sum_{\{j \notin S:\, S \cup j \in \mathcal{L}\}} a_{S \cup j}^j, \quad \text{for all } S \in \mathcal{L},\ \emptyset \neq S \neq N.
\end{aligned}$$

Therefore, it suffices to show that

$$a_S^i = \frac{c(S \setminus i)\, c([S,N])}{c(N)}, \quad \text{for all } S \in \mathcal{L} \text{ and } i \in \mathrm{ex}(S).$$

The coefficients are $a_S^i = \Phi_i(\delta_S)$, hence the chain axiom implies that for all pairs $i, j \in \mathrm{ex}(S)$, $a_S^j\, c(S \setminus i) = a_S^i\, c(S \setminus j)$. If we fix $i \in \mathrm{ex}(S)$, then we obtain

$$\begin{aligned}
\sum_{j \in \mathrm{ex}(S)} a_S^j &= a_S^i + \sum_{\{j \in \mathrm{ex}(S):\, j \neq i\}} \frac{c(S \setminus j)}{c(S \setminus i)} a_S^i \\
&= \frac{a_S^i}{c(S \setminus i)} \sum_{j \in \mathrm{ex}(S)} c(S \setminus j) \\
&= a_S^i \frac{c(S)}{c(S \setminus i)}.
\end{aligned}$$

For $S = N$ we take the first efficiency equation and hence $c(N\backslash i) = a^i_N\, c(N)$, for every $i \in \mathrm{ex}(N)$. Thus

$$a^i_N = \frac{c(N \setminus i)\, c([N, N])}{c(N)},$$

for all $i \in \mathrm{ex}(N)$. We assume the following induction hypothesis:
For every $T \in \mathcal{L}$, with $|T| = k \geq 2$ we have

$$a^i_T = \frac{c(T \setminus i)\, c([T, N])}{c(N)},$$

for all $i \in \mathrm{ex}(T)$. The case $k = n$ has just been proved. Let $S \in \mathcal{L}$, such that $|S| = k - 1 < n$. Then $S \neq N$, and the efficiency equations imply that

$$\begin{aligned}
\sum_{i\in \mathrm{ex}(S)} a^i_S &= \sum_{\{j\notin S\,:\,S\cup j\in\mathcal{L}\}} a^j_{S\cup j} \\
&= \sum_{\{j\notin S\,:\,S\cup j\in\mathcal{L}\}} \frac{c(S)\, c([S\cup j, N])}{c(N)} \\
&= \frac{c(S)}{c(N)} \sum_{\{j\notin S\,:\,S\cup j\in\mathcal{L}\}} c([S\cup j, N]) \\
&= \frac{c(S)\, c([S, N])}{c(N)},
\end{aligned}$$

where the second equation follows from the induction hypothesis for $T = S \cup j$. Finally, for every $i \in \mathrm{ex}(S)$, the identity

$$a^i_S \frac{c(S)}{c(S\setminus i)} = \frac{c(S)\, c([S, N])}{c(N)}$$

implies that the coefficients are

$$a^i_S = \frac{c(S \setminus i)\, c([S, N])}{c(N)}.$$

□

Remark 7.4.1 *Let $\{p_C : C \in \mathcal{C}(\mathcal{L})\}$ be a probability distribution such that $p_C = 1/c(N)$, for all $C \in \mathcal{C}(\mathcal{L})$. If Ω is the compatible-order associated, then Ω is the Shapley value.*

7.5 Axioms for the Banzhaf index

The analysis of power is central in political science. In general, it is difficult to define the idea of power, but for the special case of voting power there are mathematical power indices that have been used. The first such power index was proposed by Shapley and Shubik [213]. Another concept for measuring voting power was introduced by Banzhaf [8], a lawyer, whose work has appeared mainly in law journals, and whose index has been used in arguments in various legal proceedings. In this paper, we introduce the Banzhaf index for cooperative games in which only certain coalitions are allowed to form. We will study the structure of such allowable coalitions using the theory of convex geometries. We will define Banzhaf indices for simple games on convex geometries.

Definition 7.5.1 *A simple game on a convex geometry $\mathcal{L} \subseteq 2^N$ is a monotone game $v : \mathcal{L} \to \{0,1\}$, such that $v(\emptyset) = 0$.*

If $\mathcal{L} \subseteq 2^N$ is a convex geometry then there exists a closure operator on 2^N, defined by

$$\overline{A} = \bigcap_{\{S \in \mathcal{L} : S \supseteq A\}} S,$$

where $\overline{A} \in \mathcal{L}$. The collection of simple games on $\mathcal{L}$ is denoted by $\Upsilon(\mathcal{L})$.

Definition 7.5.2 *A simple game $v : \mathcal{L} \to \{0,1\}$ is called superadditive on $\mathcal{L}$ if $v\left(\overline{S \cup T}\right) \geq v(S) + v(T)$, whenever $S \cap T = \emptyset$.*

The collection of superadditive games is denoted by $\Upsilon_{sa}(\mathcal{L})$. On the set $\Upsilon(\mathcal{L})$ we define the internal operations meet $\wedge$ and join $\vee$ by

$$\begin{aligned}(v \wedge w)(S) &= \min\{v(S), w(S)\},\\ (v \vee w)(S) &= \max\{v(S), w(S)\}.\end{aligned}$$

In $\Upsilon_{sa}(\mathcal{L})$ operation join is not internal. Games $\widehat{1}$ and $\widehat{0}$ are simple games such that $\widehat{1}(S) = 1$, and $\widehat{0}(S) = 0$, for every nonempty $S \in \mathcal{L}$.

Example: Let $N = (1 < \cdots < 2k+1)$ be the linear order and let $Co\,(\mathbf{2k+1})$ be the convex geometry whose convex sets are the empty set and the intervals $[i,j] = \{i, i+1, \ldots, j-1, j\}$. Game v, defined by

$$v(S) = \begin{cases} 1, & \text{if } |S| \geq k+1 \\ 0, & \text{if } |S| \leq k, \end{cases}$$

for all $S \in Co\,(\mathbf{2k+1})$, is the *majority simple* game on the convex sets for the *policy order* (see Edelman [67]).

Convex sets $S \in \mathcal{L}$ with $v(S) = 1$ are called *winning* and convex sets with $v(S) = 0$, *losing*. In the classical theory (see Dubey and Shapley [59]) a *swing* for player i is a pair of coalitions $(S, S \setminus i)$ such that S is winning and $S \setminus i$ is not. This concept allows us to define convex swings for games on convex geometries.

Definition 7.5.3 *For a player i and a simple game $v \in \Upsilon(\mathcal{L})$, we say that the pair $(S, S \setminus i)$ is a convex swing if $S \in \mathcal{L}$, $i \in \operatorname{ex}(S)$, $v(S) = 1$, $v(S \setminus i) = 0$. The number of convex swings of player i is denoted by $cs_i(v)$, and the total number of convex swings for game v is*

$$cs(v) = \sum_{i \in N} cs_i(v).$$

We will study an *extreme model* in which the power of a player depends on his extreme position in the winning feasible coalitions (see Bilbao, Jiménez Losada and López [12]). In this model, it is obvious that the convex sets where a player is extreme are the only coalitions to take part in the game. We define the following numbers, for all $i \in N$,

$$\begin{aligned} \mathcal{E}_i &:= \left|\{S \in \mathcal{L} : i \in \operatorname{ex}(S)\}\right|, \\ \mathcal{E}_i(T) &:= \left|\{S \in \mathcal{L} : i \in \operatorname{ex}(S),\ S \supseteq T\}\right|. \end{aligned}$$

The number $\mathcal{E}_i$ will be called *extreme power* of the player i and $\mathcal{E}_i(T)$ is the *extreme power* of i on T. The ratio $\mathcal{E}_i(T)/\mathcal{E}_i$ is the *relative extreme power* of i on T.

We consider the relationship between the convex swings of a player and the set of all convex sets such that the player i is an extreme point.

Definition 7.5.4 *For every player $i \in N$, the number*

$$cs_i'(v) = \frac{cs_i(v)}{\mathcal{E}_i},$$

is called convex swing probability of i for the game $v \in \Upsilon(\mathcal{L})$. The sum of all these numbers is denoted $cs'(v)$.

The pair $(S, S \setminus i)$ is a convex swing if and only if $v(S) - v(S \setminus i) = 1$. Therefore, the following proposition holds.

Proposition 7.5.1 *Let $v \in \Upsilon(\mathcal{L})$ and $i \in N$. Then,*

$$cs_i(v) = \sum_{\{S \in \mathcal{L} : i \in \operatorname{ex}(S)\}} \left(v(S) - v(S \setminus i)\right).$$

Definition 7.5.5 *A player i with $cs_i(v) = 0$ is called a dummy in $v \in \Upsilon(\mathcal{L})$.*

Notice that a dummy player can never help a convex set to win.

Proposition 7.5.2 *If $S \in \mathcal{L}$ and $i \notin \text{ex}(S)$, then player i is a dummy in the unanimity game $u_S \in \Upsilon(\mathcal{L})$.*

Proof. If $i \notin \text{ex}(S)$, then there are two possibilities for i. If $i \notin S$ then $u_S(T) = u_S(T \setminus i)$, thus $cs_i(u_S) = 0$. If $i \in S \setminus \text{ex}(S)$ and $S \subseteq T$, then we have that $i \notin \text{ex}(T)$. Otherwise $T \setminus i$ and S are convex sets and $(T \setminus i) \cap S = S \setminus i$ is a convex set, hence $i \in \text{ex}(S)$, a contradiction. □

Proposition 7.5.3 *If $S \in \mathcal{L}$ then the number of convex swings for u_S is*

$$cs_i(u_S) = \begin{cases} 0, & \text{if } i \notin \text{ex}(S) \\ \mathcal{E}_i(S), & \text{if } i \in \text{ex}(S). \end{cases}$$

Proof. By Proposition 7.5.1 we have

$$cs_i(u_S) = \sum_{\{T \in \mathcal{L} : i \in \text{ex}(T)\}} (u_S(T) - u_S(T \setminus i)).$$

If $i \notin \text{ex}(S)$, then Proposition 7.5.2 implies the result. If $i \in \text{ex}(S)$ then we have that the marginal contribution $u_S(T) - u_S(T \setminus i) = 1$, if and only if $T \in \{S \in \mathcal{L} : i \in \text{ex}(S)\}$ and $T \supseteq S$. □

Let us define two different types of indices: the normalized and probabilistic convex Banzhaf indices.

Definition 7.5.6 *The vector whose components are the numbers of convex swings $cs_i(v)$ is called the convex Banzhaf index. When this number is normalized then we obtain the normalized convex Banzhaf index $c\beta : \Upsilon(\mathcal{L}) \to \mathbb{R}^n$,*

$$c\beta(v) = (c\beta_1(v), \ldots, c\beta_n(v)), \quad \text{where } c\beta_i(v) = \frac{cs_i(v)}{cs(v)}.$$

The interest of these numbers lies in their ratios and it has been useful to normalize them with the total number of swings (see Dubey and Shapley [59]). Another possibility is to consider the measure given by the expected number of convex swings $cs'_i(v) = cs_i(v)/\mathcal{E}_i$, for all $i \in N$.

Definition 7.5.7 *The probabilistic convex Banzhaf index is $c\beta' : \Upsilon(\mathcal{L}) \to \mathbb{R}^n$,*

$$c\beta'(v) = \left(c\beta'_1(v), \ldots, c\beta'_n(v)\right), \quad \text{where } c\beta'_i(v) = cs'_i(v).$$

If we apply Proposition 7.5.1 then these indices satisfy

$$c\beta_i(v) = \sum_{\{S\in\mathcal{L}\,:\, i\in \mathrm{ex}(S)\}} \frac{1}{cs(v)} \left(v(S) - v(S\setminus i)\right),$$

$$c\beta_i'(v) = \sum_{\{S\in\mathcal{L}\,:\, i\in \mathrm{ex}(S)\}} \frac{1}{\mathcal{E}_i} \left(v(S) - v(S\setminus i)\right),$$

and we can show that $c\beta_i'(v)$ is a probabilistic index because

$$\sum_{\{S\in\mathcal{L}: i\in \mathrm{ex}(S)\}} \frac{1}{\mathcal{E}_i} = 1.$$

The relationship between these indices and the internal operations of $\Upsilon(\mathcal{L})$ is the *transfer property.*

Proposition 7.5.4 *The convex Banzhaf index satisfies the transfer property, i.e.,* $cs_i(v\vee w) + cs_i(v\wedge w) = cs_i(v) + cs_i(w)$ *for all* $v, w \in \Upsilon(\mathcal{L})$.

Proof. In the following table we observe that $(v\vee w) + (v\wedge w) = v + w$.

v	w	$v\vee w$	$v\wedge w$	$v+w$	$(v\vee w)+(v\wedge w)$
1	1	1	1	2	2
1	0	1	0	1	1
0	1	1	0	1	1
0	0	0	0	0	0

Since Proposition 7.5.1 implies that cs_i is an additive operator, the proof is complete. □

Remark 7.5.1 *The normalized convex Banzhaf index does not satisfy this property because its denominator depends on the game. However the probabilistic convex Banzhaf index offers no problems. The above proposition is true in* $\Upsilon_{sa}(\mathcal{L})$ *when the operation join is an internal operation.*

Let $\varphi : \Upsilon(\mathcal{L}) \to \mathbb{R}^n$ be a group value with components φ_i, for all $i = 1, \ldots, n$. We use four axioms to obtain a characterization of the convex Banzhaf indices. If a player $i \in N$ is a dummy for $v \in \Upsilon(\mathcal{L})$ then $c\beta_i(v) = c\beta_i'(v) = cs_i(v) = 0$. Thus, the first axiom is:

Axiom B1 *Dummy:* If $i \in N$ is a dummy in v then $\varphi_i(v) = 0$.

Axiom B2a *Total swings:* The sum of all components of $\varphi(v)$ is equal to the total number of the convex swings, i.e. for each $v \in \Upsilon(\mathcal{L})$,

$$\sum_{i \in N} \varphi_i(v) = cs(v).$$

Axiom B2b *Total swing probabilities:* The sum of all components of $\varphi(v)$ is equal to the total number of the convex swing probabilities, i.e. for each $v \in \Upsilon(\mathcal{L})$,

$$\sum_{i \in N} \varphi_i(v) = cs'(v).$$

We will now consider the value of unanimity games in the *extreme* axioms.

Axiom B3a *Extreme power:* For any $S \in \mathcal{L}$, and $i, j \in S$, we have that $\mathcal{E}_i(S)\varphi_j(u_S) = \mathcal{E}_j(S)\varphi_i(u_S)$.

Axiom B3b *Relative extreme power:* For any $S \in \mathcal{L}$, and $i, j \in S$, we have that

$$\frac{\mathcal{E}_i(S)}{\mathcal{E}_i}\varphi_j(u_S) = \frac{\mathcal{E}_j(S)}{\mathcal{E}_j}\varphi_i(u_S).$$

The last axiom is based on the transfer property.

Axiom B4 *Transfer:* If $v, w \in \Upsilon(\mathcal{L})$ then $\varphi(v) + \varphi(w) = \varphi(v \vee w) + \varphi(v \wedge w)$.

We will now prove that the normalized and probabilistic convex Banzhaf indices can be characterized with these axioms. In the next theorem, the normalized convex Banzhaf index is not obtained directly by Axiom B4 (see Remark 7.5.1).

Theorem 7.5.5 *There exists a unique group value φ on $\Upsilon(\mathcal{L})$ that satisfies the dummy, total swings, extreme power and transfer axioms. Moreover, for all $v \in \Upsilon(\mathcal{L})$,*

$$c\beta(v) = \frac{1}{cs(v)}\varphi(v).$$

Proof. Let S be a convex set. If $j \notin \text{ex}(S)$ then j is a dummy in u_S, and Axiom B1 implies

$$\sum_{i \in N} \varphi_i(u_S) = \sum_{j \in \text{ex}(S)} \varphi_j(u_S).$$

If $i \in \text{ex}(S)$, then $S \in \{T \in \mathcal{L} : i \in \text{ex}(T)\}$ and $S \supseteq S$, hence $\mathcal{E}_i(S) \neq 0$ and by Axiom B3a we obtain that

$$\varphi_j(u_S) = \frac{\mathcal{E}_j(S)}{\mathcal{E}_i(S)}\varphi_i(u_S).$$

Therefore, φ is uniquely determined over the unanimity game since Axiom B2a leads us to

$$\begin{aligned} cs\,(v) &= \sum_{j\in \mathrm{ex}(S)} \varphi_j(u_S) \\ &= \sum_{j\in \mathrm{ex}(S)} \frac{\mathcal{E}_j(S)}{\mathcal{E}_i(S)}\varphi_i(u_S) \\ &= \frac{\varphi_i(u_S)}{\mathcal{E}_i(S)} \sum_{j\in \mathrm{ex}(S)} \mathcal{E}_j(S). \end{aligned}$$

Using Proposition 7.5.3 we show that

$$\sum_{j\in \mathrm{ex}(S)} \mathcal{E}_j(S) = cs(u_S) \text{ and } \mathcal{E}_i(S) = cs_i(u_S).$$

Thus, we obtain $\varphi_i(u_S) = cs_i\,(u_S)$. Finally, let $S_1, \ldots, S_p$ be the minimal winning convex sets for $v \in \Upsilon(\mathcal{L})$, such that $v \neq \widehat{0}$. By using the monotonicity of v and the decomposition showed by Dubey and Shapley [59], we obtain

$$v = u_{S_1} \vee u_{S_2} \vee \ldots \vee u_{S_p},$$

and this is the unique decomposition of v given by the operation join in $\Upsilon(\mathcal{L})$. By the transfer axiom,

$$\varphi\,(v) = \varphi\,(u_{S_1}) + \varphi\left(u_{S_2} \vee \ldots \vee u_{S_p}\right) - \varphi\left(u_{S_1} \wedge (u_{S_2} \vee \ldots \vee u_{S_p})\right).$$

Each game that appears in the second member is a game with fewer minimal winning convex sets than v. So, we can perform an induction on the number of minimal winning convex sets. If $v = \widehat{0}$, then all the players are dummy and therefore $\varphi\left(\widehat{0}\right) = 0$. We have proved the uniqueness and the existence is obtained because the vector of components $cs_i(v)$ satisfies the axioms. □

We obtain the probabilistic convex Banzhaf index with a similar proof.

Theorem 7.5.6 *There exists a unique group value φ on $\Upsilon(\mathcal{L})$ that satisfies the dummy, total swing probabilities, relative extreme power and transfer axioms. Moreover, $c\beta'(v) = \varphi(v)$ for all v in $\Upsilon(\mathcal{L})$.*

Proposition 7.5.7 *Let v be the majority simple game on the convex geometry $Co\,(\mathbf{2k+1})$. Then:*

1. *The components of the normalized convex Banzhaf index are*

$$c\beta_i\,(v) = \begin{cases} 1/(2k+2)\,, & \text{if } i \neq k+1 \\ 1/(k+1)\,, & \text{if } i = k+1. \end{cases}$$

2. *The components of the probabilistic convex Banzhaf index are*

$$c\beta_i'(v) = \begin{cases} 1/(2k+1), & \text{if } i \neq k+1 \\ 2/(2k+1), & \text{if } i = k+1. \end{cases}$$

Proof. (1) This formula is shown by Edelman [67]. (2) First, we obtain the extreme power $\mathcal{E}_i$ of each player $i \in N$. For all $j \geq i$, the interval $[i,j]$ is a convex set which contains i as an extreme point. Then, the number of these convex sets is $2k+1-(i-1) = 2k+2-i$. For all $j < i$, the interval $[j,i]$ is a coalition with i as an extreme point, and therefore there are $i-1$ more convex sets. Then, the number $\mathcal{E}_i$, for every player i, is $\mathcal{E}_i = 2k+1$. The number of swings is

$$cs_i(v) = \begin{cases} 1, \text{ if } i \neq k+1 \\ 2, \text{ if } i = k+1, \end{cases}$$

and the proof is complete. □

We can try to generalize the probabilistic convex Banzhaf power index to the convex Banzhaf value.

Definition 7.5.8 *The convex Banzhaf value of the game $v : \mathcal{L} \to \mathbb{R}$ is the vector whose components are*

$$c\beta_i'(v) = \sum_{\{S \in \mathcal{L} : i \in \text{ex}(S)\}} \frac{1}{\mathcal{E}_i} (v(S) - v(S \setminus i)).$$

The dummy, total swing probabilities, extreme power axioms and, in this case, also the linearity axiom, hold for this value. We leave the proofs of the following two results as an exercise for the reader.

Theorem 7.5.8 *There exists a unique group value on $\Gamma(\mathcal{L})$ that satisfies the linearity, dummy, relative extreme power and total swing probabilities axioms. This value is the convex Banzhaf value.*

Proposition 7.5.9 *Let v be a game on the convex geometry $Co(\mathbf{n})$. The components of the convex Banzhaf value are*

$$c\beta_i'(v) = \frac{1}{n} \left\{ \sum_{j=1}^{i-1} (v([j,i]) - v([j,i-1])) + \sum_{j=i+1}^{n} (v([i,j]) - v([i+1,j])) \right\}.$$

7.6 Project: Marginalist and efficient values

Young [249] introduced a marginalistic property for values which implies a representation using marginal worth vectors (see page 83), and obtained a new axiomatization of the Shapley value without the linearity axiom. In a recent work, Nowak [191] provides an axiomatization of the Banzhaf value without the additivity axiom. This approximation is now developing by Khmelnitskaya [142] for probabilistic values and it is instructive to study marginalist and efficient values for games on convex geometries.

Let $\mathcal{C}(\mathcal{L})$ be the set of all compatible orderings of a convex geometry $\mathcal{L} \subseteq 2^N$. A *marginalistic value* on $\Gamma(\mathcal{L})$ is a function $\Psi : \Gamma(\mathcal{L}) \to \mathbb{R}^n$, such that

$$\Psi_i(v) = F_i\left(\{v\left(C(i)\right) - v\left(C(i) \setminus i\right)\}_{C \in \mathcal{C}(\mathcal{L})}\right)$$

for all $i \in N$.

First, provide conditions which imply that a marginalist and efficient value is a compatible-order value, that is, for every $i \in N$,

$$\Psi_i(v) = \sum_{C \in \mathcal{C}(\mathcal{L})} p_C \left[v\left(C(i)\right) - v\left(C(i) \setminus i\right)\right],$$

where $\{p_C : C \in \mathcal{C}(\mathcal{L})\}$ is a probability distribution.

Next, axiomatizations of the compatible-order values without the linearity axiom can be investigate.

Chapter 8

Values for games on matroids

In this chapter we will consider feasible coalitions of players by using a class of combinatorial geometries called *matroids.* Let us assume that there are two rules of cooperation between players:

- If a coalition can form, then any subcoalition is also feasible. In general, the players that take part in the formation of a coalition have common interests. Therefore, any subset of these players have at least the same common interests.
- If there are two coalitions where the cardinal differs in one element, a player of the largest can join with the smallest making a feasible coalition.

The characteristic function of games on matroids is only defined on the elements of the matroid. We propose a generalization of the *Shapley value* [212] as concept of solution for games on matroids. The results was given by Jiménez Losada in his Ph. D. Thesis [135] and Bilbao, Driessen, Jiménez Losada and Lebrón [17].

8.1 The rank game

Recall that a *matroid* is a pair $(N, \mathcal{M})$ consisting of a finite set N and a collection $\mathcal{M}$ of subsets of N satisfying the following properties:

(M1) $\emptyset \in \mathcal{M}$.

(M2) If $S \in \mathcal{M}$ and $T \subseteq S$, then $T \in \mathcal{M}$.

(M3) If $T, S \in \mathcal{M}$ with $|S| = |T| + 1$, then there exists $i \in S \setminus T$ such that $T \cup i \in \mathcal{M}$.

Throughout this chapter, $\mathcal{M}$ will denote the matroid $(N, \mathcal{M})$. Some examples of matroids were presented in Section 1.5. The *rank function* $r : 2^N \to \mathbb{Z}_+$ of a matroid $\mathcal{M}$ on N is defined by

$$r(X) := \max\{|S| : S \subseteq X,\ S \in \mathcal{M}\}, \tag{8.1}$$

for all $X \subseteq N$. Notice that $S \in \mathcal{M}$ if and only if $r(S) = |S|$. The following two theorems (see [144] and [245]) axiomatize matroids in terms of their rank functions.

Theorem 8.1.1 *A function $r : 2^N \to \mathbb{Z}_+$ is the rank function of a matroid on N if and only if, for all X, $Y \subseteq N$, the following holds:*

(R1) $0 \leq r(X) \leq |X|$.

(R2) $r(X) \leq r(Y)$ *whenever* $X \subseteq Y$.

(R3) $r(X \cup Y) + r(X \cap Y) \leq r(X) + r(Y)$.

Theorem 8.1.2 *A function $r : 2^N \to \mathbb{Z}_+$ is the rank function of a matroid on N if and only if, for all $X \subseteq N$ and all i, $j \in N \setminus X$, the following holds:*

(R1') $r(\emptyset) = 0$.

(R2') $r(X) \leq r(X \cup i) \leq r(X) + 1$.

(R3') *If* $r(X \cup i) = r(X \cup j) = r(X)$ *then* $r(X \cup \{i, j\}) = r(X)$.

In the setting of the above theorems, the function r determines uniquely the corresponding matroid through $\mathcal{M} = \{S \subseteq N : r(S) = |S|\}$. Elements of a given matroid are called *independent sets* and further, a maximal independent set (with respect to inclusion of sets) is called a *basis*.

We will suppose that $\bigcup_{S \in \mathcal{M}} \{i : i \in S\} = N$. Property (M2) implies that the subsets of any basis are independent sets and thus, $2^B \subseteq \mathcal{M}$ for every basis $B \in \mathcal{B}(\mathcal{M})$. Recall that $\mathcal{B}(\mathcal{M}) = \{B \in \mathcal{M} : |B| = r(N)\}$ is the family of bases of $\mathcal{M}$.

The following theorem has been shown by Edmonds [71] and provides one interpretation of the core of a cooperative game induced by the rank function of a matroid (see Theorem 2.5.3 for a different proof). The rank function indicates the maximal feasible cooperation level between the players of a coalition. With every set $X \subseteq N$, there is associated the incidence vector $\mathbf{1}_X \in \mathbb{R}^N$ such that $(\mathbf{1}_X)_i := 1$ for all $i \in X$ and $(\mathbf{1}_X)_i := 0$ otherwise.

Theorem 8.1.3 *Let $r : 2^N \to \mathbb{Z}_+$ be the rank function of a matroid $\mathcal{M}$ and let $\mathcal{B}(\mathcal{M})$ be the family of the bases of $\mathcal{M}$. Then*

$$C(r) = \text{conv}\,\{\mathbf{1}_B : B \in \mathcal{B}(\mathcal{M})\}\,.$$

Proof. First, we establish that $\mathbf{1}_B \in C(r)$ for all $B \in \mathcal{B}(\mathcal{M})$. Indeed, for every $B \in \mathcal{B}(\mathcal{M})$, $\sum_{j\in N} (\mathbf{1}_B)_j = |B| = r(N)$ holds as well as

$$\sum_{j\in S} (\mathbf{1}_B)_j = |B \cap S| \leq r(S),$$

for all $S \subseteq N$. Since the core is a convex set, we deduce that

$$\text{conv}\,\{\mathbf{1}_B : B \in \mathcal{B}(\mathcal{M})\} \subseteq C(r).$$

To prove the reverse inclusion, it suffices to show that the vertices of the core belong to the set $\{\mathbf{1}_B : B \in \mathcal{B}(\mathcal{M})\}$. In view of Theorem 8.1.1 (R3), the rank game (N, r) is submodular and hence, by Theorem 2.6.7, the vertices of the $C(r)$ are determined by the marginal worth vectors, the components of which are composed of the marginal contributions $r(S \cup i) - r(S)$ for all $S \subseteq N \setminus i$, of player $i \in N$, in the rank game (N, r).

Together with Theorem 8.1.2 (R2'), this implies that any marginal worth vector $y = (y_i)_{i\in N}$ of the rank game (N, r) satisfies $y_i \in \{0, 1\}$ for all $i \in N$. That is, $y = \mathbf{1}_S$ for some $S \subseteq N$. From $y \in C(r)$ we deduce

$$|S| = \sum_{j\in S} y_j \leq r(S),$$

whereas, by construction, $r(S) \leq |S|$. Thus, $r(S) = |S|$, and hence $S \in \mathcal{M}$. Moreover, from the efficiency of y we obtain

$$r(N) = \sum_{j\in N} y_j = |S|\,.$$

Finally, from $S \in \mathcal{M}$ and $|S| = r(N)$, we conclude that $S \in \mathcal{B}(\mathcal{M})$ and hence $y = \mathbf{1}_B$ for some $B \in \mathcal{B}(\mathcal{M})$. □

Let $S \in \mathcal{M}$. The *contraction* $\mathcal{M}/S$ of S from $\mathcal{M}$ is the new matroid

$$\mathcal{M}/S := \{T \in \mathcal{M} : T \cap S = \emptyset \text{ and } T \cup S \in \mathcal{M}\}\,.$$

Then the contraction of an independent set S is a matroid formed with the independent sets of the initial matroid that do not include any members of S, whereas its union with S is still an independent set in the initial matroid. In the particular case of $S = \{i\}$, $i \in N$, we use $\mathcal{M}/i$ instead of $\mathcal{M}/\{i\}$.

Example: Let $N = \{1, \ldots, n\}$ be a set of players. For any $i,\ j \in N$, with $i \neq j$, we consider the opposition matroid

$$M_n(i||j) := \{S \subseteq N : \{i, j\} \nsubseteq S\}$$

being the largest matroid which excludes coalitions containing both players i and j. There are two bases of $M_n(i||j)$, namely $N \setminus i$ and $N \setminus j$. The rank function $r : 2^N \to \mathbb{Z}_+$ is given by

$$r(X) = \begin{cases} |X|, & \text{if } X \in M_n(i||j) \\ |X| - 1, & \text{otherwise.} \end{cases}$$

Two examples of contractions from $M_4(1||4)$ are shown in Figure 8.1. Players 2 and 3 are called *istmus players* since they belong to every basis of $M_4(1||4)$.

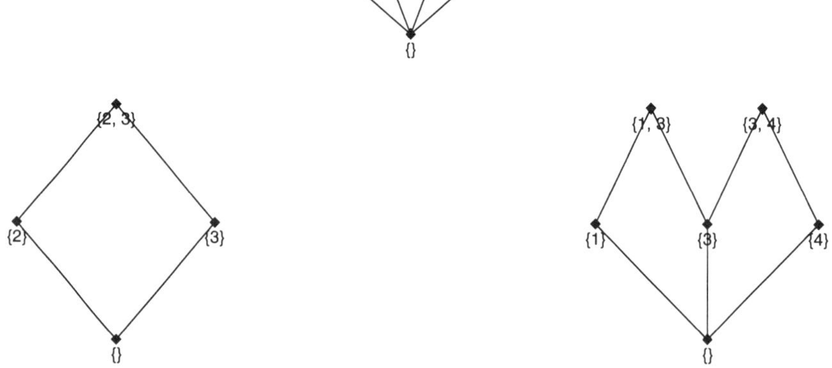

Figure 8.1 The matroid $M_4\,(1||4)$ and the contractions of 1 and 2

8.2 The Shapley value for games on matroids

Definition 8.2.1 *A game on a matroid* $\mathcal{M}$ *is a function* $v : \mathcal{M} \to \mathbb{R}$ *with* $v(\emptyset) = 0$.

We denote by $\Gamma(\mathcal{M})$ the set of all games on the matroid $\mathcal{M} \neq \{\emptyset\}$. The set $\Gamma(\mathcal{M})$ is a vector space under the operations of addition of functions and multiplication of functions by scalars. We consider now the unanimity game $u_S : \mathcal{M} \to \mathbb{R}$ and the identity game $\delta_S : \mathcal{M} \to \mathbb{R}$ defined for every nonempty $S \in \mathcal{M}$.

Proposition 8.2.1 *The dimension of the vector space* $\Gamma(\mathcal{M})$ *is* $|\mathcal{M}| - 1$, *and the collections* $\{\delta_T : T \in \mathcal{M},\ T \neq \emptyset\}$ *and* $\{u_S : S \in \mathcal{M},\ S \neq \emptyset\}$ *are basis of* $\Gamma(\mathcal{M})$.

Proof. It is easy to show the linear independence of the games δ_T and

$$v = \sum_{\{T \in \mathcal{M} : T \neq \emptyset\}} v(T)\, \delta_T, \quad \text{for all } v \in \Gamma(\mathcal{M}).$$

In order to prove that the set of the unanimity games u_S form a basis, it suffices to prove that it is an independent set or a generator system. We prove that any game on the matroid is a linear combination of the unanimity games, since it will be used further on. We can write $u_S = \sum_{\{T \in \mathcal{M} : T \supseteq S\}} \delta_T$. Applying the Möbius inversion formula [224] in the poset generated by the elements of the matroid $\mathcal{M}$ ordered by inclusion, we have

$$\delta_T = \sum_{\{S \in \mathcal{M} : S \supseteq T\}} \mu(T, S)\, u_S.$$

Therefore, we get

$$v = \sum_{\{S \in \mathcal{M} : S \neq \emptyset\}} d_S(v)\, u_S,$$

for all $v \in \Gamma(\mathcal{M})$, where the coefficients are $d_S(v) = \sum_{T \subseteq S} \mu(T, S)\, v(T)$. Note that the family $\{T \in \mathcal{M} : T \subseteq S\}$ is isomorphic to the boolean algebra 2^S. Therefore, the Möbius function of a matroid $\mathcal{M}$ is $\mu(T, S) = (-1)^{s-t}$, where $|S| = s$ and $|T| = t$. Thus, for all $v \in \Gamma(\mathcal{M})$, we have

$$v = \sum_{\{S \in \mathcal{M} : S \neq \emptyset\}} \left[\sum_{T \subseteq S} (-1)^{s-t}\, v(T) \right] u_S.$$

□

We call the bases of the matroid $\mathcal{M}$ the *basic coalitions* of $\mathcal{M}$. Recall that the influence set of $S \in \mathcal{M}$ is the set

$$\mathcal{B}_S(\mathcal{M}) = \{B \in \mathcal{B}(\mathcal{M}) : S \subseteq B\}$$

of the basic coalitions that contain S, and its influence worth is the ratio b_S/b, where $b_S = |\mathcal{B}_S(\mathcal{M})|$ and $b = |\mathcal{B}(\mathcal{M})|$. We write $\mathcal{B}_i(\mathcal{M})$ and b_i for individual coalitions $S = \{i\}$, $i \in N$.

The *Shapley value* we seek will be a group value $\Phi : \Gamma(\mathcal{M}) \to \mathbb{R}^n$, i.e., a vector valued function $v \mapsto (\Phi_1(v), \ldots, \Phi_n(v))$ that assigns to each player $i \in N$ his share $\Phi_i(v)$ relative to the game v. We will introduce some fundamental axioms that we consider are reasonable demands for the function Φ.

1. *Linearity*: For all $\alpha_1, \alpha_2 \in \mathbb{R}$, $v_1, v_2 \in \Gamma(\mathcal{M})$,

$$\Phi(\alpha_1 v_1 + \alpha_2 v_2) = \alpha_1 \Phi(v_1) + \alpha_2 \Phi(v_2).$$

Definition 8.2.2 *A player i is a* dummy *in the game $v : \mathcal{M} \to \mathbb{R}$ if for every $S \in \mathcal{M}$, such that $i \in S$, $v(S) - v(S \setminus i) = v(i)$.*

We will consider the influence of a dummy player in the following axiom.

2. *Dummy*: If i is a dummy in $v \in \Gamma(\mathcal{M})$, then

$$\Phi_i(v) = \frac{b_i}{b} v(i).$$

The classical efficiency axiom must change since the maximal feasible coalitions are the basis of N. Here, it is reasonable to suppose that the payoff vector $\Phi(v)$ may be taken as an equitable distribution of the average of the resources available for the basic coalitions.

3. *Efficiency*: For all $v \in \Gamma(\mathcal{M})$,

$$\sum_{i \in N} \Phi_i(v) = \frac{1}{b} \sum_{B \in \mathcal{B}(\mathcal{M})} v(B).$$

The last axiom stipulates that, in the unanimity game, players in the relevant coalition should be rewarded in the same way (see Nowak and Radzik [190]).

4. *Symmetry*: For every $S \in \mathcal{M}$, and $i, j \in S$, we have that

$$\Phi_i(u_S) = \Phi_j(u_S).$$

We will show in the next theorem that there is indeed a (unique) Shapley value. Moreover, we will prove that it is a generalization of the classical Shapley value.

Lemma 8.2.2 *Let $\mathcal{M}$ be a matroid with rank r and let $S \in \mathcal{M}$ such that $|S| = s$. Then, for every $t \in \mathbb{N}$ such that $s \leq t \leq r$, we have*

$$\sum_{\{T \in \mathcal{M} : T \supseteq S, |T|=t\}} b_T = b_S \binom{r-s}{t-s}.$$

Proof. Let us consider $\mathcal{B}_S(\mathcal{M})$. On the one hand, for every $B \in \mathcal{B}_S(\mathcal{M})$, there are $\binom{r-s}{t-s}$ coalitions $T \subseteq B$ that contain S and have cardinality t, since the family of coalitions contained in B forms a boolean algebra. On the other, each of the coalitions T is contained in b_T basic coalitions, hence, the number $b_S\binom{r-s}{t-s}$ is equal to

$$\sum_{\{T \in \mathcal{M} : T \supseteq S, |T|=t\}} b_T.$$

□

Theorem 8.2.3 *There exists a unique group value $\Phi : \Gamma(\mathcal{M}) \to \mathbb{R}^n$ which satisfies linearity, dummy, efficiency and symmetry axioms. Moreover, the Shapley value Φ is given by*

$$\Phi_i(v) = \sum_{\{S \in \mathcal{M} : i \in S\}} \frac{b_S}{b} \frac{(r-s)!(s-1)!}{r!} (v(S) - v(S \setminus i)),$$

for all $i \in N$.

Proof. First, it is easy to check that for any nonempty $T \in \mathcal{M}$, and $j \notin T$, the player j is a dummy player in the game u_T, and $u_T(j) = 0$. Hence, by dummy axiom, we have $\Phi_j(u_T) = 0$ for all $T \in \mathcal{M}$, $j \notin T$. Furthermore, by efficiency axiom

$$\sum_{j \in N} \Phi_j(u_T) = \sum_{i \in T} \Phi_i(u_T) = \frac{1}{b} \sum_{B \in \mathcal{B}(\mathcal{M})} u_T(B) = \frac{b_T}{b},$$

and by symmetry axiom, we conclude

$$\Phi_i(u_T) = \frac{b_T}{bt},$$

for all $T \in \mathcal{M}$, and $i \in T$, where $t = |T|$. Finally, by linearity axiom we can extend Φ to all of $\Gamma(\mathcal{M})$ uniquely via

$$\Phi_i(v) = \sum_{\{T \in \mathcal{M} : i \in T\}} d_T(v) \Phi_i(u_T)$$

$$= \sum_{\{T\in\mathcal{M}:i\in T\}} \left[\sum_{S\subseteq T} (-1)^{t-s} v(S)\right] \frac{b_T}{bt}$$

$$= \sum_{\{S\in\mathcal{M}:S\cup i\in\mathcal{M}\}} \left[\sum_{\{T\in\mathcal{M}:T\supseteq S,\, i\in T\}} (-1)^{t-s} \frac{b_T}{bt}\right] v(S).$$

Now, with respect to the coefficient of $v(S)$ in the last expression, there are two cases. If $i \in S$, then

$$\begin{aligned}
\sum_{\{T\in\mathcal{M}:T\supseteq S,\, i\in T\}} (-1)^{t-s} \frac{b_T}{bt} &= \sum_{t=s}^{r} (-1)^{t-s} \frac{1}{bt} \sum_{\{T\in\mathcal{M}:T\supseteq S,\, |T|=t\}} b_T \\
&= \frac{b_S}{b} \sum_{t=s}^{r} \binom{r-s}{t-s} (-1)^{t-s} \frac{1}{t} \\
&= \frac{b_S}{b} \sum_{h=0}^{r-s} \binom{r-s}{h} (-1)^{h} \frac{1}{h+s} \\
&= \frac{b_S}{b} \int_0^1 x^{s-1} (1-x)^{r-s}\, dx \\
&= \frac{b_S}{b} \frac{(r-s)!(s-1)!}{r!}.
\end{aligned}$$

If $i \notin S$, since $S \cup i \in \mathcal{M}$, we have

$$\begin{aligned}
\sum_{\{T\in\mathcal{M}:T\supseteq S,\, i\in T\}} (-1)^{t-s} \frac{b_T}{bt} &= \sum_{t=s+1}^{r} (-1)^{t-s} \frac{1}{bt} \sum_{\{T\in\mathcal{M}:T\supseteq S\cup i,\, |T|=t\}} b_T \\
&= \frac{b_{S\cup i}}{b} \sum_{t=s+1}^{r} \binom{r-s-1}{t-s-1} (-1)^{t-s} \frac{1}{t} \\
&= -\frac{b_{S\cup i}}{b} \frac{(r-s-1)!s!}{r!}.
\end{aligned}$$

Notice that we have used Lemma 8.2.2. There exists a bijection between the collections $\{S \in \mathcal{M} : i \in S\}$ and $\{S \in \mathcal{M} : i \notin S,\ S \cup i \in \mathcal{M}\}$. Therefore, we deduce that for all $i \in N$,

$$\Phi_i(v) = \sum_{\{S\in\mathcal{M}:i\in S\}} \frac{b_S}{b} \frac{(r-s)!(s-1)!}{r!} \left(v(S) - v(S \setminus i)\right).$$

□

Remark 8.2.1 *The Shapley value on matroids is equal to the classical Shapley value whenever* $\mathcal{M} = 2^N$*, since* $r = n$*, and* $b_S = b = 1$.

If $v : 2^N \rightarrow \mathbb{R}$ is a cooperative game, then Weber [244] proved that the Shapley value is a probabilistic value. That is, for each $i \in N$,

$$\Phi_i(v) = \sum_{\{S \subseteq N : i \in S\}} p_S^i \left(v(S) - v(S \setminus i)\right),$$

where

$$p_S^i = \frac{(n-s)!\,(s-1)!}{n!}$$

and hence $\{p_S^i : S \subseteq N,\ i \in S\}$ is a probability distribution over the collection of coalitions containing i. In our model, for every matroid $\mathcal{M}$ we have a similar formula for the Shapley value with coefficients

$$m_S = \frac{b_S}{b} \frac{(r-s)!(s-1)!}{r!}$$

Now, these coefficients satisfy $m_S > 0$ and

$$\begin{aligned} \sum_{\{S \in \mathcal{M} : i \in S\}} m_S &= \sum_{\{S \in \mathcal{M} : i \in S\}} \frac{b_S}{b} \frac{(r-s)!(s-1)!}{r!} \\ &= \sum_{s=1}^{r} \frac{b_i}{b} \binom{r-1}{s-1} \frac{(r-s)!(s-1)!}{r!} \\ &= \frac{b_i}{b}. \end{aligned}$$

Thus, the collection $\{m_S : S \in \mathcal{M},\ i \in S\}$ is not a probability distribution.

Theorem 8.2.4 *Let* $v : \mathcal{M} \rightarrow \mathbb{R}$ *be a game on a matroid. The Shapley value satisfies, for all* $i \in N$,

$$\Phi_i(v) = \frac{1}{b} \sum_{\{B \in \mathcal{B}(\mathcal{M}) : i \in B\}} \Phi_i(v_B),$$

where $v_B : 2^B \rightarrow \mathbb{R}$ *is the subgame defined in the subsets of* $B \in \mathcal{B}(\mathcal{M})$.

Proof. We compute the Shapley formula as follows,

$$\begin{aligned}\Phi_i(v) &= \sum_{\{S\in\mathcal{M}:i\in S\}} \frac{b_S}{b}\frac{(r-s)!(s-1)!}{r!}(v(S)-v(S\setminus i)) \\ &= \sum_{\{B\in\mathcal{B}(\mathcal{M}):i\in B\}}\sum_{\{S\subseteq B:i\in S\}} \frac{1}{b}\frac{(r-s)!(s-1)!}{r!}(v(S)-v(S\setminus i)) \\ &= \frac{1}{b}\sum_{\{B\in\mathcal{B}(\mathcal{M}):i\in B\}} \Phi_i(v_B).\end{aligned}$$

Notice that $\Phi_i(v_B)$ is the classical Shapley value. □

Faigle and Kern [78] introduced the concept of hierarchical strength of a player i in a coalition S, when the feasible coalitions system is a distributive lattice, generated by a hierarchy poset between the players. In the matroid model, it is possible to consider the same idea. We will prove that the hierarchical strength in a matroid only depends on the coalition.

The compatible orderings in a matroid are the orderings of its basic coalitions and corresponds to all maximal chains in the diagram of $(\mathcal{M},\subseteq)$. We denote by $c(\mathcal{M})$ the number of compatible orderings. Let $S\in\mathcal{M}$ be such that $i\in S$, and let us denote by $c(\mathcal{M},S,i)$ the number of compatible orderings of $\mathcal{M}$ in which player i is the last member of S.

Definition 8.2.3 *The hierarchical strength of the player i in a coalition $S\in\mathcal{M}$ is given by*

$$h_S(i) := \frac{c(\mathcal{M},S,i)}{c(\mathcal{M})}.$$

Proposition 8.2.5 *The hierarchical strength of a player i in $S\in\mathcal{M}$ satisfies*

$$h_S(i) = \frac{b_S}{b\,|S|}.$$

Proof. If the matroid $\mathcal{M}$ has rank r and b is the number of basic elements, then the number of maximal chains is $r!\,b$. Furthermore, the number of chains that contain S such that i is the last member of S is $r!\,b_S/|S|$. □

Proposition 8.2.6 *The Shapley value for a game $v\in\Gamma\left(U_n^k\right)$ satisfies, for all $i\in N$,*

$$\Phi_i(v) = \sum_{\{S:i\in S,\ |S|\leq k\}} \frac{(n-s)!(s-1)!}{n!}(v(S)-v(S\setminus i)).$$

Proof. It follows from

$$r = k, \; b = \binom{n}{k}, \text{ and } b_S = \binom{n-s}{k-s}.$$

□

Proposition 8.2.7 *Let v be a game on $\mathcal{M}_n(i\|j) = \{S \subseteq N : \{i,j\} \not\subseteq S\}$. Then*

$$\begin{aligned}
\Phi_i(v) &= \sum_{\{S\,:\,i\in S,\, j\notin S\}} \frac{1}{2}\frac{(n-1-s)!(s-1)!}{(n-1)!}\left(v(S)-v(S\setminus i)\right),\\
\Phi_j(v) &= \sum_{\{S\,:\,j\in S,\, i\notin S\}} \frac{1}{2}\frac{(n-1-s)!(s-1)!}{(n-1)!}\left(v(S)-v(S\setminus j)\right),\\
\Phi_k(v) &= \sum_{\{S\,:\,k\in S,\, i\notin S,\, j\notin S\}} \frac{(n-1-s)!(s-1)!}{(n-1)!}\left(v(S)-v(S\setminus k)\right)\\
&\quad + \sum_{\{S\,:\,k\in S,\, i\in S,\, j\notin S\}} \frac{1}{2}\frac{(n-1-s)!(s-1)!}{(n-1)!}\left(v(S)-v(S\setminus k)\right)\\
&\quad + \sum_{\{S\,:\,k\in S,\, i\notin S,\, j\in S\}} \frac{1}{2}\frac{(n-1-s)!(s-1)!}{(n-1)!}\left(v(S)-v(S\setminus k)\right),
\end{aligned}$$

where $k \notin \{i,j\}$.

Proof. In a matroid $\mathcal{M}_n(i\|j)$, we have $r = n-1$, $b = 2$ and

$$b_S = \begin{cases} 2, & \text{if } i \notin S \text{ and } j \notin S \\ 1, & \text{if } i \in S \text{ or } j \in S. \end{cases}$$

□

Let $N = \{1,2,3,4\}$ be a set of players and let $v : 2^N \to \{0,1\}$ be a simple game such that the *winning coalitions*, i.e., $v(S) = 1$ are

$$\mathcal{W} = \{\{1,4\},\{1,2,3\},\{1,2,4\},\{1,3,4\},\{2,3,4\},\{1,2,3,4\}\}.$$

The classical Shapley value is

$$\Phi(N,v) = \left(\frac{1}{3},\frac{1}{6},\frac{1}{6},\frac{1}{3}\right).$$

However, if we only consider the feasible coalitions belonging to the matroid

$$\mathcal{M}_4\,(1\|4) = \{S \subseteq N\, :\, \{1,4\} \not\subseteq S\}\,,$$

then the collection of coalitions which are both winning and feasible is

$$\mathcal{W} \cap \mathcal{M} = \{\{1,2,3\},\{2,3,4\}\}.$$

The Shapley value of the game $v : \mathcal{M} \to \mathbb{R}$, is given by

$$\Phi(\mathcal{M}, v) = \left(\frac{1}{6}, \frac{1}{3}, \frac{1}{3}, \frac{1}{6}\right).$$

The above example shows that strong players in the game are weak players in the game on this matroid.

A *symmetric game* $v : 2^N \to \mathbb{R}$ is characterized by a sequence of $|N|$ numbers, $\{v_1, \ldots, v_{|N|}\}$, where for any nonempty $S \in 2^N$, the worth is $v(S) = v_{|S|}$.

Definition 8.2.4 *A symmetric game v on a matroid $\mathcal{M}$ with rank r is a game such that for every nonempty $S \in \mathcal{M}$, $v(S) = v_s$, where $1 \leq s = |S| \leq r$.*

Proposition 8.2.8 *Let v be a symmetric game on a matroid $\mathcal{M}$ with rank r. If the worths are $\{0, v_1, \ldots, v_r\}$, then the Shapley value for every player $i \in N$ satisfies*

$$\Phi_i(v) = \frac{b_i}{\sum_{i\in N} b_i} v_r.$$

Proof. We compute the Shapley value for the symmetric game,

$$\begin{aligned}
\Phi_i(v) &= \sum_{\{S\in\mathcal{M}\,:\,i\in S\}} \frac{b_S}{b} \frac{(r-s)!(s-1)!}{r!} \left(v(S) - v(S\setminus i)\right) \\
&= \sum_{s=1}^{r} \frac{b_i}{b} \binom{r-1}{s-1} \frac{(r-s)!(s-1)!}{r!} (v_s - v_{s-1}) \\
&= \frac{b_i}{br} \sum_{s=1}^{r} (v_s - v_{s-1}) \\
&= \frac{b_i}{br} v_r = \frac{b_i}{\sum_{i\in N} b_i} v_r.
\end{aligned}$$

□

8.3 Quasi-probabilistic values

A cooperative game on a matroid represents an evaluation of the utility of any feasible coalition, whereas non-feasible coalitions are totally ignored since such coalitions are supposed not to be formed anyhow. For instance, in the context of the uniform matroid U_k^n, the n participants in a trip are only interested in measuring the utility of any group consisting of at most of k persons because of the limited number of seats in identical minibuses to be used for transportation during the trip.

An *individual value* for player i on the vector space $\Gamma(\mathcal{M})$ is defined to be a function $\psi_i : \Gamma(\mathcal{M}) \to \mathbb{R}$. For every game $v \in \Gamma(\mathcal{M})$, the value $\psi_i(v)$ represents an assessment by player i of his gains from participating in game v. We consider the following three axioms for an individual value ψ_i on $\Gamma(\mathcal{M})$.

1. *Linearity:* $\psi_i(\alpha v + \beta w) = \alpha\psi_i(v) + \beta\psi_i(w)$, for all $v,\ w \in \Gamma(\mathcal{M})$ and all $\alpha,\ \beta \in \mathbb{R}$.

2. λ_i*-dummy player* ($\lambda_i \in [0,1]$ is the rate of participation by player i): If player i is a dummy in the game $v \in \Gamma(\mathcal{M})$, that is $v(S \cup i) - v(S) = v(i)$ for all $S \in \mathcal{M}/i$, then $\psi_i(v) = \lambda_i v(i)$.

3. *Monotonicity*: If $v \in \Gamma(\mathcal{M})$ is monotone, that is $v(S) \leq v(T)$ for all $S,\ T \in \mathcal{M}$ with $S \subseteq T$, then $\psi_i(v) \geq 0$.

Theorem 8.3.1 *Let $\mathcal{M}$ be a matroid and $\psi_i : \Gamma(\mathcal{M}) \to \mathbb{R}$ an individual value for player $i \in N$. If $\lambda_i \in [0,1]$ is the rate of participation by player i, then the following two statements are equivalent:*

(i) *ψ_i satisfies the linearity, monotonicity, and λ_i-dummy player properties.*

(ii) *There exists a collection $\{p_S^i \geq 0 : S \in \mathcal{M}/i\}$ with $\sum_{S\in\mathcal{M}/i} p_S^i = \lambda_i$ such that for all $v \in \Gamma(\mathcal{M})$,*

$$\psi_i(v) = \sum_{S\in\mathcal{M}/i} p_S^i \left(v(S \cup i) - v(S)\right). \tag{8.2}$$

Proof. The implication $(ii) \Rightarrow (i)$ is obvious. To prove the converse implication, we consider the unanimity game $u_T : \mathcal{M} \to \mathbb{R}$ and the identity game $\delta_T : \mathcal{M} \to \mathbb{R}$, and we recall that $u_T = \sum_{\{S\in\mathcal{M}: S\supseteq T\}} \delta_S$ for all nonempty $T \in \mathcal{M}$. Suppose ψ_i satisfies the linearity, monotonicity, and λ_i-dummy player properties. Let $\Phi_i : \Gamma(\mathcal{M}) \to \mathbb{R}$ be defined by

$$\Phi_i(v) := \sum_{S\in\mathcal{M}/i} \psi_i(\delta_{S\cup i}) \left(v(S \cup i) - v(S)\right), \tag{8.3}$$

for all $v \in \Gamma(\mathcal{M})$. We claim $\psi_i = \Phi_i$. Since both values are linear on $\Gamma(\mathcal{M})$ and the collection $\{u_T : T \in \mathcal{M},\ T \neq \emptyset\}$ of unanimity games forms a basis of $\Gamma(\mathcal{M})$, it suffices to establish that $\psi_i(u_T) = \Phi_i(u_T)$ for all nonempty $T \in \mathcal{M}$.

First, let $T \in \mathcal{M}$, $T \neq \emptyset$, such that $i \notin T$. Note that $u_T(S \cup i) = u_T(S)$ for all $S \in \mathcal{M}/i$ and hence, by (8.3), $\Phi_i(u_T) = 0$ whenever $i \notin T$. Moreover, $u_T(i) = 0$, so $i \notin T$ is a dummy player in the unanimity game u_T and thus, by the λ_i-dummy player property for ψ_i, we obtain that $\psi_i(u_T) = \lambda_i u_T(i) = 0$ whenever $i \notin T$. So far, $\psi_i(u_T) = 0 = \Phi_i(u_T)$ whenever $i \notin T$.

Secondly, let $T \in \mathcal{M}$ such that $i \in T$. From (8.3) and the linearity property for ψ_i, we deduce that the following holds:

$$\begin{aligned}\Phi_i(u_T) &= \sum_{S \in \mathcal{M}/i} \psi_i(\delta_{S \cup i}) u_T(S \cup i) = \sum_{\{S \in \mathcal{M} : S \supseteq T\}} \psi_i(\delta_S) \\ &= \psi_i\left(\sum_{\{S \in \mathcal{M} : S \supseteq T\}} \delta_S\right) = \psi_i(u_T).\end{aligned}$$

We conclude that $\psi_i(u_T) = \Phi_i(u_T)$ for all nonempty $T \in \mathcal{M}$. Hence, by linearity, $\psi_i = \Phi_i$. Next, we define $p_S^i := \psi_i(\delta_{S \cup i})$ for all $S \in \mathcal{M}/i$. Further, $u_{\{i\}}(S \cup i) - u_{\{i\}}(S) = 1 = u_{\{i\}}(i)$ for all $S \in \mathcal{M}/i$. That is, player i is a dummy in the unanimity game $u_{\{i\}}$ and thus, by the λ_i-dummy player property for ψ_i and (8.3), we obtain that $\lambda_i = \lambda_i u_{\{i\}}(i) = \psi_i(u_{\{i\}}) = \sum_{S \in \mathcal{M}/i} p_S^i$.

It remains to establish that $p_S^i \geq 0$ for all $S \in \mathcal{M}/i$. For that purpose, for every $S \in \mathcal{M}$, $S \neq \emptyset$, define the monotone game $\widehat{u}_S : \mathcal{M} \to \mathbb{R}$ by $\widehat{u}_S := u_S - \delta_S$, that is $\widehat{u}_S(T) := 1$ if $T \supsetneq S$ and $\widehat{u}_S(T) := 0$ otherwise. Let $S \in \mathcal{M}/i$. By the monotonicity of game $\widehat{u}_S$, we obtain that $\widehat{u}_S(T \cup i) - \widehat{u}_S(T) \in \{0, 1\}$ for all $T \in \mathcal{M}/i$ and moreover, $\widehat{u}_S(T \cup i) - \widehat{u}_S(T) = 1$ if and only if $S = T$. Hence, by (8.3), $\psi_i(\widehat{u}_S) = p_S^i$ and by the monotonicity property for ψ_i, we conclude that $p_S^i = \psi_i(\widehat{u}_S) \geq 0$ for all $S \in \mathcal{M}/i$. □

In fact, the above theorem generalizes Weber's characterization [244, Theorem 4] of so-called *probabilistic values*, which are defined as a particular case of the new concept, as given by (8.2), applied to a free matroid.

Definition 8.3.1 *Let $\mathcal{M}$ be a matroid and let $i \in N$. An individual value $\psi_i : \Gamma(\mathcal{M}) \to \mathbb{R}$ is said to be a λ_i-quasi-probabilistic value if there exists a collection $\{p_S^i \geq 0 : S \in \mathcal{M}/i\}$ satisfying $\sum_{S \in \mathcal{M}/i} p_S^i = \lambda_i$ such that, for all $v \in \Gamma(\mathcal{M})$,*

$$\psi_i(v) = \sum_{S \in \mathcal{M}/i} p_S^i \left(v(S \cup i) - v(S)\right). \tag{8.4}$$

The following equivalence theorem states that, given a matroid $\mathcal{M}$, an individual value can be classified as a quasi-probabilistic value on $\Gamma(\mathcal{M})$ if and only if it is decomposable as a weighted sum of certain probabilistic values on free matroids induced by the basic coalitions of the given matroid.

Theorem 8.3.2 *Let $\mathcal{M}$ be a matroid. Let $\psi_i : \Gamma(\mathcal{M}) \to \mathbb{R}$ be an individual value for player $i \in N$ and $\lambda_i \in [0,1]$ the rate of participation by player i. Then the following two statements are equivalent:*

(i) *ψ_i is a λ_i-quasi-probabilistic value.*

(ii) *There exists a probability distribution P^i on the bases $\mathcal{B}(\mathcal{M})$ such that $\sum_{B \in \mathcal{B}_i(\mathcal{M})} P^i(B) = \lambda_i$ and, for every $B \in \mathcal{B}_i(\mathcal{M})$, there exists an individual probabilistic value $\psi_i^B : \Gamma(2^B) \to \mathbb{R}$ such that, for all $v \in \Gamma(\mathcal{M})$,*

$$\psi_i(v) = \sum_{B \in \mathcal{B}_i(\mathcal{M})} P^i(B) \psi_i^B(v_B), \tag{8.5}$$

where v_B is the restriction of the game v to 2^B.

Proof. Fix $i \in N$. Suppose P^i is a (yet unspecified) probability distribution on $\mathcal{B}(\mathcal{M})$ and moreover, suppose, for every $B \in \mathcal{B}_i(\mathcal{M})$, that $\psi_i^B : \Gamma(2^B) \to \mathbb{R}$ is an individual probabilistic value of the form

$$\psi_i^B(w) = \sum_{\{T \subseteq B : i \notin T\}} p_{B,T}^i \left(w(T \cup i) - w(T)\right) \quad \text{for all } w \in \Gamma(2^B), \tag{8.6}$$

where $\{p_{B,T}^i \geq 0 : T \subseteq B,\ i \notin T\}$ is a collection of (yet unspecified) numbers such that $\sum_{\{T \subseteq B \,:\, i \notin T\}} p_{B,T}^i = 1$. Therefore, we obtain, for all $v \in \Gamma(\mathcal{M})$,

$$\begin{aligned}
\sum_{B \in \mathcal{B}_i(\mathcal{M})} P^i(B) \psi_i^B(v_B) &= \sum_{B \in \mathcal{B}_i(\mathcal{M})} P^i(B) \left(\sum_{\{T \subseteq B : i \notin T\}} p_{B,T}^i \left(v(T \cup i) - v(T)\right) \right) \\
&= \sum_{T \in \mathcal{M}/i} \left(\sum_{B \in \mathcal{B}_{T \cup i}(\mathcal{M})} P^i(B) p_{B,T}^i \right) \left(v(T \cup i) - v(T)\right) \\
&= \sum_{T \in \mathcal{M}/i} p_T^i \left(v(T \cup i) - v(T)\right),
\end{aligned}$$

where the interrelationship between the coefficients is determined as follows:

$$p_T^i = \sum_{B \in \mathcal{B}_{T \cup i}(\mathcal{M})} P^i(B) p_{B,T}^i \tag{8.7}$$

for all $T \in \mathcal{M}/i$. The equality (8.7) yields the following result:

$$\begin{aligned}
\sum_{S \in \mathcal{M}/i} p_S^i &= \sum_{S \in \mathcal{M}/i} \left(\sum_{B \in \mathcal{B}_{S \cup i}(\mathcal{M})} P^i(B) p_{B,S}^i \right) \\
&= \sum_{B \in \mathcal{B}_i(\mathcal{M})} P^i(B) \left(\sum_{\{S \subseteq B : i \notin S\}} p_{B,S}^i \right) \\
&= \sum_{B \in \mathcal{B}_i(\mathcal{M})} P^i(B).
\end{aligned}$$

The implication $(ii) \Rightarrow (i)$ follows immediately from the definition of p_T^i by (8.7), provided that $P^i(B)$ and $p_{B,T}^i$ are given for all $B \in \mathcal{B}_{T \cup i}(\mathcal{M})$, $i \notin T$. In order to prove the converse implication, it suffices to solve the equation (8.7) for the variables $P^i(B)$ and $p_{B,T}^i$, provided that p_T^i are given for all $T \in \mathcal{M}/i$, and taking into account the following two additional restrictions: $\sum_{\{T \subseteq B : i \notin T\}} p_{B,T}^i = 1$ and $\sum_{B \in \mathcal{B}(\mathcal{M})} P^i(B) = 1$. Fort that purpose, let us define

$$\begin{aligned}
P^i(B) &:= \sum_{\{S \subseteq B : i \notin S\}} \frac{p_S^i}{b_{S \cup i}} \quad \text{for all } B \in \mathcal{B}_i(\mathcal{M}), \\
p_{B,T}^i &:= \frac{p_T^i}{b_{T \cup i} P^i(B)} \quad \text{for all } B \in \mathcal{B}_{T \cup i}(\mathcal{M}), \;\; i \notin T,
\end{aligned}$$

where $b_S := |\mathcal{B}_S(\mathcal{M})|$ for every $S \in \mathcal{M}$. By construction, the equation (8.7) is solved and the following restriction $\sum_{\{T \subseteq B : i \notin T\}} p_{B,T}^i = 1$ holds. Finally, it follows that the remaining restriction $\sum_{B \in \mathcal{B}(\mathcal{M})} P^i(B) = 1$ can be met due to the following chain of (in)equalities:

$$\begin{aligned}
\sum_{B \in \mathcal{B}_i(\mathcal{M})} P^i(B) &= \sum_{B \in \mathcal{B}_i(\mathcal{M})} \left(\sum_{\{S \subseteq B : i \notin S\}} \frac{p_S^i}{b_{S \cup i}} \right) \\
&= \sum_{S \in \mathcal{M}/i} p_S^i \left(\sum_{B \in \mathcal{B}_{S \cup i}(\mathcal{M})} \frac{1}{b_{S \cup i}} \right) \\
&= \sum_{S \in \mathcal{M}/i} p_S^i = \lambda_i \leq 1.
\end{aligned}$$

This completes the proof of the equivalence theorem. The reader can observe that when $P^i(B) = 0$, for some $B \in \mathcal{B}_i(\mathcal{M})$, then we can use any probabilistic value on 2^B, as definition of Ψ_i^B. □

Definition 8.3.2 *Let $\mathcal{M}$ be a matroid. A group value $\psi : \Gamma(\mathcal{M}) \to \mathbb{R}^n$ is said to be a basic value if there exists a probability distribution P on $\mathcal{B}(\mathcal{M})$ such that every component ψ_i, $i \in N$, is given by*

$$\psi_i(v) = \sum_{B \in \mathcal{B}_i(\mathcal{M})} P(B)\psi_i^B(v_B) \quad \textit{for all } v \in \Gamma(\mathcal{M}), \tag{8.8}$$

where $\psi_i^B : \Gamma(2^B) \to \mathbb{R}$ is an individual probabilistic value, for all $B \in \mathcal{B}_i(\mathcal{M})$.

Clearly, by Theorem 8.3.2, every component of a basic group value is a quasi-probabilistic value. The converse statement, however, is not true. That is, if each component of a group value is a quasi-probabilistic value, then the group value is not necessarily a basic value. For instance, we consider the matroid $M_3\,(1||2) = \{\emptyset, \{1\}, \{2\}, \{3\}, \{1,3\}, \{2,3\}\}$ and thus, $\mathcal{B}(\mathcal{M}) = \{\{1,3\}, \{2,3\}\}$. In addition, we consider any group value $\psi : \mathcal{M} \to \mathbb{R}^3$ whose components are quasi-probabilistic values such that $\psi_1(v) = p^1_{\{3\}}\,(v(13) - v(3))$ and $\psi_3(v) = p^3_{\{1\}}\,(v(13) - v(1))$, for all $v \in \Gamma(\mathcal{M})$, where $0 \le p^1_{\{3\}} \le 1$ and $0 \le p^3_{\{1\}} \le 1$. We claim that, for every $0 < p^1_{\{3\}} < 1$ and $p^3_{\{1\}} = 1$, the quasi-probabilistic group value ψ cannot be interpreted as a basic value of the form (8.8). For that purpose, suppose there exists a probability distribution P on $\mathcal{B}(\mathcal{M})$ such that (8.8) holds. In particular, for all $v \in \Gamma(\mathcal{M})$,

$$\begin{aligned}
\psi_1(v) &= P(\{1,3\})\psi_1^{\{1,3\}}(v_{\{1,3\}}) \\
&= P(\{1,3\})\left[p^1_{\{1,3\},\{3\}}\,(v(13) - v(3)) + p^1_{\{1,3\},\emptyset}v(1)\right]; \\
\psi_3(v) &= P(\{1,3\})\psi_3^{\{1,3\}}(v_{\{1,3\}}) + P(\{2,3\})\psi_3^{\{2,3\}}(v_{\{2,3\}}) \\
&= P(\{1,3\})\left[p^3_{\{1,3\},\{1\}}\,(v(13) - v(1)) + p^3_{\{1,3\},\emptyset}v(3)\right] \\
&\quad + P(\{2,3\})\left[p^3_{\{2,3\},\{2\}}\,(v(23) - v(2)) + p^3_{\{2,3\},\emptyset}v(3)\right].
\end{aligned}$$

From the two expressions for ψ_1, we deduce $p^1_{\{1,3\},\emptyset} = 0$. In fact, we apply (8.7) by looking at $0 = p^1_\emptyset = P(\{1,3\})p^1_{\{1,3\},\emptyset}$, where $P(\{1,3\}) \neq 0$ because of the assumption $p^1_{\{3\}} > 0$. Thus, $p^1_{\{1,3\},\{3\}} = 1 - p^1_{\{1,3\},\emptyset} = 1$ and consequently, (8.7) yields $p^1_{\{3\}} = P(\{1,3\})p^1_{\{1,3\},\{3\}} = P(\{1,3\})$. Finally, from the two expressions for ψ_3 (or once again (8.7)), we deduce $1 = p^3_{\{1\}} = P(\{1,3\})p^3_{\{1,3\},\{1\}}$. The resulting equality, however, is in contradiction with $0 \le p^3_{\{1,3\},\{1\}} \le 1$ and $0 < P(\{1,3\}) < 1$. Therefore, the given quasi-probabilistic group value is not a basic value.

8.4 The probabilistic Shapley value

Recall that the solution part of classical cooperative games on the free matroid $\mathcal{M} = 2^N$ is based on the assumption that the grand coalition N is formed and thus, solution concepts aim to prescribe equitable divisions of the associated worth $v(N)$ among the players of any game (N, v). In the context of any non-trivial matroid $\mathcal{M} \neq 2^N$, the fullest measure of cooperation among players is supposed to take place within feasible coalitions that are as large as possible (with respect to inclusion of sets). Thus, for every basic coalition $B \in \mathcal{B}(\mathcal{M})$, the family 2^B of feasible coalitions is called a *cooperation area* for the players of B. The basic coalitions of a matroid, however, are not necessarily disjoint and so, two basic coalitions with at least one mutual member cannot be formed at the same time.

This section is devoted to a model based on a probabilistic approach to the formation of basic coalitions. In other words, this *static model* involves probability distributions over the various cooperation areas (basic coalitions) and according to this random process, each feasible coalition interacts within its relevant cooperation areas with certain probabilities. Since it is supposed that a basic coalition is formed randomly, we will deal with an arbitrary probability distribution over the set of basic coalitions denoted by

$$\mathcal{P}(\mathcal{M}) = \left\{ P \in (\mathbb{R}_+)^{\mathcal{B}(\mathcal{M})} \ : \ \sum_{B \in \mathcal{B}(\mathcal{M})} P(B) = 1 \right\}.$$

Definition 8.4.1 *Let $\mathcal{M}$ be a matroid and let $P \in \mathcal{P}(\mathcal{M})$ a probability distribution over $\mathcal{B}(\mathcal{M})$. For every $S \in \mathcal{M}$, the probabilistic participation influence $w^P(S)$ of S within the cooperation areas of $\mathcal{M}$ (with respect to P) is given by the sum of the probabilities of basic coalitions containing S,*

$$w^P(S) := \sum_{B \in \mathcal{B}_S(\mathcal{M})} P(B) \ \ \textit{for all } S \in \mathcal{M}. \tag{8.9}$$

For all $P \in \mathcal{P}(\mathcal{M})$, the vector $w^P := (w^P(i))_{i \in N} \in \mathbb{R}^n$ is called the *probabilistic participation influence.* Their components are the probabilistic participation influences of individuals within the cooperation areas of the matroid.

The next result asserts that the set consisting of all probabilistic participation influence vectors coincides with the core of the rank game (induced by the rank function of matroid $\mathcal{M}$). In other words, every core-allocation of the rank game represents in a natural and unique manner the (probabilistic) rate of participation by individuals within the cooperation areas of the matroid. Obviously, the probabilistic rate of participation by any istmus player (who belongs to every basic coalition) equals one.

Proposition 8.4.1 *Let $\mathcal{M}$ be a matroid and $r : 2^N \to \mathbb{Z}_+$ its rank function. Then $C(r) = \{w^P : P \in \mathcal{P}(\mathcal{M})\}$.*

Proof. By Theorem 8.1.3, every $x \in C(r)$ can be written as a convex combination of the collection incidence vectors $\{\mathbf{1}_B : B \in \mathcal{B}(\mathcal{M})\}$. Then, there exist nonnegative numbers $\{P(B) : B \in \mathcal{B}(\mathcal{M})\}$, with $\sum_{B\in\mathcal{B}(\mathcal{M})} P(B) = 1$ such that $x = \sum_{B\in\mathcal{B}(\mathcal{M})} P(B)\mathbf{1}_B = w^P$, or equivalently,

$$x_i = \sum_{B\in\mathcal{B}(\mathcal{M})} P(B) = w^P(i)$$

for all $i \in N$. □

Now we are in a position to state our main theorem concerning the extension of the well-known *Shapley value* for classical games to games on matroids. In an axiomatic way we introduce a basic value on $\Gamma(\mathcal{M})$ constructed by the classical Shapley values on $\Gamma(2^B)$, for all $B \in \mathcal{B}(\mathcal{M})$, and a probability distribution $P \in \mathcal{P}(\mathcal{M})$:

$$\Phi_i^P(v) = \sum_{B\in\mathcal{B}_i(\mathcal{M})} P(B)\Phi_i^B(v_B),$$

for all $v \in \Gamma(\mathcal{M})$ and all $i \in N$, where $\Phi^B = (\Phi_i^B)_{i\in B}$ is the classical Shapley value on $\Gamma(2^B)$, $B \in \mathcal{B}(\mathcal{M})$. The axiomatic approach to the extended Shapley value involves four axioms, in which linearity is formulated in a classical manner, the substitution axiom is applied to some substitutes in unanimity games, whereas the dummy player and efficiency axioms are formulated with reference to a given probability distribution over the cooperation areas of the considered matroid.

Theorem 8.4.2 *Let $\mathcal{M}$ be a matroid and let $P \in \mathcal{P}(\mathcal{M})$ be a probability distribution over $\mathcal{B}(\mathcal{M})$. There exists a unique group value $\psi : \Gamma(\mathcal{M}) \to \mathbb{R}^n$ that satisfies the following four axioms:*

1. *Linearity: For every $i \in N$, $\psi_i(\alpha v + \beta w) = \alpha\psi_i(v) + \beta\psi_i(w)$, for all $v, w \in \Gamma(\mathcal{M})$ and $\alpha, \beta \in \mathbb{R}$.*

2. *Substitution applied to unanimity games: For each $T \in \mathcal{M}$ and for every pair $i, j \in T$, we have $\psi_i(u_T) = \psi_j(u_T)$.*

3. *P-dummy player property: $\psi_i(v) = w^P(i)\, v(i)$ for every dummy player i in the game $v \in \Gamma(\mathcal{M})$.*

4. *Probabilistic efficiency: $\sum_{i\in N} \psi_i(v) = \sum_{B\in\mathcal{B}(\mathcal{M})} P(B)v(B)$ for all $v \in \Gamma(\mathcal{M})$.*

This unique group value $\Phi^P = \left(\Phi_i^P\right)_{i\in N}$ *is called the probabilistic Shapley value on* $\Gamma(\mathcal{M})$*, and it satisfies the following two formulas:*

$$\Phi_i^P(v) = \sum_{T\in\mathcal{M}/i} \frac{w^P(T\cup i)}{(r(N)-|T|)\binom{r(N)}{|T|}}(v(T\cup i)-v(T)) \tag{8.10}$$

$$\Phi_i^P(v) = \sum_{B\in\mathcal{B}_i(\mathcal{M})} P(B)\Phi_i^B(v_B) \tag{8.11}$$

for all $v\in\Gamma(\mathcal{M})$ *and all* $i\in N$*, where* $\Phi^B = \left(\Phi_i^B\right)_{i\in B}$ *represents the classical Shapley value on* $\Gamma\left(2^B\right)$*,* $B\in\mathcal{B}(\mathcal{M})$*.*

Proof. In order to prove the uniqueness part, suppose a group value ψ satisfies the linearity, substitution applied to unanimity games, probabilistic efficiency, and P-dummy player properties. Consider, for every $T\in\mathcal{M}$, $T\neq\emptyset$, the unanimity game $u_T:\mathcal{M}\to\mathbb{R}$. Recall that every $i\notin T$ is a dummy player in game u_T and thus, by the P-dummy player property for ψ_i, it holds that $\psi_i(u_T) = w^P(i)u_T(i) = 0$ whenever $i\notin T$. Further, by the substitution property for ψ, $\psi_i(u_T)=\psi_j(u_T)$ for every pair $i,\ j\in T$. By the probabilistic efficiency property for ψ, it holds that

$$\begin{aligned}\sum_{i\in N}\psi_i(u_T) &= \sum_{B\in\mathcal{B}(\mathcal{M})} P(B)u_T(B)\\ &= \sum_{B\in\mathcal{B}_T(\mathcal{M})} P(B) = w^P(T).\end{aligned}$$

So far, we obtain that the group value ψ for every unanimity game u_T, $T\in\mathcal{M}$, $T\neq\emptyset$, is uniquely determined by

$$\psi_i(u_T) = \begin{cases} w^P(T)/|T|, & \text{if } i\in T\\ 0, & \text{if } i\notin T.\end{cases} \tag{8.12}$$

Because the set $\{u_T : T\in\mathcal{M},\ T\neq\emptyset\}$ of unanimity games forms a basis of $\Gamma(\mathcal{M})$ and ψ is supposed to be linear, we conclude that the group value ψ is uniquely determined on $\Gamma(\mathcal{M})$ by the four axioms involved.

In order to prove the existence part, we first show that formula (8.10) agrees with the alternative formula (8.11). Recall that, for every $B\in\mathcal{B}(\mathcal{M})$, the classical Shapley value $\Phi^B = \left(\Phi_i^B\right)_{i\in B}$ on $\Gamma\left(2^B\right)$ is given as follows:

$$\Phi_i^B(v') = \sum_{\{T\subseteq B: i\notin T\}} \frac{1}{(|B|-|T|)\binom{|B|}{|T|}}\left(v'(T\cup i)-v'(T)\right),$$

for all $v'\in\Gamma(2^B)$ and all $i\in B$.

By the above equation and $|B| = r(N)$ for every $B \in \mathcal{B}(\mathcal{M})$, we arrive at the following chain of equalities, for every $v \in \Gamma(\mathcal{M})$ and all $i \in N$:

$$\begin{aligned}
&\sum_{B \in \mathcal{B}_i(\mathcal{M})} P(B)\Phi_i^B(v_B) \\
&= \sum_{B \in \mathcal{B}_i(\mathcal{M})} P(B) \sum_{\{T \subseteq B : i \notin T\}} \frac{1}{(|B| - |T|)\binom{|B|}{|T|}} (v(T \cup i) - v(T)) \\
&= \sum_{T \in \mathcal{M}/i} \left(\sum_{B \in \mathcal{B}_{T \cup i}(\mathcal{M})} P(B) \right) \frac{(v(T \cup i) - v(T))}{(|B| - |T|)\binom{|B|}{|T|}} \\
&= \sum_{T \in \mathcal{M}/i} \frac{w^P(T \cup i)}{(r(N) - |T|)\binom{r(N)}{|T|}} (v(T \cup i) - v(T)) .
\end{aligned}$$

Thus, (8.10) is fully equivalent to (8.11). Clearly, by (8.10), the probabilistic Shapley value Φ^P is linear on $\Gamma(\mathcal{M})$. By (8.12) the relevant substitution property holds since the probabilistic Shapley value for a unanimity game, u_S, is the equitable allocation of the probabilistic participation influence of S between the players of S.

With the aid of (8.11) and the classical efficiency of the classical Shapley value on free matroids, we are able to establish the probabilistic efficiency property for the probabilistic Shapley value Φ^P as follows:

$$\begin{aligned}
\sum_{i \in N} \Phi_i^P(v) &= \sum_{i \in N} \left(\sum_{B \in \mathcal{B}_i(\mathcal{M})} P(B)\Phi_i^B(v_B) \right) \\
&= \sum_{B \in \mathcal{B}(\mathcal{M})} P(B) \left(\sum_{i \in B} \Phi_i^B(v_B) \right) \\
&= \sum_{B \in \mathcal{B}(\mathcal{M})} P(B) v(B),
\end{aligned}$$

for all $v \in \Gamma(\mathcal{M})$. For the sake of the P-dummy player property, suppose player i is a dummy in $v \in \Gamma(\mathcal{M})$, that is $v(S \cup i) - v(S) = v(i)$ for all $S \in \mathcal{M}/i$. In particular, for every $B \in \mathcal{B}_i(\mathcal{M})$, we have $v(S \cup i) - v(S) = v(i)$ for all $S \subseteq B$ with $i \notin S$. In other words, player i is a dummy in any subgame v_B associated with any basic coalition containing player i. Since the classical Shapley value on free matroids satisfies the classical dummy player property, we arrive at the P-dummy player property for the probabilistic Shapley value Φ_i^P as follows:

$$\Phi_i^P(v) = \sum_{B \in \mathcal{B}_i(\mathcal{M})} P(B)\Phi_i^B(v_B)$$

$$= \sum_{B \in \mathcal{B}_i(\mathcal{M})} P(B)v(i)$$
$$= w^P(i)v(i).$$

An alternative proof of the P-dummy player property for the probabilistic Shapley value Φ_i^P proceeds as follows. From (8.10), we derive that, for every dummy player i in a game $v \in \Gamma(\mathcal{M})$, the following holds:

$$\begin{aligned}\Phi_i^P(v) &= \sum_{S \in \mathcal{M}/i} \frac{w^P(S \cup i)}{(r(N) - |S|)\binom{r(N)}{|S|}} (v(S \cup i) - v(S)) \\ &= v(i) \sum_{S \in \mathcal{M}/i} \frac{w^P(S \cup i)}{(r(N) - |S|)\binom{r(N)}{|S|}} \\ &= v(i)\, w^P(i).\end{aligned}$$

In order to prove the last equality, we claim that for all $S \in \mathcal{M}$ and all $k \geq |S|$,

$$\sum_{\{T \in \mathcal{M}: T \supseteq S, |T| = k\}} w^P(T) = \binom{r(N) - |S|}{k - |S|} w^P(S). \tag{8.13}$$

Indeed, in view of (8.9), the straightforward proof of (8.13) proceeds as follows,

$$\begin{aligned}\sum_{\{T \in \mathcal{M}: T \supseteq S, |T| = k\}} w^P(T) &= \sum_{\{T \in \mathcal{M}: T \supseteq S, |T| = k\}} \sum_{B \in \mathcal{B}_T(\mathcal{M})} P(B) \\ &= \sum_{B \in \mathcal{B}_S(\mathcal{M})} P(B) \sum_{\{T \in \mathcal{M}: S \subseteq T \subseteq B, |T| = k\}} 1 \\ &= \binom{r(N) - |S|}{k - |S|} w^P(S).\end{aligned}$$

From (8.13) we deduce

$$\begin{aligned}&\sum_{S \in \mathcal{M}/i} \frac{w^P(S \cup i)}{(r(N) - |S|)\binom{r(N)}{|S|}} \\ &= \sum_{k=1}^{r(N)} \frac{1}{(r(N) - k + 1)\binom{r(N)}{k-1}} \sum_{\{T \in \mathcal{M}: |T| = k, i \in T\}} w^P(T) \\ &= \sum_{k=1}^{r(N)} \frac{\binom{r(N)-1}{k-1}}{(r(N) - k + 1)\binom{r(N)}{k-1}} w^P(i) \\ &= w^P(i) \sum_{k=1}^{r(N)} \frac{1}{r(N)} = w^P(i).\end{aligned}$$

This completes the alternative proof of the P-dummy player property for the probabilistic Shapley value Φ_i^P as given by (8.10). □

By (8.10), the probabilistic Shapley value for games on an arbitrary matroid represents some kind of expected value, in that the expected payoff to every player is composed of the player's marginal contributions in the game with respect to feasible coalitions contained in the contraction of the matroid to the player. We take into account the probabilistic participation influences of these feasible coalitions enlarged with the player as well as a (classical) probability distribution arising from the belief that the feasible coalition, which the player joins, is equally likely with probability $r(N)^{-1}$ to be of any cardinality t, where $0 \leq t \leq r(N) - 1$, and that all such coalitions of the same cardinality t are equally likely with probability $\binom{r(N)-1}{t}^{-1}$. To be exact, we claim

$$\sum_{T \in \mathcal{M}/i} \frac{w^P(T \cup i)}{(r(N) - |T|)\binom{r(N)}{|T|}} = w^P(i)$$

for every $i \in N$ (see the last stage of the proof of Theorem 8.4.2).

By the decomposition formula (8.11), the probabilistic Shapley value for games on a matroid is fully determined by classical Shapley values on free matroids induced by the basic coalitions of the given matroid, taking into account the probabilities of these basic coalitions. The impact of the decomposition formula (8.11) was already illustrated in the proof of Theorem 8.4.2 in that known properties (like efficiency and dummy player property) for the classical Shapley value are basic tools to prove similar properties for the extension of the Shapley value to games on an arbitrary matroid.

Remark 8.4.1 *For the future research on the Shapley value for games on matroids, the decomposition formula (8.11) may be an extremely helpful tool to obtain knowledge of particular properties.*

The probabilistic Shapley value for games on matroids as given by (8.10) does not satisfy the classical substitution property. For instance, consider again the matroid $M_3(1\|2)$. Given game v, defined by $v(1) = v(3) = v(2) = 1$, $v(13) = v(23) = 2$, players 1 and 3 are substitutes in the classical sense since $v(S \setminus 1) = v(S \setminus 3)$ for all $S \in M_3(1\|2)$ with $\{1,3\} \subseteq S$. By choosing the probabilities $P(\{1,3\}) = P(\{2,3\}) = 1/2$, formula (8.10) yields $\Phi_1^P(v) = 1/2$, whereas $\Phi_3^P(v) = 1$, although players 1 and 3 are substitutes in the classical sense.

Chapter 9

The core, the selectope and the Weber set

This chapter directs the attention to the games defined on a family of subsets of a finite set N and in this context, different solution concepts are considered. First, we define *imputations*, *core* and *selectope* for games defined on a family $\mathcal{L} \subseteq 2^N$ and we establish the relations between these solution concepts. If $\mathcal{L} = 2^N$ we obtain a classical cooperative game. In this case, we know that the core of a game v is included in the selectope. However, this is not true when $\mathcal{L} \neq 2^N$.

Next, some conditions on the family $\mathcal{L}$ are imposed and we prove that under those conditions, the selectope contains the core of the game. To obtain this result, closure spaces are introduced, and in particular, intersecting families. It turns out, roughly, that the latter structure of partial cooperation is necessary and sufficient for the core to be always included in the selectope. Moreover, the concept of a closure space generalizes those partial cooperation structures where the set of feasible coalitions is closed under taking intersections. The results of this chapter are due to Jiménez Jiménez [134], Bilbao, Jiménez Jiménez and Lebrón [14], and Bilbao, Jiménez Jiménez, Lebrón and Peters [15].

9.1 The core

Let $N = \{1, \ldots, n\}$ be a set of players and let $\mathcal{L} \subseteq 2^N$ be a collection of feasible coalitions. We only assume that $\emptyset, N \in \mathcal{L}$. When $\{i\} \in \mathcal{L}$ for all $i \in N$, the family $\mathcal{L}$ is called *atomic*. Sometimes, additional conditions are required on $\mathcal{L}$. A *game on* $\mathcal{L}$ is a function $v : \mathcal{L} \to \mathbb{R}$ such that $v(\emptyset) = 0$.

We denote by $\Gamma(\mathcal{L})$ the set of all games defined on the family $\mathcal{L}$. The set $\Gamma(\mathcal{L})$ is a vector space over $\mathbb{R}$, and there is a collection of games which deserves special consideration since it is a basis of $\Gamma(\mathcal{L})$. These *unanimity games* are defined for any nonempty $T \in \mathcal{L}$, and denoted by $u_T : \mathcal{L} \to \{0,1\}$, where

$$u_T(S) := \begin{cases} 1, & \text{if } T \subseteq S \\ 0, & \text{otherwise} \end{cases}$$

for every coalition $S \in \mathcal{L}$. It is left to the reader as a simple exercise to show linear independence of the collection of unanimity games. The spanning property is obvious, and the associated coefficients play a major role in this chapter and are derived below.

One of the main topics dealt with in cooperative game theory is to divide the amount $v(N)$ between the players if the grand coalition N is formed. The solution concepts that we study will prescribe distributions of the worth $v(N)$ among the players.

We introduce the following solution concepts. The set of *preimputations* of game $v : \mathcal{L} \to \mathbb{R}$ is the set

$$I^*(\mathcal{L}, v) := \{x \in \mathbb{R}^n : x(N) = v(N)\},$$

and the *imputation set* is given by

$$I(\mathcal{L}, v) := \{x \in I^*(\mathcal{L}, v) : x_i \geq v(i) \text{ for all } i \in N \text{ with } \{i\} \in \mathcal{L}\}.$$

It should be noted that $I(\mathcal{L}, v) \subseteq I^*(\mathcal{L}, v) \neq \emptyset$, but could be $I(\mathcal{L}, v) = \emptyset$. However, we may observe that if the family $\mathcal{L}$ does not contain any unitary coalition, then $I(\mathcal{L}, v) = I^*(\mathcal{L}, v)$, and hence $I(\mathcal{L}, v) \neq \emptyset$. On the other hand, if the family $\mathcal{L}$ contains at least one unitary coalition, the following cases may happen:

1. If $\mathcal{L}$ is not atomic, i.e., there exists $j \in N$ such that $\{j\} \notin \mathcal{L}$, then $I(\mathcal{L}, v) \neq \emptyset$. The hypothesis implies that vector $x = (x_i)_{i \in N}$ defined by

$$x_i = \begin{cases} v(i), & \text{if } \{i\} \in \mathcal{L} \\ v(N) - \sum_{\{k\} \in \mathcal{L}} v(k), & \text{if } i = j \\ 0, & \text{otherwise,} \end{cases}$$

satisfies $x \in I(\mathcal{L}, v)$.

2. If $\mathcal{L}$ is atomic then

$$I(\mathcal{L}, v) \neq \emptyset \quad \text{if and only if} \quad v(N) \geq \sum_{i \in N} v(i).$$

Unlike the imputation set of a cooperative game (N, v), the imputation set of a game on a family $\mathcal{L}$ may or may not be bounded. Note that the set $I(\mathcal{L}, v)$ is determined by a finite number of linear inequalities, and thus it is a polyhedron. We will use the following characterization of polyhedra (see Chapter 2):

- The polyhedron $P = \{x \in \mathbb{R}^n : Ax \geq b\}$ is bounded if and only if

$$\{x \in \mathbb{R}^n : Ax \geq 0\} = \{0\}.$$

Proposition 9.1.1 *Let $v \in \Gamma(\mathcal{L})$. The following statements are equivalent:*

(a) *The family $\mathcal{L}$ is atomic.*

(b) *The set $I(\mathcal{L}, v)$ is bounded.*

Proof. The imputation set $I(\mathcal{L}, v)$ is bounded if and only if the set

$$\left\{x \in \mathbb{R}^n : \sum_{i \in N} x_i = 0,\ x_i \geq 0, \text{ for all } \{i\} \in \mathcal{L}\right\} = \{0\}.$$

But this equality holds if and only if the family $\mathcal{L}$ is atomic. □

From the imputation definition it is immediately deduced that when the family $\mathcal{L}$ is atomic and $v(N) = \sum_{i \in N} v(i)$, the imputation set is reduced to a unique distribution.

Proposition 9.1.2 *Let $v \in \Gamma(\mathcal{L})$ be a game on an atomic family $\mathcal{L}$. Then $I(\mathcal{L}, v) = \{(v(1), \ldots, v(n))\}$ if and only if $v(N) = \sum_{i \in N} v(i)$.*

Given a cooperative game (N, v) one of the most studied solution concepts has been the *core*. For games with partial cooperation, it is natural to use, in the definition of the core, only the feasible coalitions in $\mathcal{L}$.

Definition 9.1.1 *The core of the game $v \in \Gamma(\mathcal{L})$ is the set*

$$C(\mathcal{L}, v) := \{x \in \mathbb{R}^n : x(N) = v(N),\ x(S) \geq v(S) \ \text{for all}\ \ S \in \mathcal{L}\}.$$

Obviously, all vectors of the core are imputations and thus, the core can be the empty set. If $\mathcal{L}$ is atomic, then the imputation set $I(\mathcal{L}, v)$ is bounded and hence the core is a bounded polyhedron (polytope).

Proposition 9.1.3 *Let $v \in \Gamma(\mathcal{L})$. If the family $\mathcal{L}$ is atomic, then $C(\mathcal{L}, v)$ is a polytope.*

The reverse is not true in general. Let us take $N = \{1, 2, 3\}$, the family $\mathcal{L} = \{\emptyset, \{1\}, \{2\}, \{1, 3\}, \{2, 3\}, \{1, 2, 3\}\}$, and the game v on $\mathcal{L}$ defined by $v(S) = |S| - 1$ for every nonempty coalition $S \in \mathcal{L}$. In this case,

$$C(\mathcal{L}, v) = \left\{x \in \mathbb{R}^n_+ : x_1 + x_2 + x_3 = 2,\ x_1 \leq 1,\ x_2 \leq 1,\ x_3 \leq 2\right\}$$

is a bounded set and the family $\mathcal{L}$ is not atomic.

9.2 The selectope

The central concept of this section is the *selectope*, first introduced by Hammer, Peled and Sorensen [119]. The selectope contains every way of distributing the dividends of the coalitions between their members. From a game theoretic approach, it can be seen as imposing reasonable bounds on such distributions. Indeed, for a classical cooperative game the selectope is equal to the core of a game that arises by imposing on every coalition additionally the negative dividends of all coalitions in which members of the coalition under consideration participate.

This solution concept was investigated by Derks, Haller and Peters [55] in the case of cooperative games. We extend it for games defined on an atomic family of feasible coalitions $\mathcal{L}$.

First of all, we define the *dividends* for the game $v \in \Gamma(\mathcal{L})$, recursively, by

$$d_S(v) := \begin{cases} 0, & \text{if } S = \emptyset \\ v(S) - \sum_{\{T \in \mathcal{L}: T \subset S\}} d_T(v), & \text{otherwise,} \end{cases}$$

for every coalition $S \in \mathcal{L}$, where $\subset$ denotes strict inclusion. Obviously, for every $v \in \Gamma(\mathcal{L})$ and $S \in \mathcal{L}$,

$$v(S) = \sum_{\{T \in \mathcal{L}: T \subseteq S\}} d_T(v),$$

and hence every $v \in \Gamma(\mathcal{L})$ can be uniquely represented by

$$v = \sum_{\{T \in \mathcal{L}: T \neq \emptyset\}} d_T(v)\, u_T,$$

where $u_T \in \Gamma(\mathcal{L})$ is the unanimity game for the nonempty coalition $T \in \mathcal{L}$.

Definition 9.2.1 *Let $\mathcal{L} \subseteq 2^N$ be an atomic family such that $\emptyset, N \in \mathcal{L}$. A selector on $\mathcal{L}$ is a function $\alpha : \mathcal{L} \setminus \{\emptyset\} \to N$ with $\alpha(S) \in S$ for every nonempty coalition $S \in \mathcal{L}$.*

We denote by $\mathcal{A}(\mathcal{L})$ the set of all selectors on $\mathcal{L}$. Note that the number of selectors on $\mathcal{L}$ is $\prod_{\{T \in \mathcal{L}: T \neq \emptyset\}} |T|$.

Definition 9.2.2 *Let $\alpha \in \mathcal{A}(\mathcal{L})$ be a selector. The selection corresponding to α is the vector $m^\alpha(v) \in \mathbb{R}^n$ defined by*

$$m_i^\alpha(v) = \sum_{\{S \in \mathcal{L} : \alpha(S) = i\}} d_S(v)$$

for every $i \in N$ and $v \in \Gamma(\mathcal{L})$.

Definition 9.2.3 *Let $\mathcal{L} \subseteq 2^N$ be an atomic family such that $\emptyset, N \in \mathcal{L}$. The selectope for a game $v \in \Gamma(\mathcal{L})$ is given by*

$$Sel(\mathcal{L}, v) := \text{conv}\{m^{\alpha}(v) : \alpha \in \mathcal{A}(\mathcal{L})\}.$$

Obviously, for every $v \in \Gamma(\mathcal{L})$, and $\alpha \in \mathcal{A}(\mathcal{L})$, we have $m^{\alpha}(v) \in I^*(\mathcal{L}, v)$ and hence, by convexity of $I^*(\mathcal{L}, v)$, we obtain

$$Sel(\mathcal{L}, v) \subseteq I^*(\mathcal{L}, v).$$

Example: Let $N = \{1, 2, 3\}$ and consider the family

$$\mathcal{L} = \{\emptyset, \{1\}, \{2\}, \{3\}, \{1,2\}, \{1,2,3\}\}.$$

We define $v \in \Gamma(\mathcal{L})$ by $v(1) = v(2) = v(3) = 0$, $v(12) = 2$, and $v(123) = 3$. There are six selectors on $\mathcal{L}$ which are listed below along with the corresponding selections.

α	$\{1,2\}$	$\{1,2,3\}$	$m_1^{\alpha}(v)$	$m_2^{\alpha}(v)$	$m_3^{\alpha}(v)$
1	1	1	3	0	0
2	1	2	2	1	0
3	1	3	2	0	1
4	2	1	1	2	0
5	2	2	0	3	0
6	2	3	0	2	1

Table 9.1

Note that the unitary coalitions are not in this table since all selectors satisfy $\alpha(i) = i$ for all $\{i\} \in \mathcal{L}$. For this game v, we have

$$Sel(\mathcal{L}, v) = \text{conv}\{(3,0,0), (2,1,0), (2,0,1), (1,2,0), (0,3,0), (0,2,1)\}.$$

Note that $C(\mathcal{L}, v) = Sel(\mathcal{L}, v)$ as we illustrate in Figure 9.1.

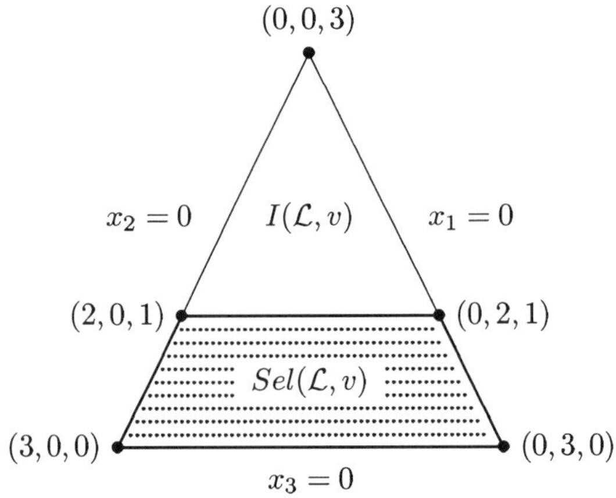

Figure 9.1

Example: Let $N = \{1, 2, 3, 4\}$ and consider the family

$$\mathcal{L} = \{\emptyset, \{1\}, \{2\}, \{3\}, \{4\}, \{1,2\}, \{2,3\}, \{1,2,3,4\}\},$$

and the game $v \in \Gamma(\mathcal{L})$ defined, for every $S \in \mathcal{L} \backslash \{\emptyset\}$, by

$$v(S) = \begin{cases} -1, & \text{if } |S| = 1 \\ 2, & \text{otherwise.} \end{cases}$$

If we calculate the dividends of the game, we get

$$d_S(v) = \begin{cases} -1, & \text{if } |S| = 1 \\ 4, & \text{if } |S| = 2 \\ -2, & \text{if } S = N, \end{cases}$$

and we show the selections in the following table:

α	$\{1,2\}$	$\{2,3\}$	$\{1,2,3,4\}$	$m_1^{\alpha}(v)$	$m_2^{\alpha}(v)$	$m_3^{\alpha}(v)$	$m_4^{\alpha}(v)$
1	1	2	1	1	3	−1	−1
2	1	2	2	3	1	−1	−1
3	1	2	3	3	3	−3	−1
4	1	2	4	3	3	−1	−3
5	1	3	1	1	−1	3	−1
6	1	3	2	3	−3	3	−1
7	1	3	3	3	−1	1	−1
8	1	3	4	3	−1	3	−3
9	2	2	1	−3	7	−1	−1
10	2	2	2	−1	5	−1	−1
11	2	2	3	−1	7	−3	−1
12	2	2	4	−1	7	−1	−3
13	2	3	1	−3	3	3	−1
14	2	3	2	−1	1	3	−1
15	2	3	3	−1	3	1	−1
16	2	3	4	−1	3	3	−3

Table 9.2

On the other hand, $C(\mathcal{L}, v)$ is the set of vectors $x \in \mathbb{R}^4$ such that

$$\begin{aligned} x_1 + x_2 + x_3 + x_4 &= 2 \\ x_1 + x_2 &\geq 2 \\ x_2 + x_3 &\geq 2 \\ x_1, x_2, x_3, x_4 &\geq -1. \end{aligned}$$

Note that we have that $Sel(\mathcal{L}, v) \nsubseteq C(\mathcal{L}, v)$ and $C(\mathcal{L}, v) \nsubseteq Sel(\mathcal{L}, v)$. Specifically, the vector $(3, 3, -3, -1) \in Sel(\mathcal{L}, v) \setminus C(\mathcal{L}, v)$ since $x_3 \ngeq -1$ and the vector $(0, 2, 0, 0) \in C(\mathcal{L}, v) \setminus Sel(\mathcal{L}, v)$ as it cannot be written as convex combination of $m^{\alpha}(v)$ with $\alpha \in \mathcal{A}(\mathcal{L})$, since $x_4 < 0$ for all $x \in Sel(\mathcal{L}, v)$ (see Table 9.2).

The inclusion $Sel(\mathcal{L}, v) \subseteq C(\mathcal{L}, v)$ characterizes the following *almost positive* games.

Definition 9.2.4 *A game $v \in \Gamma(\mathcal{L})$ is called almost positive if the dividends of all non-unitary coalitions are nonnegative, that is,*

$$d_S(v) \geq 0 \text{ for all } S \in \mathcal{L} \text{ with } |S| \geq 2.$$

Theorem 9.2.1 *Let $\mathcal{L} \subseteq 2^N$ be an atomic family which contains $\emptyset, N$ and let $v \in \Gamma(\mathcal{L})$. The following statements are equivalent:*

(a) $Sel(\mathcal{L}, v) \subseteq C(\mathcal{L}, v)$.

(b) $Sel(\mathcal{L}, v) \subseteq I(\mathcal{L}, v)$.

(c) *Game v is almost positive.*

Proof. $(a) \Rightarrow (b)$ follows from $C(\mathcal{L}, v) \subseteq I(\mathcal{L}, v)$.

For $(b) \Rightarrow (c)$, assume that there exists $S \in \mathcal{L}$ with $|S| \geq 2$ and $d_S(v) < 0$ and consider a selector $\alpha \in \mathcal{A}(\mathcal{L})$ such that $\alpha(S) = i$ and $\alpha(T) \neq i$ for all $T \in \mathcal{L}$, $T \neq S, \{i\}$. Then $m_i^\alpha(v) = v(i) + d_S(v) < v(i)$, contradicting (b).

For $(c) \Rightarrow (a)$, let v be an almost positive game. For this class of games, we assert that $m^\alpha(v) \in C(\mathcal{L}, v)$ for every selector α. Indeed, for every nonempty coalition $S \in \mathcal{L}$, we have

$$\begin{aligned} \sum_{i \in S} m_i^\alpha(v) &= \sum_{i \in S} \sum_{\{T \in \mathcal{L} : \alpha(T) = i\}} d_T(v) \\ &\geq \sum_{\{T \in \mathcal{L} : T \subseteq S\}} d_T(v) \\ &= v(S), \end{aligned}$$

where the inequality follows since all dividends of the non-unitary coalitions are nonnegative. □

Although $C(\mathcal{L}, v) \subseteq Sel(\mathcal{L}, v)$ has already been proved by Hammer *et al.* [119] in case $\mathcal{L} = 2^N$, this result is not true when the family of the feasible coalitions $\mathcal{L} \neq 2^N$ as the above example shows.

Even, $Sel(\mathcal{L}, v) \subset C(\mathcal{L}, v)$ is possible. For instance, we consider the set of players $N = \{1, 2, 3, 4\}$, the family of feasible coalitions

$$\mathcal{L} = \{\emptyset, \{1\}, \{2\}, \{3\}, \{4\}, \{1,2\}, \{2,3\}, N\},$$

and the almost positive game v given by $v(12) = v(23) = 1$, $v(N) = 4$, and $v(S) = 0$ otherwise. Then it is easily checked that $(0, 1, 0, 3) \in C(\mathcal{L}, v)$, whereas player 4 obtains at most $d_N(v) = 2$ in the selectope.

The inclusion $C(\mathcal{L}, v) \subseteq Sel(\mathcal{L}, v)$ is true if we consider the family $\mathcal{L}$ with some additional conditions, as established below. Recall that a family $\mathcal{L}$ of subsets of N such that

1. $\emptyset \in \mathcal{L}$ and $N \in \mathcal{L}$,

2. $A \in \mathcal{L}$ and $B \in \mathcal{L}$ implies that $A \cap B \in \mathcal{L}$,

is called a *closure space* over N, and its elements are called *closed sets*. We will study conditions on the family of feasible coalitions which imply that the inclusion $C(\mathcal{L}, v) \subseteq Sel(\mathcal{L}, v)$ holds. The selectope of a game $v \in \Gamma(\mathcal{L})$ is defined as the convex hull of a finite set of vectors and it is always a polytope. So that the above inclusion makes sense we note the core of a game $v \in \Gamma(\mathcal{L})$ is a bounded set if the family $\mathcal{L}$ is atomic. Although it is not relevant in the relation between core and selectope, we can observe that, if $\mathcal{L}$ is a closure space, this sufficient condition is also made necessary.

Theorem 9.2.2 *Let $\mathcal{L} \subseteq 2^N$ be a closure space and let $v \in \Gamma(\mathcal{L})$ be a game such that $C(\mathcal{L}, v)$ is nonempty. Then, the following statements are equivalent:*

(a) *The core of the game v is a polytope.*

(b) *The family $\mathcal{L}$ is atomic.*

Proof. To prove that $(a) \Rightarrow (b)$, suppose that (b) is false. Then there exists $j \in N$ such that $\{j\} \notin \mathcal{L}$ we consider $\overline{\{j\}} \in \mathcal{L}$. Let $k \in \overline{\{j\}}$ with $k \neq j$, and we define the vector $x \in \mathbb{R}^n$,

$$x_i = \begin{cases} -1, & \text{if } i = j \\ 1, & \text{if } i = k \\ 0, & \text{otherwise.} \end{cases}$$

This vector $x \neq 0$ satisfies $x(N) = 0$, and $x(S) \geq 0$, for all $S \in \mathcal{L}$, since if $j \in S$ then $\overline{\{j\}} \subseteq S$, hence $k \in S$. This implies that

$$\{x \in \mathbb{R}^n : x(N) = 0, \ x(S) \geq 0, \ \text{for all } S \in \mathcal{L}\} \neq \{0\}.$$

which is a contradiction with (a).

$(b) \Rightarrow (a)$ follows from Proposition 9.1.3. □

Two special cases of closure spaces are *intersecting families* and *convex geometries*. The closure space $\mathcal{L}$ is called an intersecting family when

$$\text{If } S, T \in \mathcal{L} \ \text{ with } S \cap T \neq \emptyset, \ \text{ then } S \cup T \in \mathcal{L}.$$

Note that all the closure spaces on $N = \{1, 2, 3\}$ are intersecting families. In order to prove the next theorem, it is necessary to first define some concepts. Let $\mathcal{L} \subseteq 2^N$ be an atomic intersecting family and let $S \subseteq N$. We consider the set of all closed sets of $\mathcal{L}$ contained in S, i.e., $\mathcal{L}_S = \{T \in \mathcal{L} : T \subseteq S\}$.

Obviously, $\mathcal{L}_S \neq \emptyset$ since $\emptyset, \{i\} \in \mathcal{L}_S$ for all $i \in S$. Moreover, $\mathcal{L}_S$ is an atomic intersecting family on S if and only if $S \in \mathcal{L}$. If we consider $\mathcal{L}_S$ ordered by inclusion, then $(\mathcal{L}_S, \subseteq)$ is a finite partially ordered set where the first element is $\emptyset$. Let $\Pi(S)$ be the set of all maximal elements of $(\mathcal{L}_S, \subseteq)$. Every one of these elements is called a *maximal closed set* of S. Under these conditions we have the following property.

Proposition 9.2.3 *The coalitions of* $\Pi(S)$ *form a partition of every* $S \subseteq N$.

Proof. Note that if $S \in \mathcal{L}$, then $\Pi(S) = \{S\}$ and therefore, the proposition is trivially true. If $S \notin \mathcal{L}$ and we write $\Pi(S) = \{S_1, \ldots, S_k\}$, then $k \geq 2$ and $\bigcup_{i=1}^k S_i \subseteq S$. Also, $S \subseteq \bigcup_{i=1}^k S_i$ since $\mathcal{L}$ is an atomic family. In order to prove that the coalitions in $\Pi(S)$ form a partition of S, we claim that $S_i \cap S_j = \emptyset$ for all $1 \leq i, j \leq k, i \neq j$. Indeed, if $S_i \cap S_j \neq \emptyset$, then $S_i \cup S_j \in \mathcal{L}$ since $\mathcal{L}$ is an intersecting family and also $S_i \cup S_j \subseteq S$, and hence $S_i \cup S_j \in \mathcal{L}_S$ but this is a contradiction since S_i and S_j are maximal closed sets of S. □

Theorem 9.2.4 *Let* $\mathcal{L} \subseteq 2^N$ *be an atomic intersecting family. If* $v \in \Gamma(\mathcal{L})$ *then* $C(\mathcal{L}, v) \subseteq Sel(\mathcal{L}, v)$.

Proof. Assume that there exists $x \in C(\mathcal{L}, v)$ such that $x \notin Sel(\mathcal{L}, v)$. By convexity and closedness of $Sel(\mathcal{L}, v)$ and applying the Separation Theorem (see Aubin [6, pages 29-32]), there exists $y \in \mathbb{R}^n$ such that $z \cdot y > x \cdot y$ for every $z \in Sel(\mathcal{L}, v)$. In particular, this holds for every $m^\alpha(v)$ with $\alpha \in \mathcal{A}(\mathcal{L})$. If the components of vector y are ordered in nonincreasing order

$$y_{i_1} \geq y_{i_2} \geq \cdots \geq y_{i_{n-1}} \geq y_{i_n},$$

then we have that $x \cdot y$

$$\begin{aligned}
&= \sum_{j=1}^{n} x_{i_j} y_{i_j} = y_{i_n} \sum_{j=1}^{n} x_{i_j} + \sum_{k=1}^{n-1} \left(y_{i_k} - y_{i_{k+1}}\right) \sum_{j=1}^{k} x_{i_j} \\
&\geq y_{i_n} v(N) + \sum_{k=1}^{n-1} \left(y_{i_k} - y_{i_{k+1}}\right) \sum_{S \in \Pi(\{i_1, i_2, \ldots, i_k\})} v(S) \\
&= y_{i_1} v(\{i_1\}) + \sum_{j=2}^{n} y_{i_j} \left(\sum_{S \in \Pi(\{i_1, \ldots, i_j\})} v(S) - \sum_{S \in \Pi(\{i_1, \ldots, i_{j-1}\})} v(S) \right)
\end{aligned}$$

$$
\begin{aligned}
&= y_{i_1} d_{\{i_1\}}(v) + \\
&\quad \sum_{j=2}^{n} y_{i_j} \left(\sum_{S \in \Pi(\{i_1,\dots,i_j\})} \sum_{T \subseteq S} d_T(v) - \sum_{S \in \Pi(\{i_1,\dots,i_{j-1}\})} \sum_{T \subseteq S} d_T(v) \right) \\
&= y_{i_1} d_{\{i_1\}}(v) + \\
&\quad \sum_{j=2}^{n} y_{i_j} \left(\sum_{\{T \in \mathcal{L} : T \subseteq \{i_1,\dots,i_j\}\}} d_T(v) - \sum_{\{T \in \mathcal{L} : T \subseteq \{i_1,\dots,i_{j-1}\}\}} d_T(v) \right) \\
&= \sum_{j=1}^{n} y_{i_j} \left(\sum_{\{T \in \mathcal{L} : T \subseteq \{i_1,i_2,\dots,i_j\},\, i_j \in T\}} d_T(v) \right),
\end{aligned}
$$

where the inequality follows since $x \in C(\mathcal{L}, v)$, the y_{i_j} are nonincreasing, and by Proposition 9.2.3, and the last but one equality also follows by Proposition 9.2.3. Obviously, it is sufficient to take the selector $\alpha \in \mathcal{A}(\mathcal{L})$ such that

$$
m_{i_j}^{\alpha}(v) = \sum_{\{T \in \mathcal{L} : T \subseteq \{i_1,i_2,\dots,i_j\},\, i_j \in T\}} d_T(v)
$$

for all $1 \leq j \leq n$ in order to have a contradiction. This selector $\alpha \in \mathcal{A}(\mathcal{L})$ is defined by $\alpha(S) = i_k$ for all $S \in \mathcal{L}$, where $k = \max\{p : i_p \in S\}$. □

It should be noted that the proof of Theorem 9.2.4 is closely related to the proof in [54] of the inclusion of the core in the Weber set for games on 2^N. There is also a converse to Theorem 9.2.4.

Theorem 9.2.5 *Let $\mathcal{L} \subseteq 2^N$ be an atomic closure space such that for every game $v \in \Gamma(\mathcal{L})$ we have $C(\mathcal{L}, v) \subseteq Sel(\mathcal{L}, v)$. Then $\mathcal{L}$ is an intersecting family.*

Proof. Suppose $\mathcal{L}$ is not an intersecting family. Then there are $S, T \in \mathcal{L}$ with $S \cap T \neq \emptyset$ but $S \cup T \notin \mathcal{L}$. Let the number of players in $N \backslash (S \cup T)$ be equal to $k > 0$. Define v by $v(M) = 1$ if $S \subseteq M$ or $T \subseteq M$ but $S \cup T \not\subseteq M$, $v(M) = 2$ if $S \cup T \subseteq M$ but $M \neq N$, $v(N) = k + 1$, and $v(M) = 0$ otherwise. Then the dividends of S and T are equal to 1, the dividend of N is equal to $k - 1$, and all other dividends are equal to 0.

Fix a player $i \in S \cap T$. Then it is easily seen that the vector x with $x_i = 1$, $x_j = 1$ for all $j \in N \backslash (S \cup T)$, and $x_j = 0$ otherwise, is a core element. In this core element the players outside $S \cup T$ receive k together. In the selectope, however, they receive at most the dividend $k - 1$ of N together.

Hence, the core is not contained in the selectope. Observe that in this proof it is indeed essential that $S \cup T \notin \mathcal{L}$, since otherwise x as above would not be a core element. □

As has already been indicated, the game $v \in \Gamma(\mathcal{L})$ can be written as a linear combination of unanimity games, where the coefficients are the dividends of the coalitions in $\mathcal{L}$.

Definition 9.2.5 *Let $v \in \Gamma(\mathcal{L})$ be a game. We define*

$$\begin{aligned} v^+ &= \sum_{\{S \in \mathcal{L} : d_S(v) > 0\}} d_S(v)\, u_S, \\ v^- &= \sum_{\{S \in \mathcal{L} : d_S(v) < 0\}} -d_S(v)\, u_S. \end{aligned}$$

Note that for every game v we obtain the decomposition $v = v^+ - v^-$.

Theorem 9.2.6 *Let $\mathcal{L} \subseteq 2^N$ be an atomic intersecting family and $v \in \Gamma(\mathcal{L})$. Then*

(a) *$Sel(\mathcal{L}, v^+) = C(\mathcal{L}, v^+)$ and $Sel(\mathcal{L}, v^-) = C(\mathcal{L}, v^-)$.*

(b) *$Sel(\mathcal{L}, v) = C(\mathcal{L}, v^+) - C(\mathcal{L}, v^-)$.*

Proof. (a) This is a direct consequence of Theorem 9.2.1 and Theorem 9.2.4. (b) For every $\alpha \in \mathcal{A}(\mathcal{L})$ we have $m^\alpha(v) = m^\alpha(v^+) - m^\alpha(v^-)$, and (a) implies that $m^\alpha(v) \in C(\mathcal{L}, v^+) - C(\mathcal{L}, v^-)$. Therefore,

$$Sel(\mathcal{L}, v) \subseteq C(\mathcal{L}, v^+) - C(\mathcal{L}, v^-).$$

In order to prove the converse inclusion, in view of part (a), it is sufficient to prove that for any two selectors $\alpha, \beta \in \mathcal{A}(\mathcal{L})$, we have

$$m^\alpha(v^+) - m^\beta(v^-) \in Sel(\mathcal{L}, v).$$

We define $\gamma \in \mathcal{A}(\mathcal{L})$ by

$$\gamma(S) = \begin{cases} \alpha(S), & \text{if } d_S(v) \geq 0 \\ \beta(S), & \text{if } d_S(v) < 0, \end{cases}$$

for every nonempty $S \in \mathcal{L}$, and thus $m^\gamma(v) = m^\alpha(v^+) - m^\beta(v^-)$. Therefore, $m^\alpha(v^+) - m^\beta(v^-) \in Sel(\mathcal{L}, v)$. □

Remark 9.2.1 *Note that if $\mathcal{L}$ is an atomic family such that $\emptyset, N \in \mathcal{L}$, but it is not an intersecting family, then the relations in Theorem 9.2.6 hold only with $\subseteq$ instead of equality.*

9.3 The Weber set

The *convex geometry* structure generalizes many forms of partial cooperation studied in the literature. Moreover, it facilitates the extension of the definition of *marginal worth vectors* to games with partial cooperation since the role of permutations of the player set in this definition is played by maximal chains. We introduce the *Weber set* as the convex hull of the marginal worth vectors, and we show that for games on convex geometries the Weber set is contained in the selectope. In classical cooperative games the Weber set plays a role as a core catcher. Moreover, for the important class of convex games it coincides with the core and it is possible to implement the greedy algorithm to compute core elements. Here, as we will see, it singles out distributions according to consistent selectors from the selectope.

We introduce the Weber set for games on convex geometries and investigate its relation with the selectope. A closure space $\mathcal{L}$ is a *convex geometry* on N if it satisfies the augmentation property: if $A \neq N$ is a closed set, then $A \cup i$ is closed for some $i \in N \setminus A$.

Edelman and Jamison defined a *compatible ordering* of a convex geometry $\mathcal{L} \subseteq 2^N$ as a total ordering of the elements of N, $i_1 < i_2 < \cdots < i_n$ such that

$$\{i_1, i_2, \ldots, i_k\} \in \mathcal{L} \text{ for all } 1 \leq k \leq n.$$

A compatible ordering of $\mathcal{L}$ corresponds to a *maximal chain* of $\mathcal{L}$. Here, a maximal chain C of $\mathcal{L}$ is an ordered collection of convex sets

$$\emptyset = C_0 \subset C_1 \subset \cdots \subset C_{n-1} \subset C_n = N.$$

We denote by $\mathcal{C}(\mathcal{L})$ the set of all maximal chains of $\mathcal{L}$. Note that if $\mathcal{L} = 2^N$, then there are $n!$ maximal chains.

Then, in every maximal chain of $\mathcal{L}$, there is a minimal convex set that contains a player i. Moreover, i is extreme in this convex set. Thus, for every $i \in N$ and $C \in \mathcal{C}(\mathcal{L})$, we denote by $C(i)$ the minimal convex set in chain C which contains player i,

$$C(i) = \{j \in N : j \leq i \text{ in the chain } C\}.$$

In other words, $C(i)$ is the set of the predecessors of player i with respect to chain C together with player i. It is clear that $i \in \text{ex}(C(i))$ since $C(i) \setminus i \in \mathcal{L}$.

In the following result, we will use (see Chapter 2) that:

- The polyhedron $P = \{x \in \mathbb{R}^n : Ax \geq b\}$ is pointed if and only if

$$\{x \in \mathbb{R}^n : Ax = 0\} = \{0\}.$$

Proposition 9.3.1 *Let $\mathcal{L} \subseteq 2^N$ be a convex geometry and let $v \in \Gamma(\mathcal{L})$ be a game. The $C(\mathcal{L}, v)$ is either empty or a polyhedron pointed.*

Proof. For all $i \in N$, we know that $i \in \text{ex}\left(\overline{\{i\}}\right)$. If $S = \overline{\{i\}}$ then $S \setminus i \in \mathcal{L}$. Since $x_i = x(S) - x(S \setminus i)$, we obtain

$$\{x \in \mathbb{R}^n : x(S) = 0 \text{ for all } S \in \mathcal{L}\} = \{0\}.$$

It follows that the core is pointed, and hence each minimal face consists of just one point. □

Let $\mathcal{L}$ be a convex geometry on N, let $C \in \mathcal{C}(\mathcal{L})$ and let $v \in \Gamma(\mathcal{L})$. The *marginal worth vector* with respect to chain C in v is the vector $a^C(v) \in \mathbb{R}^n$ given by $a_i^C(v) = v(C(i)) - v(C(i) \setminus i)$, for all $i \in N$.

The i-th coordinate $a_i^C(v)$ represents the marginal contribution of player i to the coalition of his predecessors with respect to the chain C.

Definition 9.3.1 *The Weber set of a game $v \in \Gamma(\mathcal{L})$ is the convex hull of the marginal worth vectors, that is, $Weber(\mathcal{L}, v) := \text{conv}\left\{a^C(v) : C \in \mathcal{C}(\mathcal{L})\right\}$.*

For an atomic convex geometry, we will investigate the relation between the Weber set and the selectope. The following result establishes connections between selections corresponding to selectors on $\mathcal{L}$ and the marginal worth vectors associated with the maximal chains of $\mathcal{L}$.

Proposition 9.3.2 *Let $C \in \mathcal{C}(\mathcal{L})$ be a maximal chain and let $\alpha : \mathcal{L} \setminus \{\emptyset\} \to N$ be defined for every nonempty coalition $S \in \mathcal{L}$, by $\alpha(S) = j$, where $j \in S$ and $S \subseteq C(j)$. Then α is a selector and $m^\alpha(v) = a^C(v)$ for every $v \in \Gamma(\mathcal{L})$.*

Proof. First of all, we prove that α is well defined. For every nonempty coalition $S \in \mathcal{L}$, it is clear that there exists a unique $j \in S$ such that $S \subseteq C(j)$, where j is the last element of S that is incorporated in the maximal chain C. This element $j \in S$ is such that $C(k) \subseteq C(j)$ for all $k \in S$. Moreover, $j \in \text{ex}(S)$ since

$$C(j) \setminus j \in \mathcal{L} \text{ and } S \in \mathcal{L} \text{ imply } (C(j) \setminus j) \cap S = S \setminus j \in \mathcal{L}.$$

Therefore, α is well defined. Now, for every $i \in N$ and $v \in \Gamma(\mathcal{L})$, we have

$$\begin{aligned} a_i^C(v) &= v(C(i)) - v(C(i) \setminus i) \\ &= \sum_{\{T \in \mathcal{L} : T \subseteq C(i)\}} d_T(v) - \sum_{\{T \in \mathcal{L} : T \subseteq C(i) \setminus i\}} d_T(v) \\ &= \sum_{\{T \in \mathcal{L} : T \subseteq C(i),\, i \in T\}} d_T(v) \end{aligned}$$

$$= \sum_{\{T\in\mathcal{L}\,:\,\alpha(T)=i\}} d_T(v)$$
$$= m_i^\alpha(v).$$

□

As the selectope for game $v \in \Gamma(\mathcal{L})$ is a convex set, one consequence of Proposition 9.3.2 is that $Weber(\mathcal{L}, v) \subseteq Sel(\mathcal{L}, v)$.

A condition on the selector which will ensure the converse of Proposition 9.3.2 is consistency.

Definition 9.3.2 *A selector $\alpha \in \mathcal{A}(\mathcal{L})$ is called consistent if it satisfies the following conditions:*

1. $\alpha(S) \in \mathrm{ex}(S)$ *for all* $S \in \mathcal{L} \setminus \{\emptyset\}$.
2. *For all $S, T \in \mathcal{L}$, if $S \subset T$ and $\alpha(T) \in S$, then $\alpha(S) = \alpha(T)$.*

Note that if we take a maximal chain $C \in \mathcal{C}(\mathcal{L})$ and we define the selector α as in Proposition 9.3.2, then α is consistent. Moreover, different chains correspond to different selectors.

Theorem 9.3.3 *Let $\mathcal{L}$ be an atomic convex geometry and let $\alpha \in \mathcal{A}(\mathcal{L})$. Then α is consistent if and only if there exists a maximal chain $C \in \mathcal{C}(\mathcal{L})$ such that $m^\alpha(v) = a^C(v)$ for every $v \in \Gamma(\mathcal{L})$. In this case, the maximal chain*

$$C : \emptyset \subset \{i_1\} \subset \{i_1, i_2\} \subset \cdots \subset \{i_1, \ldots, i_{n-1}\} \subset \{i_1, \ldots, i_n\} = N$$

is unique and it is recursively defined by

$$i_n = \alpha(N),$$
$$i_k = \alpha(N \setminus \{i_n, i_{n-1}, \ldots, i_{k+1}\}), \quad \text{for all } 1 \le k < n.$$

Proof. First of all, note that for every unanimity game $u_S \in \Gamma(\mathcal{L})$,

$$d_S(u_S) = 1 \text{ and } d_R(u_S) = 0 \text{ for } R \in \mathcal{L},\ R \neq S,$$

since $\{u_T : T \in \mathcal{L},\ T \neq \emptyset\}$ is a basis of $\Gamma(\mathcal{L})$ and, for every $v \in \Gamma(\mathcal{L})$, we have that

$$v = \sum_{\{T\in\mathcal{L}\,:\,T\neq\emptyset\}} d_T(v)\, u_T.$$

($\Leftarrow$) Let $\alpha \in \mathcal{A}(\mathcal{L})$ be a selector such that there exists a maximal chain $C \in \mathcal{C}(\mathcal{L})$ satisfying $m^\alpha(v) = a^C(v)$ for all $v \in \Gamma(\mathcal{L})$. We assert that α is consistent. In order to prove the first condition of consistency, we consider the unanimity game $u_T \in \Gamma(\mathcal{L})$. If $\alpha(T) = i$, we have

$$m_i^{\alpha}(u_T) = \sum_{\{R \in \mathcal{L} : \alpha(R) = i\}} d_R(u_T) = 1$$

and by hypothesis, $m_i^{\alpha}(u_T) = a_i^C(u_T)$. Therefore,

$$u_T(C(i)) - u_T(C(i) \setminus i) = 1.$$

Hence, $T \subseteq C(i)$ but $T \nsubseteq C(i) \setminus i$, i.e., i is the last element of T in the order of C. So $\alpha(T) \in T$ and $T \subseteq C(\alpha(T))$, and thus $\alpha(T) \in \operatorname{ex}(T)$ since $\alpha(T) \in \operatorname{ex}(C(\alpha(T)))$.

Now, we prove the second condition of consistency. Let $S \subset T$ and $\alpha(T) = i$ with $i \in S$. Since $i \in \operatorname{ex}(T)$ and $S \subset T$ we have $i \in \operatorname{ex}(S)$. Then $S \subset C(i)$ since $T \subseteq C(i)$ and we have

$$a_i^C(u_S) = u_S(C(i)) - u_S(C(i) \setminus i) = 1,$$

and thus

$$m_i^{\alpha}(u_S) = \sum_{\{R \in \mathcal{L} : \alpha(R) = i\}} d_R(u_S) = 1.$$

Hence, $i = \alpha(S)$ and so, α is consistent.

($\Rightarrow$) Let $\alpha \in \mathcal{A}(\mathcal{L})$ be a consistent selector. First, we prove that there exists a maximal chain $C \in \mathcal{C}(\mathcal{L})$ which satisfies $a^C(v) = m^{\alpha}(v)$ for every $v \in \Gamma(\mathcal{L})$. We consider the following maximal chain

$$C : \emptyset \subset \{i_1\} \subset \{i_1, i_2\} \subset \cdots \subset \{i_1, \ldots, i_{n-1}\} \subset \{i_1, \ldots, i_n\} = N,$$

or equivalently

$$C : \emptyset \subset C(i_1) \subset C(i_2) \subset \cdots \subset C(i_{n-1}) \subset C(i_n) = N,$$

where

$$\begin{aligned} i_n &= \alpha(N), \\ i_k &= \alpha(N \setminus \{i_n, i_{n-1}, \ldots, i_{k+1}\}), \quad \text{for all } 1 \leq k < n. \end{aligned}$$

Note that the sets $N, N \setminus \{i_n\}, \ldots, N \setminus \{i_n, i_{n-1}, \ldots, i_{k+1}\} \in \mathcal{L}$ since α is consistent. Moreover, the selector α coincides with the selector defined in Proposition 9.3.2 for this chain, i.e., if $\beta \in \mathcal{A}(\mathcal{L})$ is such that for all $S \in \mathcal{L}$, $S \neq \emptyset$,

$$\beta(S) = j \text{ where } j \in S \text{ and } S \subseteq C(j),$$

then $\alpha = \beta$. Indeed, it is clear that $\alpha(S) = \beta(S)$ for every nonempty coalition $S \in C$; if $S \notin C$ then $\beta(S) = i \in \operatorname{ex}(S)$ and $S \subset C(i)$.

Since $S \subset C(i)$ and $\alpha(C(i)) = i \in \mathrm{ex}(S)$, the consistency of the selector α implies that $\alpha(S) = i$. Thus $\beta(S) = \alpha(S)$ for every nonempty coalition $S \in \mathcal{L}$. Applying Proposition 9.3.2 for the selector α, we obtain $a^C(v) = m^\alpha(v)$ for every $v \in \Gamma(\mathcal{L})$.

In order to prove that chain C is the unique chain such that $a^C(v) = m^\alpha(v)$ for all $v \in \Gamma(\mathcal{L})$, we consider the unanimity game $u_N \in \Gamma(\mathcal{L})$. For every $j \in N$, we have

$$m_j^\alpha(u_N) = \sum_{\{R \in \mathcal{L} : \alpha(R) = j\}} d_R(u_N) = \begin{cases} 1, & \text{if } \alpha(N) = j \\ 0, & \text{otherwise.} \end{cases}$$

On the other hand,

$$a_j^C(u_N) = u_N(C(j)) - u_N(C(j) \setminus j) = \begin{cases} 1, & \text{if } C(j) = N \\ 0, & \text{otherwise,} \end{cases}$$

so that, as $\alpha(N) = i_n$, the vectors coincide if and only if $C(i_n) = N$.

Next consider $u_{N \setminus \{i_n\}} \in \Gamma(\mathcal{L})$. For every $j \in N$, we have

$$m_j^\alpha\left(u_{N \setminus \{i_n\}}\right) = \sum_{\{R \in \mathcal{L} : \alpha(R) = j\}} d_R\left(u_{N \setminus \{i_n\}}\right) = \begin{cases} 1, & \text{if } \alpha(N \setminus \{i_n\}) = j \\ 0, & \text{otherwise,} \end{cases}$$

and

$$\begin{aligned} a_j^C\left(u_{N \setminus \{i_n\}}\right) &= u_{N \setminus \{i_n\}}(C(j)) - u_{N \setminus \{i_n\}}(C(j) \setminus j) \\ &= \begin{cases} 1, & \text{if } C(j) \supseteq N \setminus \{i_n\} \text{ and } C(j) \setminus j \not\supseteq N \setminus \{i_n\} \\ 0, & \text{otherwise.} \end{cases} \end{aligned}$$

Note that if $C(j) \supseteq N \setminus \{i_n\}$ and $C(j) \setminus j \not\supseteq N \setminus \{i_n\}$, then $C(j)$ is a convex set with n or $n-1$ elements, but in every maximal chain there is only one convex set with k elements, for all $1 \leq k \leq n$, and the convex set of n elements is excluded. Therefore,

$$a_j^C\left(u_{N \setminus \{i_n\}}\right) = \begin{cases} 1, & \text{if } C(j) = N \setminus \{i_n\} \\ 0, & \text{otherwise.} \end{cases}$$

Thus, since $\alpha(N \setminus \{i_n\}) = i_{n-1}$, we have $a^C\left(u_{N \setminus \{i_n\}}\right) = m^\alpha\left(u_{N \setminus \{i_n\}}\right)$ if and only if $C(i_{n-1}) = N \setminus \{i_n\}$. By repeating this argument, the maximal chain C is obtained. □

Example: Let $N = \{1, 2, 3\}$ and consider the convex geometry

$$\mathcal{L} = \{\emptyset, \{1\}, \{2\}, \{3\}, \{1,2\}, \{2,3\}, \{1,2,3\}\},$$

and the game $v \in \Gamma(\mathcal{L})$ given by

$$v(1) = v(2) = v(3) = 0, \ v(12) = v(23) = 2, v(123) = 3.$$

There are 12 selectors and the selections are:

α	$\{1,2\}$	$\{2,3\}$	$\{1,2,3\}$	$m_1^\alpha(v)$	$m_2^\alpha(v)$	$m_3^\alpha(v)$
1	1	2	3	2	2	−1
2	2	2	3	0	4	−1
3	1	3	3	2	0	1
4	2	3	3	0	2	1
5	1	2	1	1	2	0
6	2	2	1	−1	4	0
7	1	3	1	1	0	2
8	2	3	1	−1	2	2
9	1	2	2	2	1	0
10	2	2	2	0	3	0
11	1	3	2	2	−1	2
12	2	3	2	0	1	2

Table 9.3

Note that in this example, selectors 3, 4, 5, and 7 are consistent and the selections corresponding to these selectors coincide with the marginal worth vectors with respect to the four maximal chains in $\mathcal{L}$. Therefore,

$$Weber(\mathcal{L}, v) = \text{conv}\{(2,0,1), (0,2,1), (1,2,0), (1,0,2)\}.$$

PORTA (Polyhedron Representation Transformation Algorithm) by Christof [37] is a collection of C codes to analize polytopes and polyhedra. The polyhedra are either given as the convex hull of a set of vertices plus the convex cone of a set of vectors, or as a system of linear equations an inequalities. The basic function `traf` performs the transformation between these representations by using Gaussian and Fourier-Motzkin elimination (see Christof and Reinelt [36]).

Files with name suffix `.ieq` contain a representation of a polyhedron as a system of linear equations an inequalities. The input file for the

$$C(\mathcal{L}, v) = \left\{x \in \mathbb{R}^3_+ : x_1 + x_2 + x_3 = 3,\ x_1 + x_2 \geq 2,\ x_2 + x_3 \geq 2\right\},$$

named `core.ieq` is given by

```
DIM = 3
VALID
1 1 1

ELIMINATION_ORDER
1 2 3
INEQUALITIES_SECTION
x1 + x2 + x3 == 3
x1 >= 0
x2 >= 0
x3 >= 0
x1 + x2 >= 2
x2 + x3 >= 2

END
```

The command `traf core.ieq` produces the following output, in a file named `core.ieq.poi`

```
DIM = 3

CONV_SECTION
(  1) 0 3 0
(  2) 0 2 1
(  3) 1 2 0
(  4) 1 1 1

END
```

That is, $C(\mathcal{L}, v) = \operatorname{conv}\{(0,3,0), (0,2,1), (1,2,0), (1,1,1)\}$.

In Figure 9.2, we show the position of the Weber set, the selectope and the core of game v in the above example.

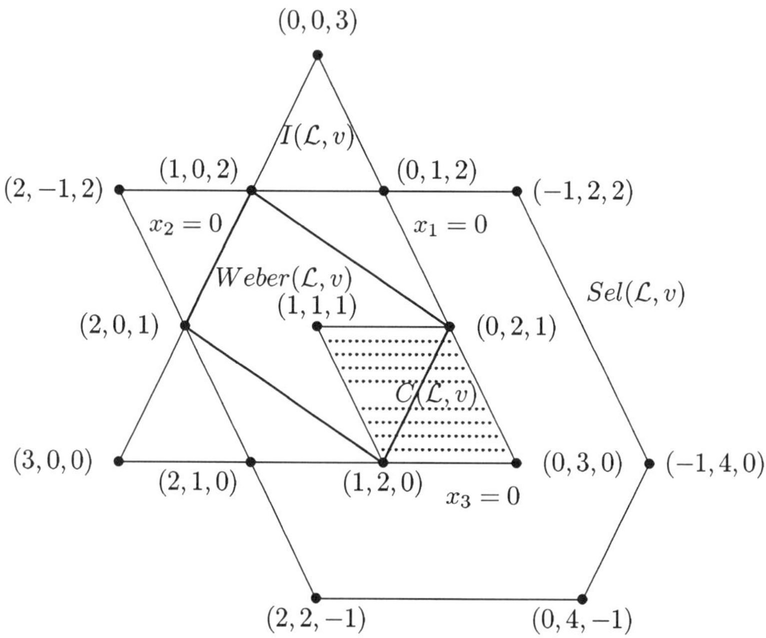

Figure 9.2

In this example, note that the Weber set is strictly contained in the selectope. In general, $Weber\,(\mathcal{L},v) \neq Sel\,(\mathcal{L},v)$. Also, $C\,(\mathcal{L},v) \subseteq Sel\,(\mathcal{L},v)$.

It is well worth noting that the inclusion of $C\,(\mathcal{L},v) \subseteq Weber\,(\mathcal{L},v)$ (see [244] and [54]) is not true if (the convex geometry) $\mathcal{L} \neq 2^N$, as the above example shows. Therefore, the proof of $C\,(\mathcal{L},v) \subseteq Sel\,(\mathcal{L},v)$ cannot be supported by $Weber\,(\mathcal{L},v) \subseteq Sel\,(\mathcal{L},v)$.

9.4 Convex and quasi-convex games

The classic result states that if $v : 2^N \to \mathbb{R}$, is a *convex* (*supermodular*) game then $C(v) = Weber(v)$, where the *Weber set* is the convex hull of the marginal worth vectors for v (Shapley [217]). Weber [244] has shown that any game satisfies $C(v) \subseteq Weber(v)$. Ichiishi [129] proved that if $Weber(v) \subseteq C(v)$ then v is a convex game. Thus, these results imply the following characterization of convex games:

$v : 2^N \to \mathbb{R}$ is a convex game if and only if $C(v) = Weber(v)$.

If $\mathcal{L}$ is the Boolean algebra 2^N and v is a cooperative game, then the core of v is contained in its Weber set. However, the inclusion $C(\mathcal{L}, v) \subseteq Weber(\mathcal{L}, v)$ does not hold when v is a game on a convex geometry $\mathcal{L} \neq 2^N$.

In the following examples, we will use the convex geometry $Co(\mathbf{3}) \subseteq 2^N$, where $N = \{1, 2, 3\}$ and $Co(\mathbf{3}) = \{\emptyset, \{1\}, \{2\}, \{3\}, \{1, 2\}, \{2, 3\}, \{1, 2, 3\}\}$, is the collection of the convex subsets of the poset $\mathbf{3} = \{1 < 2 < 3\}$. There are four maximal chains in $\mathcal{L}$

$$C_1 : \emptyset \subset \{1\} \subset \{1, 2\} \subset \{1, 2, 3\},$$
$$C_2 : \emptyset \subset \{2\} \subset \{1, 2\} \subset \{1, 2, 3\},$$
$$C_3 : \emptyset \subset \{2\} \subset \{2, 3\} \subset \{1, 2, 3\},$$
$$C_4 : \emptyset \subset \{3\} \subset \{2, 3\} \subset \{1, 2, 3\}.$$

The Hasse diagram of the lattice $Co(\mathbf{3})$ is

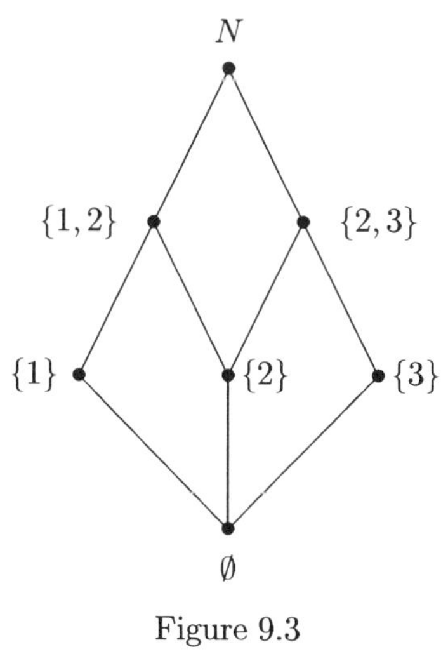

Figure 9.3

Example: Let $v : Co(\mathbf{3}) \to \mathbb{R}$ be a game, given by

$$v(1) = v(2) = v(3) = 0,\ v(12) = v(23) = 2,\ v(N) = 3.$$

The marginal worth vectors in the game v are

$$\begin{aligned}
a^{C_1}(v) &= (v(1) - v(\emptyset),\ v(12) - v(1),\ v(N) - v(12)) = (0,2,1), \\
a^{C_2}(v) &= (v(12) - v(2),\ v(2) - v(\emptyset),\ v(N) - v(12)) = (2,0,1), \\
a^{C_3}(v) &= (v(N) - v(23),\ v(2) - v(\emptyset),\ v(23) - v(2)) = (1,0,2), \\
a^{C_4}(v) &= (v(N) - v(23),\ v(23) - v(3),\ v(3) - v(\emptyset)) = (1,2,0),
\end{aligned}$$

and hence

$$Weber(\mathcal{L}, v) = \operatorname{conv}\{(0,2,1), (2,0,1), (1,0,2), (1,2,0)\}.$$

On the other hand, the core of v is

$$\begin{aligned}
C(\mathcal{L}, v) &= \left\{x \in \mathbb{R}^3_+ : x_1 + x_2 + x_3 = 3,\ x_3 \leq 1,\ x_1 \leq 1\right\} \\
&= \operatorname{conv}\{(0,2,1), (1,2,0), (0,3,0), (1,1,1)\}.
\end{aligned}$$

Then $C(\mathcal{L}, v) \nsubseteq Weber(\mathcal{L}, v)$ and $Weber(\mathcal{L}, v) \nsubseteq C(\mathcal{L}, v)$. These solution sets are depicted in Figure 9.4.

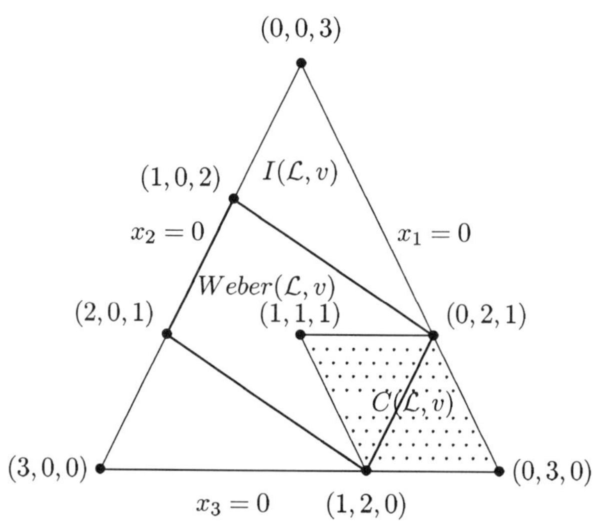

Figure 9.4

Example: Let $v : Co(\mathbf{3}) \to \mathbb{R}$ be a game, given by

$$v(1) = v(2) = v(3) = 0,\ v(12) = 1,\ v(23) = 0,\ v(N) = 3.$$

The marginal worth vectors in this game v are

$$a^{C_1}(v) = (0,1,2),\ a^{C_2}(v) = (1,0,2),\ a^{C_3}(v) = (3,0,0),\ a^{C_4}(v) = (3,0,0),$$

and the Weber set is

$$Weber(\mathcal{L}, v) = \text{conv}\{(0,1,2), (1,0,2), (3,0,0)\}.$$

Now, we get that the core is the set

$$\begin{aligned} C(\mathcal{L}, v) &= \left\{x \in \mathbb{R}^3_+ : x_1 + x_2 + x_3 = 3,\ x_3 \leq 2,\ x_1 \leq 3\right\} \\ &= \text{conv}\{(0,1,2), (1,0,2), (3,0,0), (0,3,0)\}. \end{aligned}$$

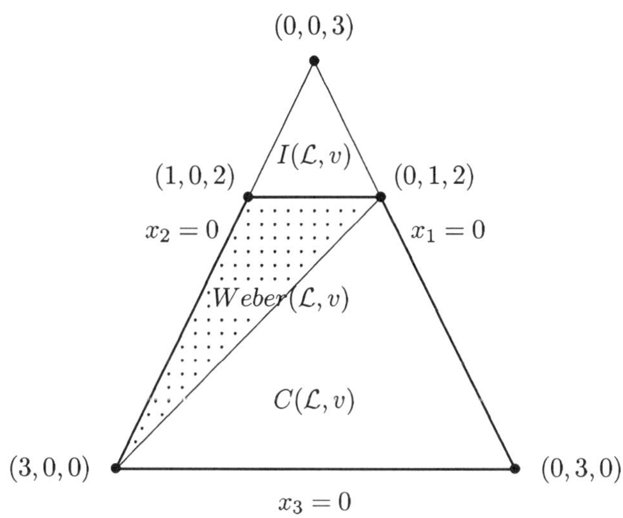

Figure 9.5

In this game, each marginal worth vector coincides with some vertex of the core and hence $Weber(\mathcal{L}, v) \subseteq C(\mathcal{L}, v)$.

The following two propositions provides a special type of vertex of the core: the marginal worth vectors $a^C(v) \in \mathbb{R}^n$ with respect to the maximal chains of $\mathcal{L}$ in game v.

Proposition 9.4.1 *Let $v \in \Gamma(\mathcal{L})$ and $C \in \mathcal{C}(\mathcal{L})$. For all $S \in C$, we have*

$$\sum_{j \in S} a_j^C(v) = v(S).$$

Proof. Let $v \in \Gamma(\mathcal{L})$ and $C \in \mathcal{C}(\mathcal{L})$. We denote by S_k the coalition of C such that $|S_k| = k$ for any $k \in N$. Put $S_0 = \emptyset$ and $S_k = \{i_1, i_2, \ldots, i_k\}$ for all $1 \leq k \leq n$. By definition, we have that $a_i^C(v) = v(C(i)) - v(C(i) \setminus i)$ and therefore it follows that

$$\begin{aligned}
\sum_{j \in S_k} a_j^C(v) &= \sum_{j=1}^{k} a_{i_j}^C(v) \\
&= \sum_{j=1}^{k} [v(C(i_j)) - v(C(i_j) \setminus i_j)] \\
&= \sum_{j=1}^{k} [v(S_j) - v(S_{j-1})] \\
&= v(S_k),
\end{aligned}$$

for all $k \in N$. Note that $S_n = N$ and hence $\sum_{j \in N} a_j^C(v) = v(N)$. □

Proposition 9.4.2 *Let $v \in \Gamma(\mathcal{L})$ and $C \in \mathcal{C}(\mathcal{L})$. If the marginal worth vector $a^C(v) \in C(\mathcal{L}, v)$, then $a^C(v)$ is a vertex of $C(\mathcal{L}, v)$.*

Proof. By Proposition 9.4.1, we have

$$\sum_{j \in S} a_j^C(v) = v(S)$$

for all $S \in C$. Since each maximal chain in $\mathcal{L}$ has n nonempty coalitions, we get n equations and due to their linear independence, we may conclude that $a^C(v)$ is a vertex of the polyhedron $C(\mathcal{L}, v)$. □

Faigle and Kern [82] introduced the concept of a convex game for a distributive lattice $(\mathcal{A}, \vee, \wedge)$. In this section, we consider a convex geometry $\mathcal{L}$ which is a lattice of subsets with the join and meet operations defined by

$$S \vee T = \overline{S \cup T}, \quad S \wedge T = S \cap T \text{ for all } S, T \in \mathcal{L}.$$

Definition 9.4.1 *Let $\mathcal{L}$ be a convex geometry. A game $v \in \Gamma(\mathcal{L})$ is said to be convex on $\mathcal{L}$ if $v(S \vee T) + v(S \wedge T) \geq v(S) + v(T)$, for all $S, T \in \mathcal{L}$. A game is called concave on a convex geometry if the reverse inequality holds.*

We now introduce the concept of quasi-convexity for games on a convex geometry $\mathcal{L}$ and we investigate the relation $Weber(\mathcal{L}, v) \subseteq C(\mathcal{L}, v)$.

Definition 9.4.2 *A game $v \in \Gamma(\mathcal{L})$ is quasi-convex if for all $S, T \in \mathcal{L}$ with $S \cup T \in \mathcal{L}$, we have $v(S \cup T) + v(S \cap T) \geq v(S) + v(T)$. A game is called quasi-concave if the reverse inequality holds.*

It is obvious that convex implies quasi-convex.

Remark 9.4.1 *Note that if there is only a maximal chain in $\mathcal{L}$, all games defined on $\mathcal{L}$ are convex.*

Proposition 9.4.3 *A game $v \in \Gamma(\mathcal{L})$ is quasi-convex if and only if for all $S, T \in \mathcal{L}$ such that $T \subset S$ and for all $i \in \operatorname{ex}(S) \cap T$, we have*

$$v(S) - v(S \setminus i) \geq v(T) - v(T \setminus i).$$

Proof. Let $S, T \in \mathcal{L}$ such that $T \subset S$ and let $i \in \operatorname{ex}(S) \cap T$. Consider $S' = S \setminus i$ and $T' = T$. We have

$$\begin{aligned} S' \cap T' &= (S \setminus i) \cap T = T \setminus i \in \mathcal{L}, \\ S' \cup T' &= (S \setminus i) \cup T = S \in \mathcal{L}. \end{aligned}$$

Applying the definition of quasi-convexity to S' and T', it follows that

$$v(S) + v(T \setminus i) \geq v(S \setminus i) + v(T).$$

Conversely, let $S, T \in \mathcal{L}$ such that $S \cup T \subset \mathcal{L}$. If $T \subset S$ or $S \subset T$, then the equality is clear. Consider $S \cap T \neq S$ and $S \cap T \neq T$ and let $C \in \mathcal{C}(\mathcal{L})$ be a maximal chain such that $S \cap T$, T, $S \cup T \in C$. As $S \setminus T \neq \emptyset$, put $|S \setminus T| = k$ and write $S \setminus T = \{i_1, i_2, \ldots, i_k\}$ with $C(i_1) \subset C(i_2) \subset \cdots \subset C(i_k)$, i.e., the chain C is given by

$$\cdots \subset T \subset T \cup \{i_1\} \subset T \cup \{i_1, i_2\} \subset \cdots \subset T \cup \{i_1, i_2, \ldots, i_k\} = T \cup S \subset \cdots$$

Let $R = S \cap T$ and let us denote $S_j = \{i_1, i_2, \ldots, i_j\}$ for all $1 \leq j \leq k$ with $S_0 = \emptyset$. Then we have that for all $1 \leq j \leq k$, $T \cup S_j \in \mathcal{L}$ and

$$R \cup S_j = (S \cap T) \cup S_j = (S \cup S_j) \cap (T \cup S_j) = S \cap (T \cup S_j) \in \mathcal{L}.$$

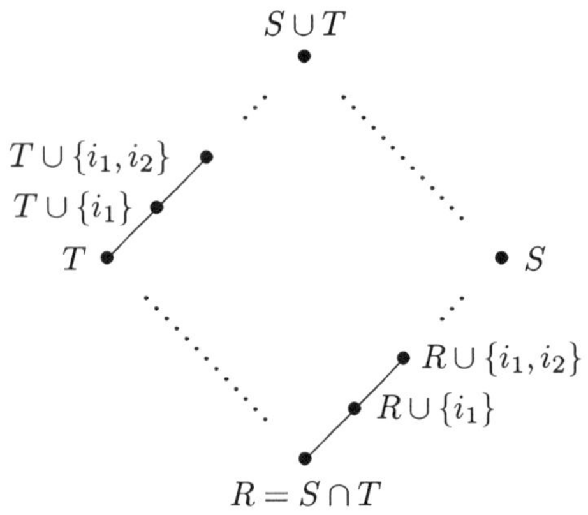

Now, considering the hypothesis, we get

$$v(R \cup S_j) - v(R \cup S_{j-1}) \leq v(T \cup S_j) - v(T \cup S_{j-1}),$$

since $R \cup S_j \subset T \cup S_j$ and $i_j \in \text{ex}(R \cup S_j) \cap (T \cup S_j)$, and now it follows that

$$\begin{aligned} v(S) - v(S \cap T) &= v(R \cup S_k) - v(R) \\ &= \sum_{j=1}^{k} [v(R \cup S_j) - v(R \cup S_{j-1})] \\ &\leq \sum_{j=1}^{k} [v(T \cup S_j) - v(T \cup S_{j-1})] \\ &= v(T \cup S) - v(T). \end{aligned}$$

□

Theorem 9.4.4 *A game v on a convex geometry $\mathcal{L}$ is quasi-convex if and only if $a^C(v) \in C(\mathcal{L}, v)$, for all $C \in \mathcal{C}(\mathcal{L})$.*

Proof. ($\Rightarrow$) Let $a^C(v)$ be the marginal worth vector associated to the maximal chain $C \in \mathcal{C}(\mathcal{L})$. By Proposition 9.4.1, $\sum_{j \in S} a_j^C(v) = v(S)$ for all $S \in C$. It remains to prove that $\sum_{j \in S} a_j^C(v) \geq v(S)$ for all $S \notin C$. Let $S \in \mathcal{L}$ such that $S \notin C$ and put $|S| = s \geq 1$. Write $S = \{i_1, i_2, \ldots, i_s\}$ where

$$C(i_1) \subset C(i_2) \subset \cdots \subset C(i_s).$$

Let us denote $S_j = \{i_1, i_2, \ldots, i_j\}$ for all $1 \leq j \leq s$ and $S_0 = \emptyset$. Then for all $1 \leq j \leq s$ we have $S_j = S \cap C(i_j) \in \mathcal{L}$, and also $i_j \in \text{ex}(C(i_j))$. Proposition 9.4.3 implies that for all $1 \leq j \leq s$, we have

$$v(C(i_j)) - v(C(i_j) \setminus i_j) \geq v(S_j) - v(S_{j-1}),$$

and hence

$$\begin{aligned}\sum_{j\in S} a_j^C(v) &= \sum_{j=1}^{s} a_{i_j}^C(v) \\ &= \sum_{j=1}^{s} \left[v\left(C\left(i_j\right)\right) - v\left(C\left(i_j\right) \setminus i_j\right)\right] \\ &\geq \sum_{j=1}^{s} \left[v\left(S_j\right) - v\left(S_{j-1}\right)\right] \\ &= v(S).\end{aligned}$$

($\Leftarrow$) For any $S, T \in \mathcal{L}$ with $S \cup T \in \mathcal{L}$, let $C \in \mathcal{C}(\mathcal{L})$ be a maximal chain containing $S \cap T$ and $S \cup T$. The marginal worth vector $a^C(v) \in \mathbb{R}^n$ belongs to $C(\mathcal{L}, v)$, hence $\sum_{j\in S} a_j^C(v) \geq v(S)$ and $\sum_{j\in T} a_j^C(v) \geq v(T)$. By construction, $\sum_{j\in S\cup T} a_j^C(v) = v(S\cup T)$ and $\sum_{j\in S\cap T} a_j^C(v) = v(S\cap T)$. Therefore,

$$\begin{aligned}v(S) + v(T) &\leq \sum_{j\in S} a_j^C(v) + \sum_{j\in T} a_j^C(v) \\ &= \sum_{j\in S\cup T} a_j^C(v) + \sum_{j\in S\cap T} a_j^C(v) \\ &= v(S\cup T) + v(S\cap T).\end{aligned}$$

□

Corollary 9.4.5 *A game v on a convex geometry $\mathcal{L}$ is quasi-convex if and only if $Weber(\mathcal{L}, v) \subseteq C(\mathcal{L}, v)$.*

Proof. It is a direct consequence of Theorem 9.4.4 since the core of a game $v \in \Gamma(\mathcal{L})$ is a convex set. □

Remark 9.4.2 *A similar characterization of submodular games on distributive lattices is given in Fujishige [96, Theorem 3.18], who considered the greedy algorithm for obtaining the result.*

Example: Let $v : Co(\mathbf{3}) \to \mathbb{R}$ be a game, defined by

$$v(1) = 1,\ v(2) = -1,\ v(3) = 1,\ v(12) = v(23) = 0,\ v(N) = 1.$$

This game is quasi-convex but it is not convex since for $S = \{1\}$ and $T = \{3\}$, we have $S \wedge T = \emptyset$, $S \vee T = \overline{\{1,3\}} = N$, and

$$1 = v(N) + v(\emptyset) < v(1) + v(3) = 1 + 1 = 2.$$

The marginal worth vectors for v are:

$$a^{C_1}(v) = a^{C_2}(v) = a^{C_3}(v) = a^{C_4}(v) = (1, -1, 1),$$

and hence

$$Weber(\mathcal{L}, v) = C(\mathcal{L}, v) = \{(1, -1, 1)\}.$$

In the above example, we obtain that the core and the Weber set are equal for a non convex game. Then, property $Weber(\mathcal{L}, v) \subseteq C(\mathcal{L}, v)$ is not a characterization of the convex games on convex geometries.

Recall that a game $v \in \Gamma(\mathcal{L})$ is monotone if $\emptyset \subseteq S \subseteq T \subseteq N$ implies $v(S) \leq v(T)$. Note that if v is monotone then $0 = v(\emptyset) \leq v(S) \leq v(N)$, for all $S \in \mathcal{L}$.

Theorem 9.4.6 *Let $v \in \Gamma(\mathcal{L})$ be a monotone game. Then, the game v is convex on the convex geometry $\mathcal{L}$ if and only if $Weber(\mathcal{L}, v) \subseteq C(\mathcal{L}, v)$.*

Proof. If v is convex then it is quasi-convex and Corollary 9.4.5 implies the property. Conversely, for all $S, T \in \mathcal{L}$, let C be a maximal chain containing $S \vee T = \overline{S \cup T}$ and $S \wedge T = S \cap T$. We observe that $\sum_{j \in S} a_j^C(v) \geq v(S)$ and $\sum_{j \in T} a_j^C(v) \geq v(T)$, since the marginal worth vector $a^C(v) \in C(\mathcal{L}, v)$. On the other hand, $a_i^C(v) = v(C(i)) - v(C(i) \setminus i) \geq 0$ since v is monotone and therefore $a^C \geq 0$. By Proposition 9.4.1,

$$\sum_{j \in \overline{S \cup T}} a_j^C(v) = v\left(\overline{S \cup T}\right) \quad \text{and} \quad \sum_{j \in S \cap T} a_j^C(v) = v(S \cap T).$$

We get the result from the following inequalities

$$\begin{aligned} v(S) + v(T) &\leq \sum_{j \in S} a_j^C(v) + \sum_{j \in T} a_j^C(v) \\ &= \sum_{j \in S \cup T} a_j^C(v) + \sum_{j \in S \cap T} a_j^C(v) \\ &\leq \sum_{j \in \overline{S \cup T}} a_j^C(v) + \sum_{j \in S \cap T} a_j^C(v) \\ &= v\left(\overline{S \cup T}\right) + v(S \cap T). \end{aligned}$$

□

9.5 Marginal operators

In this section, we analyze *marginal operators* for games on convex geometries. This concept generalizes the model developed by Curiel and Tijs for cooperative games [50]. We introduce two operators on the vector space of the games defined on a convex geometry $\mathcal{L}$. One of them reflects an optimistic view on part of each coalition with regard to the order of formation of the grand coalition. The other operator reflects a pessimistic view on part of each coalition with regard to the order of formation of the grand coalition.

Let $v \in \Gamma(\mathcal{L})$ be a game, and let $C \in \mathcal{C}(\mathcal{L})$ be a maximal chain. We define the game $a^C(v) : \mathcal{L} \to \mathbb{R}$, given by

$$a^C(v)(S) := \sum_{i \in S} a_i^C(v),$$

which is called the *marginal game* associated to chain C. We now define two operators: one of them assigns the minimum of these marginal games while the other assigns the maximum.

The *minimal marginal operator* $Mi : \Gamma(\mathcal{L}) \to \Gamma(\mathcal{L})$ is defined by

$$Mi(v) := \min_{C \in \mathcal{C}(\mathcal{L})} a^C(v).$$

For every $S \in \mathcal{L}$, we have $Mi(v)(S) = \min_{C \in \mathcal{C}(\mathcal{L})} a^C(v)(S)$. The *maximal marginal operator* $Ma : \Gamma(\mathcal{L}) \to \Gamma(\mathcal{L})$ is defined by

$$Ma(v) := \max_{C \in \mathcal{C}(\mathcal{L})} a^C(v).$$

Note that if the convex geometry has a unique maximal chain, it holds that $Mi(v) = Ma(v) = v$. It is also verified if all marginal worth vectors coincide.

Theorem 9.5.1 *Let $v \in \Gamma(\mathcal{L})$ be a game on a convex geometry. Then*

1. *$Mi(v)(S) \leq v(S) \leq Ma(v)(S)$, for all $S \in \mathcal{L}$ and the equality holds for $S = N$. Therefore, $C(\mathcal{L}, Ma(v)) \subseteq C(\mathcal{L}, v) \subseteq C(\mathcal{L}, Mi(v))$.*

2. *$Weber(\mathcal{L}, v) \subseteq C(\mathcal{L}, Mi(v))$.*

3. *If $\mathcal{L}$ is atomic, then $C(\mathcal{L}, Ma(v))$ is empty or contains a unique vector. Moreover, $C(\mathcal{L}, Ma(v)) \subseteq Weber(\mathcal{L}, v)$.*

Proof. 1. We have that

$$Mi(v)(S) = \min_{C \in \mathcal{C}(\mathcal{L})} a^C(v)(S) \leq v(S),$$

since for each $S \in \mathcal{L}$ there is a maximal chain C that contains S, and for this chain, we have $a^C(v)(S) = v(S)$. A similar argument shows that $Ma(v)(S) \geq v(S)$, for all $S \in \mathcal{L}$. Taking $S = N$, we obtain that $a^C(v)(N) = v(N)$, for every maximal chain C. Then

$$Mi(v)(N) = Ma(v)(N) = v(N).$$

We therefore conclude that

$$C(\mathcal{L}, Ma(v)) \subseteq C(\mathcal{L}, v) \subseteq C(\mathcal{L}, Mi(v)).$$

2. For every maximal chain $C \in \mathcal{C}(\mathcal{L})$ we have $a^C(v)(S) \geq Mi(v)(S)$, for all $S \in \mathcal{L}$ and $a^C(v)(N) = v(N)$. Therefore, $a^C(v) \in C(\mathcal{L}, Mi(v))$. Since the core is a convex set in $\mathbb{R}^n$, we obtain $Weber(\mathcal{L}, v) \subseteq C(\mathcal{L}, Mi(v))$.

3. Let $\mathcal{L}$ be an atomic convex geometry. We first prove that if there exist two maximal chains $C_1, C_2 \in \mathcal{C}(\mathcal{L})$ such that $a^{C_1}(v) \neq a^{C_2}(v)$ then $C(\mathcal{L}, Ma(v))$ is empty. Assume that $a^{C_1}(v) = (y_1, \ldots, y_n)$ and $a^{C_2}(v) = (z_1, \ldots, z_n)$ are not equal. Since these vectors satisfy $a^{C_1}(v)(N) = a^{C_2}(v)(N) = v(N)$, then there exist $i, j \in N$ such that $y_i < z_i$, and $y_j > z_j$.

If $x \in C(\mathcal{L}, Ma(v))$, then we have

$$\begin{aligned} x_i &\geq Ma(v)(i) \geq z_i > y_i, \\ x_j &\geq Ma(v)(j) \geq y_j, \\ x_k &\geq Ma(v)(k) \geq y_k, \quad \text{for } k \neq i, j. \end{aligned}$$

Thus, $v(N) = \sum_{i \in N} x_i > \sum_{i \in N} y_i = v(N)$. This contradiction proves that $C(\mathcal{L}, Ma(v))$ is empty.

On the other hand, if the $C(\mathcal{L}, Ma(v))$ is nonempty, then all marginal worth vectors coincide. Hence $Mi(v) = v = Ma(v)$, and the game v is defined by $v(S) = a^C(v)(S)$, where $a_i^C(v) = v(i)$, for all $1 \leq i \leq n$. □

Definition 9.5.1 *A game $v \in \Gamma(\mathcal{L})$ is called quasi-superadditive if*

$$v(S \cup T) \geq v(S) + v(T),$$

for all $S, T \in \mathcal{L}$ such that $S \cap T = \emptyset$ and $S \cup T \in \mathcal{L}$. A game $v \in \Gamma(\mathcal{L})$ is called quasi-subadditive if the reverse inequality holds.

Proposition 9.5.2 *Let $v \in \Gamma(\mathcal{L})$ be a game on a convex geometry. Then*

1. *$Mi(v)$ is quasi-superadditive and $Ma(v)$ is quasi-subadditive.*

2. *If $n \leq 3$, $Mi(v)$ is quasi-convex and $Ma(v)$ is quasi-concave.*

Proof. 1. These properties follow from the definitions of $Mi(v)$ and $Ma(v)$.

2. In case $n = 2$, we clearly have that $Mi(v)$ is quasi-superadditive. When $n = 3$, that is, $N = \{i, j, k\}$, let $S, T \in \mathcal{L}$ such that $S \cup T \in \mathcal{L}$. We have to prove that

$$Mi(v)(S \cup T) + Mi(v)(S \cap T) \geq Mi(v)(S) + Mi(v)(T).$$

If $S \cap T = \emptyset$, then the above inequality follows from the quasi-superadditivity of $Mi(v)$. If $S \cap T \neq \emptyset$, then suppose that $S \not\subseteq T$ and $T \not\subseteq S$ since in other case it is obvious. Therefore, we have $|S| = |T| = 2$, and $|S \cap T| = 1$. We will suppose that $S \cap T = \{i\} \in \mathcal{L}$, $S = \{i, j\}$, and $T = \{i, k\}$. Let $C \in \mathcal{C}(\mathcal{L})$ be a maximal chain that contains $\{i\}$ such that $Mi(v)(i) = a^C(v)(i)$. We deduce that

$$\begin{aligned} Mi(v)(\{i,j,k\}) + Mi(v)(i) &= a^C(v)(\{i,j,k\}) + a^C(v)(i) \\ &= a^C(v)(\{i,j\}) + a^C(v)(\{i,k\}) \\ &\geq Mi(v)(\{i,j\}) + Mi(v)(\{i,k\}). \end{aligned}$$

A similar proof gives that $Ma(v)$ is quasi-concave. □

Definition 9.5.2 *A game $v \in \Gamma(\mathcal{L})$ is called symmetric if $v(S) = v(T)$ for all $S, T \in \mathcal{L}$ such that $|S| = |T|$.*

Theorem 9.5.3 *If $v \in \Gamma(\mathcal{L})$ is symmetric, then $Mi(v)$ is a quasi-convex game.*

Proof. Since the game $v \in \Gamma(\mathcal{L})$ is symmetric, we have that $v(S) = f(|S|)$, for all $S \in \mathcal{L}$, where $f : \{0, 1, \ldots, n\} \to \mathbb{R}$. Let $d_1, d_2, \ldots, d_n$ be the sequence of differences $f(k) - f(k-1)$, $k = 1, 2, \ldots, n$, ordered in nondecreasing order $d_1 \leq d_2 \leq \cdots \leq d_n$. Note that if $\mathcal{L}$ is a convex geometry, there is at least one coalition of cardinal k for all $k = 1, \ldots, n$. By definition of $Mi(v)$, we have

$$Mi(v)(S) = \sum_{i=1}^{s} d_i,$$

where $s = |S|$. We observe that if $S \subseteq T$ and $i \in \text{ex}(T) \cap S$, then

$$\begin{aligned} Mi(v)(S) - Mi(v)(S \setminus i) &= \sum_{j=1}^{s} d_j - \sum_{j=1}^{s-1} d_j = d_s \leq d_t \\ &= Mi(v)(T) - Mi(v)(T \setminus i). \end{aligned}$$

The result therefore follows from Proposition 9.4.3. □

Theorem 9.5.4 *Let $v \in \Gamma(\mathcal{L})$ be a game on an atomic convex geometry $\mathcal{L}$. If we inductively define the games*

$$v^{(1)} := v, \quad v^{(k)} := Mi\left(v^{(k-1)}\right), \quad k = 2, 3, \ldots,$$

then $\lim_{k\to\infty} v^{(k)}$ exists.

Proof. Let $S \in \mathcal{L}$, by Theorem 9.5.1, we have that

$$v^{(1)}(S) \geq v^{(2)}(S) \geq \cdots \geq v^{(k)}(S) \geq \cdots$$

Moreover, if w is a quasi-superadditive game, then $a^C(w)(i) \geq w(i)$, for all $C \in \mathcal{C}(\mathcal{L})$ and hence $Mi(w)(S) \geq \sum_{i\in S} w(i)$. By Proposition 9.5.2, $v^{(k)}$ is a quasi-superadditive game for $k = 2, \ldots$ Then, for every $S \in \mathcal{L}$,

$$v^{(k)}(S) \geq \sum_{i\in S} Mi(v)(i),$$

and therefore, the sequence $v^{(k)}(S)$ has a lower bound, and hence it converges to a limit $z(S) = \lim_{k\to\infty} v^{(k)}(S)$. Further, Mi is a continuous operator on $\Gamma(\mathcal{L})$, and therefore

$$z = \lim_{k\to\infty} v^{(k)} = \lim_{k\to\infty} Mi\left(v^{(k-1)}\right) = Mi(z).$$

□

Theorem 9.5.5 *A game $v \in \Gamma(\mathcal{L})$ is quasi-convex if and only if $Mi(v) = v$.*

Proof. Assume that $Mi(v) = v$. Then $C(\mathcal{L}, v) = C(\mathcal{L}, Mi(v))$ and by Theorem 9.5.1, $Weber(\mathcal{L}, v) \subseteq C(\mathcal{L}, Mi(v))$. Hence, by using the characterization of Corollary 9.4.5, game v is quasi-convex.

Conversely, if game v is quasi-convex, then vector $a^C(v) \in C(\mathcal{L}, v)$, for all $C \in \mathcal{C}(\mathcal{L})$. Therefore, $a^C(v)(N) = v(N)$, and $a^C(v)(S) \geq v(S)$, for all $S \in \mathcal{L}$. On the other hand, for each $S \in \mathcal{L}$, there exists a maximal chain $C_S \in \mathcal{C}(\mathcal{L})$ such that $Mi(v)(S) = a^{C_S}(v)(S)$. Thus, we can write

$$v(S) \geq Mi(v)(S) = a^{C_S}(v)(S) \geq v(S),$$

and this implies that $Mi(v)(S) = v(S)$ for all $S \in \mathcal{L}$. □

Chapter 10

Simple games on closure spaces

In this chapter, we study the different solution concepts already defined in the previous chapters for an interesting and particular class of games: the *simple games*. This type of game arises, for instance, from modeling voting situations in which the result reflects two possibilities (win-lose, accept-reject, etc.) and they have been the object of a specific study like the initial ones by Isbell [130] and Shapley [214], and Einy and Wettstein [74], which have a particular incidence in this chapter, and the work of Carreras [33], who studies the simple games restricted by cooperation graphs. The results were given by Jiménez Jiménez in her Ph. D. Thesis [134].

10.1 Minimal winning coalitions

In the following, we assume that the family of feasible coalitions $\mathcal{L} \subseteq 2^N$ is such that $\emptyset, N \in \mathcal{L}$, and only in certain occasions, other requirements are needed for this family.

Definition 10.1.1 *A game $v \in \Gamma(\mathcal{L})$ is called simple if it satisfies the following conditions:*

1. *For every coalition $S \in \mathcal{L}$, $v(S) \in \{0,1\}$ and $v(\emptyset) = 0$.*
2. *If $S,T \in \mathcal{L}$ with $S \subseteq T$, then $v(S) \leq v(T)$.*

We denote by $\Omega(\mathcal{L})$ the class of all simple games defined on the family $\mathcal{L}$. In a simple game, a coalition $S \in \mathcal{L}$ is *winning* if $v(S) = 1$, otherwise, is *losing*. A winning coalition S is *minimal* if it is winning, and there does not exist any winning coalition contained in S. We denote by $\mathcal{W}$ the collection of all minimal winning coalitions. This collection is nonempty and, in general, we write

$$\mathcal{W} = \{S_1, \ldots, S_r\} \quad \text{where } r \geq 1.$$

Note that every simple game is totally characterized by the minimal winning coalitions. If a game has only one minimal winning coalition $\mathcal{W} = \{S_1\}$, then v is the unanimity game u_{S_1}.

There is a certain type of players that, in the simple games, perform a very important role, and which are so-called veto players. A player $i \in N$ is a *veto player* in the game v if he belongs to each winning coalition $S \in \mathcal{L}$, that is,

$$v(S) = 1 \text{ implies } i \in S.$$

We denote by $\mathcal{V}$ the set of all veto players in the game $v \in \Omega(\mathcal{L})$, i.e.,

$$\mathcal{V} = \bigcap \{S \in \mathcal{L} : v(S) = 1\}.$$

If there exist $\{i\}, \{j\} \in \mathcal{L}$, $i \neq j$ such that $v(i) = v(j) = 1$, then $\mathcal{V} = \emptyset$. This means that there are no veto players in all simple games. Because of this, to distinguish between the simple games which have veto players, and those games that do not, we introduce the following definition.

Definition 10.1.2 *A simple game $v \in \Omega(\mathcal{L})$is called weak if it has at least one veto player, that is, $\mathcal{V} \neq \emptyset$.*

Each unanimity game u_T is weak and the set of veto players is $\mathcal{V} = T$.

Let $\mathcal{L}$ be a closure space. A game $v \in \Gamma(\mathcal{L})$ is said to be *convex on* $\mathcal{L}$ if $v(S) + v(T) \leq v\left(\overline{S \cup T}\right) + v(S \cap T)$, for all $S, T \in \mathcal{L}$.

Recall that a game $v \in \Gamma(\mathcal{L})$ is *quasi-convex* if for all $S, T \in \mathcal{L}$ with $S \cup T \in \mathcal{L}$, we have $v(S) + v(T) \leq v(S \cup T) + v(S \cap T)$.

The following characterization of the unanimity games is the key to the proofs of some theorems in the last two sections of this chapter.

Proposition 10.1.1 *If $\mathcal{L}$ is a closure space and $v \in \Omega(\mathcal{L})$, the following statements are equivalent:*

(a) *The game v is convex on $\mathcal{L}$.*

(b) *The game v is a unanimity game.*

Moreover, if $\mathcal{L}$ is a convex geometry, then (a) and (b) are equivalent to

(c) *The game v is quasi-convex.*

Proof. First of all, we show that if v is a unanimity game, then v is a convex game. Indeed, consider the unanimity game u_T, for any nonempty coalition $T \in \mathcal{L}$, and let $A, B \in \mathcal{L}$. We distinguish three cases:

1. If $A \supseteq T$ and $B \supseteq T$, then

$$u_T(A) = 1,\ u_T(B) = 1,\ u_T\left(\overline{A \cup B}\right) = 1 \text{ and } u_T(A \cap B) = 1.$$

2. If $A \supseteq T$ and $B \not\supseteq T$, then

$$u_T(A) = 1,\ u_T(B) = 0,\ u_T\left(\overline{A \cup B}\right) = 1 \text{ and } u_T(A \cap B) = 0.$$

3. If $A \not\supseteq T$ and $B \not\supseteq T$, then

$$u_T(A) = 0,\ u_T(B) = 0,\ u_T\left(\overline{A \cup B}\right) \geq 0 \text{ and } u_T(A \cap B) = 0.$$

Therefore, the convexity condition holds in all cases.

Conversely, let $v \in \Omega(\mathcal{L})$ be a convex game and let $T \in \mathcal{L}$ such that

$$|T| = \min\{|S| : v(S) = 1\} = \alpha.$$

This coalition T is unique because if there are two coalitions $A, T \in \mathcal{L},\ A \neq T$ which satisfy $v(T) = v(A) = 1$ and $|T| = |A| = \alpha$ then, the inequality

$$v(T) + v(A) \leq v\left(\overline{T \cup A}\right) + v(T \cap A),$$

lead to $v(T \cap A) = 1$. This contradicts the election of T.

Now, taking the coalition T, for any other $A \in \mathcal{L}$ such that $A \not\supseteq T$, as it holds that

$$v(A) + v(T) \leq v\left(\overline{A \cup T}\right) + v(A \cap T) = 1,$$

we deduce that $v(A) = 0$. Thus, the game v is the unanimity game corresponding to the coalition T, that is, $v = u_T$.

Finally, note that if the game is monotone on a convex geometry, then the equivalence between (a) and (c) is already proved in Theorem 9.4.6. □

The equivalence of (a) and (b) with the affirmation (c) is not true when $\mathcal{L}$ is not a convex geometry. If we consider $N = \{1, 2, 3, 4\}$, the family

$$\mathcal{L} = \{\emptyset, \{1\}, \{2\}, \{3\}, \{4\}, \{3, 4\}, \{1, 2, 3\}, N\}$$

and the game $v : \mathcal{L} \to \mathbb{R}$ which has the coalitions $\{1\}, \{2\}, \{3\}$ as minimal winning coalitions, this game is quasi-convex but is not a unanimity game.

10.2 The core

Throughout this section, we assume that the family of feasible coalitions $\mathcal{L}$ is atomic. If we denote by $\{e_i\}_{i=1}^n$ the vectors of the canonical basis of $\mathbb{R}^n$, it is easy to check that there are three possibilities for the imputation set $I(\mathcal{L}, v)$ in a game $v \in \Omega(\mathcal{L})$:

1. $I(\mathcal{L}, v) = \emptyset$, if there are two winning coalitions of cardinality one.
2. $I(\mathcal{L}, v) = \{e_k\}$, if there exists $k \in N$, with $v(k) = 1$, and $v(i) = 0$ for all $i \in N$, $i \neq k$.
3. $I(\mathcal{L}, v) = \operatorname{conv}\{e_1, \ldots, e_n\}$, if $v(i) = 0$ for all $i \in N$.

Thus, the imputation set is a larger set when no other minimal winning coalition is a unitary coalition.

The core of a simple game is completely determined by the veto players. Recall that the core is defined by

$$C(\mathcal{L}, v) := \{x \in \mathbb{R}^n : x(N) = v(N),\ x(S) \geq v(S) \text{ for all } S \in \mathcal{L}\},$$

and note that, if $v \in \Omega(\mathcal{L})$, there are some inequalities $x(S) \geq v(S)$ which are redundant. To be exact, the inequalities corresponding to non-unitary coalitions that are losing coalitions or winning coalitions but not minimal are not needed in the description of the core. Indeed, the vectors of the core must verify the inequalities $x_i \geq 0$ for all $i \in N$, and therefore, the corresponding restrictions to non-unitary losing coalitions are redundant. Moreover, if $S \in \mathcal{L}$ is winning but not minimal, there exists a minimal winning coalition S^* such that $S^* \subset S$. In this case, we have

$$x(S) \geq x(S^*) \geq v(S^*) = v(S)$$

and hence, the condition $x(S) \geq v(S)$ is a consequence of $x(S) \geq v(S^*)$ and therefore, it is redundant. Thus, we can conclude that

$$C(\mathcal{L}, v) = \{x \in \mathbb{R}^n : x \geq 0,\ x(N) = x(S) = 1 \text{ for all } S \in \mathcal{W}\}.$$

Theorem 10.2.1 *Let $v \in \Omega(\mathcal{L})$ be a simple game on an atomic family. A necessary and sufficient condition so that $C(\mathcal{L}, v) \neq \emptyset$ is that game v is weak. Furthermore, in this case,*

$$C(\mathcal{L}, v) = \{x \in \mathbb{R}^n : x \geq 0,\ x(N) = x(\mathcal{V}) = 1\},$$

where $\mathcal{V}$ is the set of veto players in v.

Proof. *Sufficient condition.* Assume that v is weak, then $\mathcal{V} \neq \emptyset$ and for each $i \in \mathcal{V}$, we consider the corresponding vector e_i of the canonical basis of $\mathbb{R}^n$. Then, we have that $e_i(N) = 1 = v(N)$ and $e_i(S) = v(S)$ for all $S \in \mathcal{W}$ and therefore, $e_i \in C(\mathcal{L}, v)$.

Necessary condition. We construct an $r \times n$ matrix $A = (a_{ij})$ such that

$$a_{ij} = \begin{cases} 1, & \text{if } j \in S_i \\ 0, & \text{otherwise,} \end{cases}$$

where $S_1, \ldots, S_r$ are the minimal winning coalitions in game v. Then $C(\mathcal{L}, v)$ is the set of vectors satisfying

$$\begin{aligned} \sum_{j=1}^{n} x_j &= 1, \\ \sum_{j=1}^{n} a_{ij} x_j &= 1, \quad 1 \leq i \leq r, \\ x_j &\geq 0, \quad 1 \leq j \leq n. \end{aligned}$$

If this set is nonempty, there must be one element $x \neq 0$. Then, if $\mathcal{V} = \emptyset$, each column of matrix A has at least one entry equal to 0, and taking the sum of the equations corresponding to the rows of Ax we obtain

$$\alpha_1 x_1 + \cdots + \alpha_n x_n = |\mathcal{W}|, \quad \text{with } \alpha_j < |\mathcal{W}| \quad \text{for } 1 \leq j \leq n.$$

Therefore, $(|\mathcal{W}| - \alpha_1) x_1 + \cdots + (|\mathcal{W}| - \alpha_n) x_n = 0$, and this is a contradiction because $x_j \geq 0$ for $1 \leq j \leq n$ and any other $x_k > 0$ with $1 \leq k \leq n$. Thus, game v is weak. Finally, note that if the game is weak and we take $i \notin \mathcal{V}$, then $x_i = 0$ for every vector of the core, and hence

$$C(\mathcal{L}, v) = \{x \in \mathbb{R}^n : x \geq 0, \ x(N) = x(\mathcal{V}) = 1\}.$$

□

The previous theorem generalizes an analogous result of Curiel [48] and permits the identification of the extreme points of the polyhedron $C(\mathcal{L}, v)$. As a consequence of this, the following proposition is immediate.

Proposition 10.2.2 *For every weak game $v \in \Omega(\mathcal{L})$ on an atomic family,*

$$C(\mathcal{L}, v) = \text{conv}\{e_i : i \in \mathcal{V}\}$$

where $\mathcal{V}$ is the set of veto players in v.

Although in this section, we have considered simple games on atomic families, it is easy to check that every weak simple game on any family of feasible coalitions has a nonempty core.

However, if we consider weak games on non-atomic families, the Proposition 10.2.2 does not hold in general. Indeed, if we consider $N = \{1,2,3\}$, the family $\mathcal{L} = \{\emptyset, \{1\}, \{1,2\}, \{1,2,3\}\}$ and the unanimity game $u_{\{1,2\}}$, the core is not bounded.

On the other hand, if the family is not atomic, then the core can be nonempty and the game not weak. For example, by taking $N = \{1,2,3,4,5\}$,

$$\mathcal{L} = \{\emptyset, \{1\}, \{2\}, \{3\}, \{1,2\}, \{1,2,3,4,5\}\}$$

and the simple game v for which its minimal winning coalitions are $\{1,2\}$ and $\{3\}$. In this case, the vector $(1/2, 1/2, 1, -1, 0) \in C(\mathcal{L}, v)$ and v is not weak.

10.3 The Weber set

By considering the Weber set in this section, it is necessary to require, in the following, that the family of feasible coalitions $\mathcal{L}$ is a convex geometry on N, not necessarily atomic. First of all, we prove that the Weber set is a convex set where its extreme points are determined by the players who by joining a losing coalition convert it into a winning coalition.

Proposition 10.3.1 *If $v \in \Omega(\mathcal{L})$ is a game on a convex geometry, then*

$$Weber(\mathcal{L}, v) = \operatorname{conv}\left\{ e_i : i \in \bigcup_{l=1}^{r} \operatorname{ex}(S_l) \right\}.$$

where $\{S_1, \ldots, S_r\}$ is the set of minimal winning coalitions in v.

Proof. Let $C \in \mathcal{C}(\mathcal{L})$ be a maximal chain and let $a^C(v) \in \mathbb{R}^n$ be the marginal worth vector associated with this chain. For every $i \in N$ we have that

$$a_i^C(v) = v(C(i)) - v(C(i) \setminus i) \in \{0,1\}$$

since v is monotone. Moreover $a^C(v)$ is efficient and hence $\sum_{i \in N} a_i^C(v) = 1$. Therefore, the vector $a^C(v) \in \mathbb{R}^n$ has only one of its components equal to 1 and the rest of them equal to 0. If we assume that the component j is equal to 1, i.e., $v(C(j)) = 1$ and $v(C(j) \setminus j) = 0$, then $a^C(v) = e_j$.

On the other hand, $C(j)$ is a winning coalition and so, there exists a minimal winning coalition S_k such that $S_k \subseteq C(j)$. Note that $j \in S_k$, since if $S_k \subseteq C(j) \setminus j$ then $C(j) \setminus j$ is a winning coalition but it is not possible as $v(C(j) \setminus j) = 0$. Moreover, $j \in \operatorname{ex}(S_k)$ because $S_k \setminus j = (C(j) \setminus j) \cap S_k \in \mathcal{L}$.

Since $Weber(\mathcal{L}, v)$ is the convex hull of marginal worth vectors, we have proved that $Weber(\mathcal{L}, v) \subseteq \operatorname{conv}\{e_i : i \in \bigcup_{l=1}^{r} \operatorname{ex}(S_l)\}$.

In order to prove the inverse inclusion, we suppose that $i \in \bigcup_{l=1}^{r} \operatorname{ex}(S_l)$. Then there exists a minimal winning coalition S_l such that $i \in \operatorname{ex}(S_l)$. Let $C \in \mathcal{C}(\mathcal{L})$ be a maximal chain such that $S_l = C(i)$. For this chain, we obtain $a^C(v) = e_i$. □

Note that if $\mathcal{L} = 2^N$ then $Weber(\mathcal{L}, v) = \operatorname{conv}\{e_i : i \in \bigcup_{l=1}^{r} S_l\}$ and it contains the core. However, this inclusion is not always verified for simple games on convex geometries $\mathcal{L} \neq 2^N$. For example, let $N = \{1, 2, 3, 4\}$, let $\mathcal{L} = Co(\{1 < 2 < 3 < 4\})$, and let $v : \mathcal{L} \to \mathbb{R}$ be the simple game which has only one minimal winning coalition $\{1, 2, 3\}$. Then $C(\mathcal{L}, v) = \operatorname{conv}\{e_1, e_2, e_3\}$ and $Weber(\mathcal{L}, v) = \operatorname{conv}\{e_1, e_3\}$.

Definition 10.3.1 *A nonempty set $J \subseteq \mathbb{R}^n$ satisfies the dummy player property when for all $i \in N$ such that i is a dummy player in a game v and for all $x \in J$, then $x_i = v(i)$ if $\{i\} \in \mathcal{L}$, and 0 otherwise.*

The following result shows that each efficient distribution that satisfies the individual rationality principle and the dummy axiom is an element of the Weber set.

Theorem 10.3.2 *Let $v \in \Omega(\mathcal{L})$ be a simple game on an atomic convex geometry. If $J \subseteq I(\mathcal{L}, v)$ satisfies the dummy player property, then $J \subseteq Weber(\mathcal{L}, v)$. The converse is true when $I(\mathcal{L}, v) = \operatorname{conv}\{e_1, \ldots, e_n\}$.*

Proof. If $I(\mathcal{L}, v) = \{e_i\}$ the result follows easily. Thus, we assume that $I(\mathcal{L}, v) = \operatorname{conv}\{e_1, \ldots, e_n\}$, i.e., there is no winning coalition of cardinal one and so $v(i) = 0$ for all $i \in N$.

If $x \in J$, then x is an imputation and hence $x \in \operatorname{conv}\{e_1, \ldots, e_n\}$, so that $x = \sum_{i=1}^{n} \mu_i e_i$ with $\sum_{i=1}^{n} \mu_i = 1$ and $\mu_i \geq 0$ for $i = 1, \ldots, n$.

By the above theorem, to prove that $x \in Weber(\mathcal{L}, v)$ is equivalent to prove that $\mu_k = 0$ when k is not extreme for any minimal winning coalition. Then, let us take $k \in N \setminus \bigcup_{j=1}^{r} \operatorname{ex}(S_j)$. As game v is monotone, we deduce that $v(S) - v(S \setminus k) \geq 0$ for all $S \in \mathcal{L}$ with $k \in \operatorname{ex}(S)$. Moreover, in this case, we assert that $v(S) - v(S \setminus k) = 0$. Indeed, if $v(S) - v(S \setminus k) = 1$, then necessarily $v(S) = 1$ and $v(S \setminus k) = 0$, and hence, there would be a minimal winning coalition $S_j \subseteq S$ with $k \in S_j$, because if $S_j \subseteq S \setminus k$ we would obtain $v(S \setminus k) = 1$, but it is not possible. Therefore, k is a dummy player in v and so, $\mu_k = x_k = v(k) = 0$.

Conversely, let $v \in \Omega(\mathcal{L})$ such that $I(\mathcal{L}, v) = \operatorname{conv}\{e_1, \ldots, e_n\}$ and let $J \subseteq Weber(\mathcal{L}, v)$. Since $Weber(\mathcal{L}, v) \subseteq I(\mathcal{L}, v)$, then we have $J \subseteq I(\mathcal{L}, v)$. Moreover, if $k \in N$ is a dummy player in v and $x \in J$, then $x_k = v(S) - v(S \setminus k)$ for any $S \in \mathcal{L}$ with $k \in \operatorname{ex}(S)$, because $x \in Weber(\mathcal{L}, v)$. Therefore, we show that $x_k = v(k)$. □

10.4 Stable sets

We introduce the notion of dominance and the concept of stable set. Let $v \in \Gamma(\mathcal{L})$ and $x, y \in I(\mathcal{L}, v)$. The imputation x *dominates* the imputation y (we denote $x \operatorname{dom} y$) if there exists a nonempty $S \in \mathcal{L}$ such that $x_i > y_i$ for all $i \in S$, and $x(S) \leq v(S)$. A set $E \subseteq I(\mathcal{L}, v)$ is *stable* if the imputations in E are undominated by other elements of E, and every imputation not in E is dominated by an imputation in E. First of all, we show that every simple game in which the imputation set is nonempty has at least one stable set.

Theorem 10.4.1 *Every simple game $v \in \Omega(\mathcal{L})$ on an atomic family with a nonempty imputation set has at least one stable set.*

Proof. As $\mathcal{L}$ is atomic and the imputation set is nonempty, we have only two possibilities for $I(\mathcal{L}, v)$. When the imputation set is reduced to a vector, it can be immediately deduced that this set is the unique stable set of the game. For this, we assume that $I(\mathcal{L}, v) = \operatorname{conv}\{e_1, \ldots, e_n\}$. In this case, for each minimal winning coalition S in v, the set formed by the imputations that distribute the unit among the players of S is a stable set, as we prove in the following. If we denote by $E_S = \left\{x \in \mathbb{R}^n_+ : \sum_{i \in S} x_i = 1,\ x_j = 0 \text{ for all } j \notin S\right\}$, then we have the following properties:

(*i*) *Internal stability*: No imputation in E_S dominates another. Assume that there are $x, y \in E_S$ such that $x \operatorname{dom} y$. Then, there exists a nonempty coalition $T \in \mathcal{L}$, such that $x_i > y_i$ for all $i \in T$, and $\sum_{j \in T} x_j \leq v(T)$. These conditions can only be verified if $T \subset S$ and $v(T) = 1$, since $x, y \in E_S$. But it is impossible because S is a minimal winning coalition.

(*ii*) *External stability*: If $x \in I(\mathcal{L}, v) \setminus E_S$, then there exists $y \in E_S$ such that $y \operatorname{dom} x$. Indeed, taking into account that $v(S) - x(S) > 0$, upon taking the vector y where its components are

$$y_i = \begin{cases} 0, & \text{if } i \notin S \\ x_i + \dfrac{v(S) - x(S)}{|S|}, & \text{if } i \in S. \end{cases}$$

□

The following result affirms that the unanimity games are the unique simple games on atomic convex geometries for which the core is a stable set.

Theorem 10.4.2 *Let $v \in \Omega(\mathcal{L})$ be a simple game on an atomic convex geometry. Then, $C(\mathcal{L}, v)$ is stable if and only if v is a unanimity game.*

Proof. If $C(\mathcal{L}, v)$ is a stable set, then it must be nonempty and we also know that $C(\mathcal{L}, v) = \operatorname{conv}\left\{e_i : i \in \bigcap_{j=1}^r S_j\right\}$, where $\{S_1, \ldots, S_r\}$ is the set of all minimal winning coalitions with $|S_1| \leq \cdots \leq |S_r|$. We consider two cases:

1. If $|S_1| = 1$ or $|S_1| = n$, then $v = u_{S_1}$.
2. Let $1 < |S_1| < n$, and assume that $Weber(\mathcal{L}, v) \not\subseteq C(\mathcal{L}, v)$. Then there exists $k \in \bigcup_{j=1}^{r} \text{ex}(S_j) \setminus \bigcap_{j=1}^{r} S_j$ and so $e_k \notin C(\mathcal{L}, v)$. As $C(\mathcal{L}, v)$ is a stable set, there exists $x \in C(\mathcal{L}, v)$ such that $x \operatorname{dom} e_k$ using a coalition $T \in \mathcal{L}$. Since $x(T) = v(T)$, $x_i > 0$ for all $i \in T$, $i \neq k$, and $x_k > 1$ if $k \in T$, there must be $v(T) = 1$ and $k \notin T$. Thus, there exists a minimal winning coalition $S_{j^*} \in \mathcal{L}$ such that $S_{j^*} \subseteq T$ and $x(S_{j^*}) = v(S_{j^*}) = 1$. So, $x_i > 0$ for all $i \in S_{j^*}$ and as $x(N \setminus S_{j^*}) = 0$, it implies that $x(S_p \setminus S_{j^*}) = 0$ for every minimal winning coalition $S_p \in \mathcal{L}$, $S_p \neq S_{j^*}$. However, as $x \in C(\mathcal{L}, v)$, $v(S_p) = 1$ holds and hence $x(S_p \cap S_{j^*}) = 1$ for every minimal winning coalition $S_p \in \mathcal{L}$. Then, we assert that $S_{j^*} \subseteq \bigcap_{j=1}^{r} S_j$. Suppose this inclusion fails to holds. Then there exists a minimal winning coalition $S_p \in \mathcal{L}$ such that $S_{j^*} \not\subseteq S_p$ and hence

$$x(N) = x(S_{j^*} \cap S_p) + x(S_{j^*} \setminus S_p) + \ldots > 1.$$

But this is a contradiction. Therefore, $\bigcap_{j=1}^{r} S_j = S_{j^*}$ and so $v = u_{S_{j^*}}$.

In order to prove the converse, let $v = u_T$ be a unanimity game. Then, $C(\mathcal{L}, v) = \text{conv}\{e_i : i \in T\}$ because T is the unique minimal winning coalition. By the above theorem, $E_T = C(\mathcal{L}, v)$ is a stable set. □

Example: Consider $N = \{1, 2, 3\}$, and the game $v : 2^N \to \mathbb{R}$ given by $v(S) = 1$ if $|S| \geq 2$ and $v(S) = 0$, otherwise. Then we have $I(v) = \text{conv}\{e_1, e_2, e_3\}$ and $C(v) = \emptyset$. Furthermore (see Driessen [57]), there is an infinity of stable sets. Every set formed by imputations that assign a constant value c to a player i, and determine all the possible distributions of the amount $1 - c$ among the other two players, are stables sets. Each one of these stable sets contains infinite imputations and the following finite set is also stable

$$E = \left\{ \left(\frac{1}{2}, \frac{1}{2}, 0 \right), \left(\frac{1}{2}, 0, \frac{1}{2} \right), \left(0, \frac{1}{2}, \frac{1}{2} \right) \right\}.$$

In this example, we observe that every stable set can be interpreted by a model of behavior for which the imputations are selected. Furthermore, it suggests that each stable set delimits forms of distribution that the player must negotiate, or indicates a smaller game between the coalitions.

If we now consider $\mathcal{L} = Co(\{1 < 2 < 3\})$, and the game $v \in \Omega(\mathcal{L})$ defined by $v(S) = 1$ if $|S| \geq 2$ and $v(S) = 0$, otherwise. Taking into account that the sets E_S, with $S \in \mathcal{W}$, are stable, in this case, the sets $E_{\{1,2\}} = \text{conv}\{e_1, e_2\}$ and $E_{\{2,3\}} = \text{conv}\{e_2, e_3\}$ are stable. However, the other stable sets for $\mathcal{L} = 2^N$ are not stable in this case. Furthermore, we can observe that, although the core is not empty, this neither offers reasonable solutions to the game, nor is it a stable set.

10.5 Bargaining sets

Although in the literature several concepts of bargaining sets exist (for a survey see Maschler [169]), we consider only three of them in the context of simple games, and we show some results that relate these bargaining sets with the core and the Weber set. The different solution concepts called bargaining sets have in common the study of how to distribute the profits of cooperation (it is supposed that cooperation has already been decided) taking into account the negotiation processes that can be established among the different coalitions. That is, it bears in mind that certain players can offer pacts to others with the aim of obtaining more advantageous distributions and at the same time other pacts that could put the ones that do this in danger.

We consider that the family of feasible coalitions $\mathcal{L}$ is an atomic closure space on N. One of the three different bargaining sets considered is the Aumann-Maschler bargaining set. Its definition is based on the *objections* and *counter-objections* of some players against others, and they are introduced when the imputation set is nonempty.

- *The Aumann-Maschler bargaining set* (1964) [7].

Let $i, j \in N$ with $i \neq j$. We denote by Λ_{ij} the set of all feasible coalitions of $\mathcal{L}$ that contains player i but not player j, i.e.,

$$\Lambda_{ij} = \{S \in \mathcal{L} : i \in S,\ j \notin S\}.$$

Since the family $\mathcal{L}$ is atomic, the set Λ_{ij} is nonempty for all $i, j \in N$.

An *objection* of player i against player j with respect to the imputation x in game v is a pair (y, S), where $S \in \Lambda_{ij}$ and the vector $y = (y_k)_{k \in S}$ satisfy

$$\begin{aligned} y(S) &= v(S), \\ y_k &> x_k \text{ for all } k \in S. \end{aligned}$$

A *counterobjection* of j against the pair (y, S) is a pair (z, T), where $T \in \Lambda_{ji}$ and $z = (z_k)_{k \in T}$ satisfy

$$\begin{aligned} z(T) &= v(T), \\ z_k &\geq y_k \text{ for all } k \in T \cap S, \\ z_k &\geq x_k \text{ for all } k \in T \setminus S. \end{aligned}$$

An objection is said to be *justified* if there does not exist a counterobjection. An imputation $x \in I(\mathcal{L}, v)$ is said to belong to $B(\mathcal{L}, v)$ if for each objection of a player against another with respect to x, there exists a counterobjection, that is,

$$B(\mathcal{L}, v) = \{x \in I(\mathcal{L}, v) : \text{ there does not exist a justified objection to } x\}.$$

Note that $C(\mathcal{L}, v) \subseteq B(\mathcal{L}, v)$ as there cannot be an objection with respect to an element of the core.

Example: Consider the game (N, v) defined in the previous section. For this game, we have $I(v) = \text{conv}\{e_1, e_2, e_3\}$ and $C(v) = \emptyset$. From this, in the case that the cooperation among the players is already decided, how do the players share out the worth of the grand coalition $v(N) = 1$? In this case, the core does not determine any distribution. Furthermore, it seems logical to think that not any imputation is accepted as a distribution by all players. For example, the imputation $(1, 0, 0)$ must not satisfy the players 2 and 3, since their coalition is evaluated by 1 in the game. On the other hand, taking into account the symmetry of the players in this game, it is reasonable to expect that the distribution that is adopted treats each one equally, that is, $(1/3, 1/3, 1/3)$. In this case, $B(v) = \{(1/3, 1/3, 1/3)\}$. To check it, observe that each player i can object against another player j with respect to this imputation using the coalition that he forms with the other player k. However, every objection presented has a counterobjection by player j using the coalition $\{j, k\}$.

If imputation set is empty, then the Aumann-Maschler bargaining set is empty. Consider the game (N, v) where $N = \{1, 2, 3\}$ and v is given by

$$v(S) = \begin{cases} 0, & \text{if } S \in \{\emptyset, \{3\}\} \\ 1, & \text{otherwise.} \end{cases}$$

In cases like this, in which the imputation set is empty, how should the sharing out be done when the players have decided to form the grand coalition? The following bargaining sets can offer solutions.

- *The Mas-Colell bargaining set* (1989) [170].

Consider the set $X(\mathcal{L}, v) = \left\{x \in \mathbb{R}^n_+ : x(N) = v(N)\right\}$ and let $x \in X(\mathcal{L}, v)$. An *objection to* x *in the sense of Mas-Colell* is a pair (y, S) such that $y \in \mathbb{R}^n_+$ and $S \in \mathcal{L}$, satisfying

$$\begin{aligned} y(S) &\leq v(S), \\ y_k &\geq x_k \text{ for all } k \in S, \text{ with at least one strict inequality.} \end{aligned}$$

A *counterobjection* to (y, S) is a pair (z, T), where $z \in \mathbb{R}^n_+$, $T \in \mathcal{L}$, satisfy

$$\begin{aligned} z(T) &\leq v(T), \\ z_k &\geq y_k \text{ for all } k \in T \cap S, \\ z_k &\geq x_k \text{ for all } k \in T \setminus S, \end{aligned}$$

with at least one strict inequality.

An objection (y, S) to x is said *to be justified in the sense of Mas-Colell* if there does not exist any counterobjection to (y, S). The *Mas-Colell bargaining set* of v, $MB(\mathcal{L}, v)$, consists of all vectors $x \in X(\mathcal{L}, v)$ for which there does not exist (y, S) which is a justified objection to x in the sense of Mas-Colell.

- *The Greenberg bargaining set* (1992) [114].

Consider the set $X^*(\mathcal{L}, v) = \{x \in \mathbb{R}^n_+ : x(N) \leq v(N)\}$, and the vector $x \in X^*(\mathcal{L}, v)$. An *objection* to x *in the sense of Greenberg* is a pair (y, S) with $y \in \mathbb{R}^n_+$, $S \in \mathcal{L}$, satisfying

$$\begin{aligned} y(S) &= v(S), \\ y_k &> x_k \text{ for all } k \in S, \\ y_k &= x_k \text{ for all } k \in N \setminus S. \end{aligned}$$

A *counterobjection* to the pair (y, S) is a pair (z, T), where $z \in \mathbb{R}^n_+$, $T \in \mathcal{L}$, satisfy

$$\begin{aligned} z(T) &= v(T), \\ z_k &> y_k \text{ for all } k \in T, \\ z_k &= y_k \text{ for all } k \in N \setminus T. \end{aligned}$$

An objection (y, S) is said *to be justified in the sense of Greenberg* if there does not exist any counterobjection to (y, S). The *Greenberg bargaining set*, $GB(\mathcal{L}, v)$, consists of all vectors $x \in X^*(\mathcal{L}, v)$ for which there exist no justified objections in the sense of Greenberg.

We can check that for the last example the vector $(1/2, 1/2, 0)$ belongs to the two last bargaining sets. Moreover, this vector is a reasonable distribution since, in this case, we can understand that player 3 does not contribute anything to the coalitions in which he participates, but however, players 1 and 2 have the same role in the game.

The three bargaining sets introduced contain the core, as with these definitions, there does not exist objection to vectors of the core. Furthermore, Einy, Holzman and Monderer [74] showed that the least core of a game (N, v) satisfies $LC(v) \subseteq MB(v)$. Since the least core is always nonempty (see page 4) this result provides a direct proof for the existence of the Mas-Colell bargaining set.

We now study a certain class of simple games in which these sets coincide with the core.

Theorem 10.5.1 *Let $\mathcal{L}$ be an atomic closure space and $v \in \Omega(\mathcal{L})$ a weak simple game. Then $B(\mathcal{L}, v) = MB(\mathcal{L}, v) = C(\mathcal{L}, v)$.*

Proof. If $\mathcal{V}$ is the set of veto players, we know, by Theorem 10.2.1 that

$$C(\mathcal{L},v)=\{x\in\mathbb{R}^n : x\geq 0,\ x(N)=x(\mathcal{V})=1\}.$$

First, we show that $B(\mathcal{L},v)=C(\mathcal{L},v)$. Since one inclusion is evident, assume that there is $x\in B(\mathcal{L},v)\setminus C(\mathcal{L},v)$. Then $x(\mathcal{V})<1$ and hence, $\mathcal{V}\neq N$. So, there exists $i\in N\setminus\mathcal{V}$ such that $x_i>0$ because $x(N)=1$. If $j\in\mathcal{V}$, then j can object against i in x. Indeed, if $S\in\mathcal{L}$ a winning coalition such that $j\in S$ and $i\notin S$, then $x(S)<1$. We define $y=(y_k)_{k\in S}$ by

$$y_k=x_k+\frac{1-x(S)}{|S|},\quad \text{for all } k\in S.$$

This vector y satisfies that $y(S)=v(S)$ and $y_k>x_k$ for each $k\in S$. Therefore, (y,S) is an objection of j against i in x. We suppose that there exists a counterobjection (z,T) to (y,S). Then $z(T)=v(T)$, $v(T)=0$ and $i\in T$, and hence $z_i=0$. But this is a contradiction because $z_i\geq x_i>0$.

We now prove that $MB(\mathcal{L},v)=C(\mathcal{L},v)$. If there exists $x\in MB(\mathcal{L},v)$ but $x\notin C(\mathcal{L},v)$, then $x(\mathcal{V})<1$ and hence $\mathcal{V}\neq N$. Therefore, there exists $i\in N\setminus\mathcal{V}$ such that $x_i>0$ and $x(N\setminus i)<1$. Let $T\in\mathcal{L}$ be a winning coalition such that $x(T)=\min\{x(S):v(S)=1\}$. Then $x(T)<1$. We define $y\in\mathbb{R}^n_+$ by

$$y_k=\begin{cases} x_k+\dfrac{1-x(T)}{|\mathcal{V}|}, & \text{if } k\in\mathcal{V}\\ x_k, & \text{if } k\in N\setminus\mathcal{V}.\end{cases}$$

Thus $y_k>x_k$ for each $k\in\mathcal{V}$ and

$$y(T)=x(\mathcal{V})+1-x(T)+x(T\setminus\mathcal{V})=1=v(T).$$

Therefore (y,T) is an objection to x in the sense of Mas-Colell. Assume that (y,T) is not justified. Then there exists a counterobjection (z,R) such that $z(R)\leq v(R)=1$, and $z_k\geq y_k$ for each $k\in R\cap T$, and $z_k\geq x_k$ for $k\in R\setminus T$ with at least one strict inequality. Therefore,

$$\begin{aligned} z(R) &= z(R\setminus T)+z(R\cap T)>x(R\setminus T)+y(R\cap T)\\ &= y(R\setminus T)+y(R\cap T)=y(R)\\ &= 1+x(R)-x(T)\geq 1, \end{aligned}$$

which is a contradiction. □

The above result is not true if the game $v\in\Omega(\mathcal{L})$ is not weak. It is enough to see the example of Section 10.4.

Theorem 10.5.2 *Let $\mathcal{L}$ be an atomic convex geometry and let $v \in \Omega(\mathcal{L})$ be a weak simple game. Then $GB(\mathcal{L}, v) = C(\mathcal{L}, v)$ if and only if v is a unanimity game.*

Proof. Let $v = u_T$ for some nonempty $T \in \mathcal{L}$ and let $x \in GB(\mathcal{L}, v)$ be such that $x \notin C(\mathcal{L}, v)$. Then $x(T) < 1$. Define $y \in \mathbb{R}^n_+$ by

$$y_i = \begin{cases} x_i + \dfrac{1 - x(T)}{|T|}, & \text{if } i \in T \\ x_i, & \text{if } i \notin T. \end{cases}$$

Then (y, T) is an objection to x in the sense of Greenberg and this objection is justified. Indeed, if (z, S) is a counterobjection to (y, T), then $v(S) = 1$ and hence $S \supset T$. Therefore, $1 = z(S) > y(S) \geq y(T) = 1$, which is a contradiction.

To obtain the converse, we will show that $GB(\mathcal{L}, v) = C(\mathcal{L}, v)$ implies that v is a quasi-convex game, since the Proposition 10.1.1 affirms that v is a unanimity game. If v is not a quasi-convex game then Theorem 9.4.4 implies that there exists $C \in \mathcal{C}(\mathcal{L})$, such that $a^C(v) \notin C(\mathcal{L}, v)$. Note that $a^C_j(v) \in \{0, 1\}$ for all $j \in N$ and $\sum_{i \in N} a^C_i(v) = 1$. Thus, the vector $a^C(v) \in \mathbb{R}^n$ is equal to a canonical vector of $\mathbb{R}^n$.

On the other hand, if $\mathcal{V}$ is the set of veto players in v, then $\mathcal{V} \neq \emptyset, N$. Note that $\mathcal{V} \neq \emptyset$ because the game is weak and $\mathcal{V} \neq N$ since we assume that $v \neq u_N$. Furthermore, $C(\mathcal{L}, v) = \operatorname{conv}\{e_j : j \in \mathcal{V}\}$ and as $a^C(v) \notin C(\mathcal{L}, v)$ then there exists a player $i \in N \setminus \mathcal{V}$ such that $a^C(v) = e_i$. As $a^C(v) \notin GB(\mathcal{L}, v)$, there exists a justified objection (x, S) to $a^C(v)$ in the sense of Greenberg, where $x \in \mathbb{R}^n$, $S \in \mathcal{L}$ satisfy $x \geq 0$, $x(S) = v(S)$, $x_k > a^C_k(v)$ for each $k \in S$ and $x_k = a^C_k(v)$ for $k \in N \setminus S$. Therefore, $x(S) = v(S) = 1$ and $i \notin S$. Let

$$A = \{G \in \mathcal{L} : i \notin G \text{ and } S \nsubseteq G\}.$$

This set is nonempty, and we can distinguish two cases:

1. There exists $G \in A$ such that $v(G) = 1$. In this case

$$\begin{aligned} x(G) &= x(G \cap S) + x(G \cap (N \setminus S)) = x(G \cap S) \\ &= x(S) - x(S \setminus G) < 1. \end{aligned}$$

If we define $y \in \mathbb{R}^n$ by

$$y_k = \begin{cases} x_k + \dfrac{x(S \setminus G)}{|G|}, & \text{if } k \in G \\ x_k, & \text{if } k \in N \setminus G, \end{cases}$$

then $y \geq 0$, $y_k > x_k$ for $k \in G$, $y_k = x_k$ for $k \in N \setminus G$, and

$$y(G) = x(G) + x(S \setminus G) = x(S) = 1 = v(G).$$

Thus, (y, G) is a counterobjection to (x, S) and this is a contradiction.

2. For all $G \in A$, $v(G) = 0$. In this case $v(T) = 1$ implies $i \in T$ or $S \subseteq T$. Since $v \in \Omega(\mathcal{L})$ is not a quasi-convex game, then $v(\mathcal{V}) = 0$. Let $j \in S \setminus \mathcal{V}$. Let us define $y \in \mathbb{R}^n_+$ by

$$y_k = \begin{cases} 1/(2|\mathcal{V}|), & \text{if } k \in \mathcal{V} \\ 1/4, & \text{if } k \in \{i, j\} \\ 0, & \text{otherwise.} \end{cases}$$

Then $y \notin C(\mathcal{L}, v)$. We show that $y \in GB(\mathcal{L}, v)$ and will get a contradiction. Let (z, T) be an objection to (y, S). Therefore, $z \in \mathbb{R}^n$ and $T \in \mathcal{L}$ satisfy $z \geq 0$, $z(T) = v(T)$, $z_k > y_k$ for $k \in T$, $z_k = y_k$ for $k \in N \setminus T$. Hence $z(T) = v(T) = 1$ and $\mathcal{V} \subset T$. Consider two possibilities:
(a) If $i \in T$ then $j \notin T$ because $z(T) = 1$. We have $z(N \setminus T) = y(N \setminus T) = 1/4$, $z(N) = 5/4$ and $z_i > y_i = 1/4$ and then $z(N \setminus i) < 1$. Thus, we can construct a counterobjection to (z, T) using a winning coalition T^* such that $i \notin T^*$.
(b) If $i \notin T$ then $S \subseteq T$ and so $j \in T$, hence $z_j > y_j = 1/4$. Therefore, it is a counterobjection to (z, T) using a winning coalition T^* such that $j \notin T^*$. □

The unanimity games on atomic convex geometries are weak games. If we consider the unanimity game $v = u_T$, with $T \in \mathcal{L}$, $T \neq \emptyset$, then:

1. $\mathcal{V} = T$,
2. $C(\mathcal{L}, v) = \operatorname{conv}\{e_i : i \in T\} = B(\mathcal{L}, v) = MB(\mathcal{L}, v) = GB(\mathcal{L}, v)$,
3. $Weber(\mathcal{L}, v) = \operatorname{conv}\{e_i : i \in \operatorname{ex}(T)\}$,
4. $C(\mathcal{L}, v) = Weber(\mathcal{L}, v) \Longleftrightarrow T = \operatorname{ex}(T) \Longleftrightarrow 2^T \subseteq \mathcal{L}$.

Theorem 10.5.3 *Let $v \in \Omega(\mathcal{L})$ be a simple game on an atomic convex geometry. If $B(\mathcal{L}, v) = Weber(\mathcal{L}, v)$, then v is a unanimity game.*

Proof. If v is not a unanimity game, then it is not quasi-convex and so $Weber(\mathcal{L}, v) \nsubseteq C(\mathcal{L}, v)$. Let

$$k \in \bigcup_{j=1}^{r} \operatorname{ex}(S_j) \setminus \bigcap_{j=1}^{r} S_j,$$

or equivalently $e_k \in Weber(\mathcal{L}, v) \setminus C(\mathcal{L}, v)$. Then, there exists a minimal winning coalition $S_{j^*} \in \mathcal{L}$ such that $k \notin S_{j^*}$. A player $i \in S_{j^*}$ can object against player k with respect to e_k using the coalition S_{j^*}. Define $y = (y_k)_{k \in S_{j^*}}$ by

$$y_t = \frac{1}{|S_{j^*}|} \qquad \text{if } t \in S_{j^*}.$$

It is clear that $y(S_{j^*}) = v(S_{j^*}) = 1$ and $y_t > 0$ for all $t \in S_{j^*}$. Then, the pair (y, S_{j^*}) is an objection of i against k with respect to

$$e_k \in Weber(\mathcal{L}, v) = B(\mathcal{L}, v).$$

We show that there exists no counterobjection to this objection. Indeed, player k cannot make a counterobjection using $R = \{k\}$ because, in this case, he/she needs $v(k) = 1$ and then

$$I(\mathcal{L}, v) = Weber(\mathcal{L}, v) = B(\mathcal{L}, v)$$

and it implies that $v = u_{\{k\}}$.

Moreover, k cannot make a counterobjection using $R \in \mathcal{L}$ with $|R| > 1$ and such that $k \in R$, $i \notin R$ because in this case, there must be a vector $z = (z_t)_{t \in R}$ such that $z(R) = v(R)$, $z_t \geq y_t > 0$ for $t \in R \cap S_{j^*}$, $z_t \geq 0$ for $t \in R \setminus S_{j^*}$, $t \neq k$ and $z_k \geq 1$ but it is impossible. □

Theorem 10.5.4 *Let $v \in \Omega(\mathcal{L})$ be a simple game on an atomic convex geometry. If $C(\mathcal{L}, v) \neq \emptyset$ and $MB(\mathcal{L}, v) = Weber(\mathcal{L}, v)$, then v is a unanimity game.*

Proof. If v is not a unanimity game, then $Weber(\mathcal{L}, v) \not\subseteq C(\mathcal{L}, v)$ and since $C(\mathcal{L}, v) \neq \emptyset$, then $C(\mathcal{L}, v) = MB(\mathcal{L}, v)$. Hence, $Weber(\mathcal{L}, v) \neq MB(\mathcal{L}, v)$. □

The converses of Theorems 10.5.3 and 10.5.4 are not true. The unanimity game $v = u_T$, with $T \in \mathcal{L}$, $T \neq \emptyset$ such that $2^T \not\subseteq \mathcal{L}$, satisfies the strict inequalities $Weber(\mathcal{L}, v) \subset B(\mathcal{L}, v)$ and $Weber(\mathcal{L}, v) \subset MB(\mathcal{L}, v)$.

Chapter 11

Voting power

The analysis of power is central in political science. In general, it is difficult to define the idea of power, but for the special case of voting power there are mathematical power indices that have been used. The first such power index was proposed by Shapley and Shubik [213] who apply the Shapley value [212] to the case of simple games. Another concept for measuring voting power was introduced by Banzhaf [8], a lawyer, whose work has appeared mainly in law journals, and whose index has been used in arguments in various legal proceedings. For a comprehensive work on the problem of measuring voting power, see Felsenthal and Machover [87]. In this chapter, we present several results about generating functions and time complexity obtained by Fernández in his Ph. D. Thesis [89].

11.1 Weighted voting games

Throughout this chapter a *simple game* is a game $v : 2^N \to \{0,1\}$, such that $v(N) = 1$ and v is nondecreasing, i.e., $v(S) \subseteq v(T)$ whenever $S \subseteq T \subseteq N$. A coalition is *winning* if $v(S) = 1$, and *losing* if $v(S) = 0$. The collection of all winning coalitions is denoted by $\mathcal{W}$. The *Shapley value* for the player $i \in N$ is defined by

$$\Phi_i(v) = \sum_{\{S \subseteq N \,:\, i \notin S\}} \frac{s!(n-s-1)!}{n!} \left(v(S \cup i) - v(S)\right),$$

where $n = |N|$, $s = |S|$. This value is an average of *marginal contributions* $v(S \cup i) - v(S)$ of the player i to all to all coalitions $S \subseteq N \setminus i$. In this value, the sets S of different size get different weight. For simple games, Shapley and Shubik [213] introduced the following power index, which is a specialization of the Shapley value.

Definition 11.1.1 *The Shapley-Shubik index of the simple game (N, v) is the vector $\Phi(v) = (\Phi_1(v), \ldots, \Phi_n(v))$, given by*

$$\Phi_i(v) = \sum_{\{S \notin \mathcal{W} : S \cup i \in \mathcal{W}\}} \frac{s!(n-s-1)!}{n!}.$$

We now define the normalized Banzhaf index. A *swing* for player i is a pair of coalitions $(S \cup i, S)$ such that $S \cup i$ is winning and S is not. For each $i \in N$, we denote by $\eta_i(v)$ the number of swings for i in game v, and the total number of swings is given by

$$\overline{\eta}(v) = \sum_{i \in N} \eta_i(v).$$

Definition 11.1.2 *The normalized Banzhaf index of the simple game (N, v) is the vector $\beta(v) = (\beta_1(v), \ldots, \beta_n(v))$, where*

$$\beta_i(v) = \frac{\eta_i(v)}{\overline{\eta}(v)}.$$

Coleman [42] considered two indices to measure the power to prevent action and the power to initiate action. In the above notation, these two Coleman indices are

$$\gamma_i(v) = \frac{\eta_i(v)}{\omega}, \quad \gamma_i^*(v) = \frac{\eta_i(v)}{\lambda},$$

where ω and λ are the total number of winning and losing coalitions, respectively.

We introduce a special class of simple games called *weighted voting games.* The symbol $[q; w_1, \ldots, w_n]$ will be used, where q and $w_1, \ldots, w_n$ are positive integers with

$$0 < q \le \sum_{i=1}^{n} w_i.$$

Here there are n players, w_i is the number of votes of player i, and q is the quota needed for a coalition to win. Then, the above symbol represents the simple game v defined by

$$v(S) = \begin{cases} 1, & \text{if } w(S) \ge q \\ 0, & \text{if } w(S) < q, \end{cases}$$

where $S \subseteq \{1, \ldots, n\}$ and $w(S) = \sum_{i \in S} w_i$. Dubey and Shapley [59] studied an axiomatization of the Banzhaf index and its behavior in weighted voting games when the number of small voters tends to infinity.

11.2 Generating functions

In order to obtain the Shapley-Shubik and the normalized Banzhaf power indices exactly, we present a combinatorial method based in the generating functions. The most useful method for counting the number of elements $f(k)$ of a finite set is to obtain its generating function. The ordinary *generating function* of $f(k)$ is the formal power series $\sum_{k\geq 0} f(k)x^k$.

This power series is called formal because we ignore the evaluation on particular values and problems on convergence (see Stanley [224] [225]). We can also work with generating functions of several variables

$$\sum_{k\geq 0}\sum_{j\geq 0}\sum_{l\geq 0} f(k,j,l)x^k x^j x^l.$$

For each $n \in \mathbb{N}$, the number of subsets of k elements of $N = \{1,\ldots,n\}$ is given by the explicit formula of the binomial coefficients

$$\binom{n}{k} = \frac{n(n-1)\cdots(n-k+1)}{k!}.$$

A generating function approach to binomial coefficients may be obtained as follows. Let $S = \{x_1, x_2, \ldots, x_n\}$ be an n-element set. Let us regard the elements $x_1, x_2, \ldots, x_n$ as independent indeterminates. It is a consequence of the process of multiplication (one could also give a rigorous proof by induction) that

$$(1+x_1)(1+x_2)\cdots(1+x_n) = \sum_{T\subseteq S}\prod_{x_i\in T} x_i.$$

Note that if $T = \emptyset$ then we obtain 1. If we put each $x_i = x$, we obtain

$$(1+x)^n = \sum_{T\subseteq S}\prod_{x\in T} x = \sum_{T\subseteq S} x^{|T|} = \sum_{k\geq 0}\binom{n}{k}x^k.$$

We now present generating functions for computing the Shapley-Shubik and the Banzhaf power indices in weighted voting games $[q;\ w_1,\ldots,w_n]$. David G. Cantor used generating functions for computing the Shapley-Shubik index for large voting games exactly. As related by Mann and Shapley [162], Cantor's contribution was the following result (see Lucas [161]): The Shapley-Shubik index of the player $i \in N$, satisfies

$$\Phi_i(v) = \sum_{\{S\notin\mathcal{W}\,:\,S\cup i\in\mathcal{W}\}} \frac{s!(n-s-1)!}{n!} = \sum_{j=0}^{n-1}\frac{j!(n-j-1)!}{n!}\left(\sum_{k=q-w_i}^{q-1} A^i(k,j)\right),$$

where $A^i(k,j)$ is the number of ways in which j players, other than i, can have a sum of weights equal to k.

Proposition 11.2.1 (Cantor) *Let $[q;\, w_1, \ldots, w_n]$ be a weighted voting game. Then the generating function of the number $A^i(k,j)$ of coalitions S of j players with $i \notin S$ and $w(S) = k$, is given by*

$$ShG_i(x,z) = \prod_{j \neq i} (1 + z\, x^{w_j})\,.$$

Proof. Let $W = \{w_1, \ldots, w_n\}$ be the set of the weights of all the players. We consider the following generating function

$$\begin{aligned}(1 + z\, x^{w_1}) \cdots (1 + z\, x^{w_n}) &= \sum_{T \subseteq W} \left(z^{|T|}\, x^{\sum_{w_i \in T} w_i} \right) \\ &= \sum_{k \geq 0} \sum_{j \geq 0} A(k,j)\, x^k z^j ,\end{aligned}$$

where the coefficient $A(k,j)$ is the number of coalitions of weight k and size j. To obtain the numbers $A^i(k,j)$, we drop the factor $(1 + z\, x^{w_i})$. □

The above approach was applied by Brams and Affuso [31] for computing the normalized Banzhaf index. The number of swings for player i satisfies

$$\eta_i(v) = |\{S \notin \mathcal{W} \,:\, S \cup i \in \mathcal{W}\}| = \sum_{k=q-w_i}^{q-1} b^i(k),$$

where $b^i(k)$ is the number of coalitions with weight k that do not include i.

Proposition 11.2.2 (Brams-Affuso) *Let $[q;\, w_1, \ldots, w_n]$ be a weighted voting game. Then the generating function of the number $b^i(k)$ of coalitions S such that $i \notin S$, and $w(S) = k$, is given by*

$$BG_i(x) = \prod_{j \neq i} (1 + x^{w_j})\,.$$

Proof. For the weights $W = \{w_1, \ldots, w_n\}$, we consider the generating function

$$\begin{aligned}(1 + x^{w_1}) \cdots (1 + x^{w_n}) &= \sum_{V \subseteq W} \prod_{w_i \in V} x^{w_i} \\ &= \sum_{V \subseteq W} \left(x^{\sum_{w_i \in V} w_i} \right) \\ &= \sum_{k \geq 0} b(k)\, x^k ,\end{aligned}$$

where $b(k)$ is the number of coalitions with weight k. To obtain the numbers $b^i(k)$, we delete the factor $(1+\,x^{w_i})$. $\square$

On the collection of simple games, we define the operation meet $\wedge$ by

$$(v_1 \wedge v_2)(S) := \min\{v_1(S), v_2(S)\}.$$

Let $v_1 = [q;\ w_1, \ldots, w_n]$, $v_2 = [p;\ p_1, \ldots, p_n]$ be weighted voting games. Then the *meet game* satisfies

$$(v_1 \wedge v_2)(S) = \begin{cases} 1, & \text{if } w(S) \geq q \text{ and } p(S) \geq p \\ 0, & \text{if } w(S) < q \text{ or } \ \ p(S) < p. \end{cases}$$

Proposition 11.2.3 *Let $v_1 = [q;\ w_1, \ldots, w_n]$, $v_2 = [p;\ p_1, \ldots, p_n]$ be weighted voting games. Then the generating function of the number $b^i(k,r)$ of coalitions S such that $i \notin S$, and $w(S) = k$, $p(S) = r$, is given by*

$$BG_i(x,y) = \prod_{j \neq i} (1+\,x^{w_j} y^{p_j}).$$

Proof. For the weights $W = \{w_1, \ldots, w_n\}$ and $P = \{p_1, \ldots, p_n\}$, we consider the generating function

$$\begin{aligned} (1+x^{w_1}y^{p_1})\cdots(1+x^{w_n}y^{p_n}) &= \sum_{V\subseteq W}\sum_{R\subseteq P}\prod_{w_i\in V} x^{w_i} \prod_{p_j\in R} y^{p_i} \\ &= \sum_{V\subseteq W}\sum_{R\subseteq P}\left(x^{\sum_{w_i\in V} w_i}\right)\left(y^{\sum_{p_j\in R} p_j}\right) \\ &= \sum_{k\geq 0}\sum_{r\geq 0} b(k,r)\, x^k y^r, \end{aligned}$$

where $b(k,r)$ is the number of coalitions $S \subseteq N$ such that $w(S) = k$, and $p(S) = r$. To obtain the numbers $b^i(k,r)$, we delete the factor $(1+\,x^{w_i}y^{p_i})$. $\square$

Proposition 11.2.4 *Let $v_1 = [q;\ w_1, \ldots, w_n]$, $v_2 = [p;\ p_1, \ldots, p_n]$ be weighted voting games. Then the generating function of the number $A^i(k,r,j)$ of coalitions S of j players such that $i \notin S$, $w(S) = k$, and $p(S) - r$, is given by*

$$ShG_i(x,y,z) = \prod_{j \neq i} (1+ z\, x^{w_j} y^{p_j}).$$

11.3 Computational complexity

The classical procedures for computing the power indices are based in the enumeration of all coalitions. Thus, if the input size of the problem is n, then the function which measures the worst case running time for computing the indices is in $O(2^n)$. In this section, we will give the algorithms based in the generating functions to obtain these power indices and we study its computational complexity. Throughout the remainder of this chapter, we will assume the algebraic or unit cost model (see page 90).

Let $f(n)$ be a function from $\mathbb{Z}_+$ to $\mathbb{Z}_+$. Recall that we denote $O(f(n))$ for the set of all functions g such that $f(n) \leq cg(n)$ for all $n \geq n_0$. With this definition a polynomial $\sum_{i=0}^{d} a_i n^i$ is in $O(n^d)$ and this means that only the asymptotic behavior of the function as $n \to \infty$ is being considered.

The programs of our language contain only *assignments* and a **for**-*loop construct*. We use the symbol $\leftarrow$ for assignments, for example, $g(x) \leftarrow 1$ denotes setting the value of $g(x)$ to 1. A **for**-loop to calculate $\sum_{i \in I} a_i$, can be defined by

$h \leftarrow 0$
for $i \in I$ **do**
  $h \leftarrow h + a_i$
endfor

Theorem 11.3.1 *Let $[q;\ w_1, \ldots, w_n]$ be a weighted voting game. If C is the number of nonzero coefficients of the function $BG(x)$, then the time complexity of the generating algorithm for the Banzhaf indices is $O(n^2 C)$.*

Proof. Let i be a player, the function $BG_i(x) = \prod_{j \neq i} (1 + x^{w_j})$ is given by
$BG(x) \leftarrow 1$
for $j \in \{1, \ldots, n\}$ **with** $j \neq i$ **do**
$BG(x) \leftarrow BG(x) + BG(x)\, x^{w_j}$
endfor

The time to compute the line in the loop is in $O(C)$ for every player. Thus the time to compute this function is $O(nC)$. We take $BG_i(x) = \sum_{k \geq 0} b^i(k)\, x^k$, for every player $i \in N$, and consider the **for** loop

$w \leftarrow w_i$
$s \leftarrow 0$
for $k \in \{q - w, \ldots, q - 1\}$ **do**
  $s \leftarrow s + b^i(k)$
endfor

The time spent in the above loop is $O(C)$, since in the sum we only consider the nonzero coefficients and the total time in the procedure is $O(nC)$. If this procedure is executed n times, we obtain the indices of the n players. □

Corollary 11.3.2 *Let $v_1 = [q;\ w_1, \ldots, w_n]$, $v_2 = [p;\ p_1, \ldots, p_n]$ be weighted voting games. If C is the number of nonzero coefficients of $BG(x, y)$, then the time complexity of the generating algorithm for the Banzhaf indices of the meet game is $O(n^2 C)$.*

Remark 11.3.1 *If the weighted n-voting game satisfies $w_i = w$, for every player $i \in N$ then the number of nonzero coefficients of $BG(x)$ is $n+1$. For weighted n-voting games such that all the sums of the weights are different, the number of nonzero coefficients of $BG(x)$ is 2^n.*

Theorem 11.3.3 *Let $[q;\ w_1, \ldots, w_n]$ be a weighted voting game. If C is the number of nonzero coefficients of $ShG(x, z)$, then the time complexity of the generating algorithm for the Shapley-Shubik indices is $O(n^2 C)$.*

Proof. The time to compute the function

$$ShG_i(x, z) = \prod_{j \neq i} (1 + z\, x^{w_j}),$$

with a **for** loop is $O(nC)$, for every player. Also, there are two independent **for** loops:

$w \leftarrow w_i$
$gg(z) \leftarrow 0$
for $k \in \{q - w, \ldots, q - 1\}$ **do**
  $gg(z) \leftarrow gg(z) + A^i(k, j)\, z^j$
endfor

Thus we obtain the polynomial $gg(z) = \sum_{j=0}^{n-1} b_j z^j$, whose coefficients appear in the next sum,

$t \leftarrow 0$
for $j \in \mathbb{Z}$ **with** $0 \leq j \leq n-1$ **do**
  $t \leftarrow t + b_j (n - j - 1)! j!$
endfor
$t/n!$

Note that the factorial function takes $O(n)$, and $n \leq C$. Thus we can calculate the index for player i in time $O(nC)$. For the n players, this procedure is executed n times. □

Corollary 11.3.4 *Let $v_1 = [q;\ w_1, \ldots, w_n]$, $v_2 = [p;\ p_1, \ldots, p_n]$ be weighted voting games. If C is the number of nonzero coefficients of $ShG(x, y, z)$, then the time complexity of the generating algorithm for the Shapley-Shubik indices of the meet game is $O(n^2 C)$.*

11.4 Algorithms with Mathematica

We present several algorithms for computing the power indices with the system Mathematica by Wolfram [248]. The procedure is similar to that presented by Tannenbaum [229], but in our algorithms we delete Expand in the definition of the generating function and we use Apply[Plus,]. Furthermore, we compute the voting power in the Council of Ministers of the European Union with the generating function algorithms. The power game of the countries in the EU Council is defined by

$$N = \{\text{GE, UK, FR, IT, SP, NE, GR, BE, PO, SW, AU, DE, FI, IR, LU}\},$$
$$v = [q; 10, 10, 10, 10, 8, 5, 5, 5, 5, 4, 4, 3, 3, 3, 2],$$

where $q = 62$ or $q = 65$. Of course the classical procedures run in time exponential 2^n, where n is the number of players. In general, we cannot hope for a polynomial time complexity for the generating function algorithms, but in many problems we obtain polynomial time whenever the number of coefficients and the maximum of the weights are polynomial in n.

The notebook of Mathematica for computing the classical index power is the following.

In[1]:=

```
votesUE={10,10,10,10,8,5,5,5,5,4,4,3,3,3,2};
```

In[2]:=

```
VUE=N[%/Plus @@ %,3]
```

Out[2]=

```
{0.115, 0.115, 0.115, 0.115, 0.092, 0.0575, 0.0575, 0.0575,
 0.0575, 0.046, 0.046, 0.0345, 0.0345, 0.0345, 0.023}
```

The function BanzhafG computes the generating function for computing the Banzhaf index of a weighted voting game, given by a list of integer weights.

In[3]:=

```
BanzhafG[weights_List]:=Times @@ (1+x^weights)
```

We can find the complexity bound C for the function BanzhafG in the EU-game, as follows:

In[4]:=

```
Length[BanzhafG[votesUE]//Expand]
```

Out[4]=

86

The function BanzhafIndexPlus computes the normalized Banzhaf index of player i by summing the appropriate coefficients in this generating function. Dividing the index of each player by the sum of all the indices gives the BanzhafPowerPlus distribution.

In[5]:=

```
BanzhafIndexPlus[i_,weights_List,q_Integer]:=
Module[{delw,sw,g,coefi},
delw=Delete[weights,i];
sw=Apply[Plus,delw];
g=BanzhafG[delw];
coefi=CoefficientList[g,x];
Apply[Plus,coefi[[Range[Max[1,q-weights[[i]]+1],Min[q,sw]]]]]]
```

In[6]:=

```
BanzhafPowerPlus[weights_List,q_Integer]:= # /(Plus @@ #)& @
Table[BanzhafIndexPlus[i,weights,q],{i,Length[weights]}]
```

In[7]:=

```
Timing[BanzhafPowerPlus[votesUE,62]]
```

Out[7]=

```
{0.824*Second, {1849/16565, 1849/16565, 1849/16565, 1849/16565,
 1531/16565, 973/16565, 973/16565, 973/16565, 973/16565,
 793/16565, 793/16565, 119/3313, 119/3313, 119/3313, 75/3313}}
```

In[8]:=

```
Ban62=N[%[[2]],3]
```

Out[8]=

```
{0.112, 0.112, 0.112, 0.112, 0.0924, 0.0587, 0.0587, 0.0587,
 0.0587, 0.0479, 0.0479, 0.0359, 0.0359, 0.0359, 0.0226}
```

In[9]:=

```
Timing[BanzhafPowerPlus[votesUE,65]]
```

Out[9]=

```
{0.824*Second, {1227/11149, 1227/11149, 1227/11149, 1227/11149,
 1033/11149, 671/11149, 671/11149, 671/11149, 671/11149,
 507/11149, 507/11149, 411/11149, 411/11149, 411/11149,
 277/11149}}
```

In[10]:=

```
Ban65=N[%[[2]],3]
```

Out[10]=

```
{0.11, 0.11, 0.11, 0.11, 0.0927, 0.0602, 0.0602, 0.0602, 0.0602,
 0.0455, 0.0455, 0.0369, 0.0369, 0.0369, 0.0248}
```

The number of coalitions of weight k and size j is the coefficient of $x^k z^j$ in the generating function ShG for the Shapley-Shubik index. The function ShPowerPlus computes the Shapley-Shubik power distribution with the implementation in Mathematica of Tannenbaum and the modifications mentioned above.

In[11]:=

```
ShG[weights_List]:=Times @@ (1+z x^weights)
```

The complexity bound C for the function ShG in the EU-game is

In[12]:=

```
Length[ShG[votesUE]//Expand]
```

Out[12]=

338

In[13]:=

```
ShPowerPlus[weights_List,q_Integer]:=
Module[{n=Length[weights],delw,sw,g,coefi,gg},
Table[delw=Delete[weights,i];
sw=Apply[Plus,delw]+1;
g=ShG[delw];
coefi=CoefficientList[g,x];
gg=Apply[Plus,coefi[[
Range[Max[1,q-weights[[i]]+1],Min[q,sw]]]]];
Sum[Coefficient[gg,z,j] j! (n-j-1)!,{j,n-1}],{i,n}]/n!]
```

In[14]:=

```
Timing[ShPowerPlus[votesUE,62]]
```

Out[14]=

```
{4.174*Second, {7/60, 7/60, 7/60, 7/60, 860/9009, 19883/360360,
 19883/360360, 19883/360360, 19883/360360, 743/16380, 743/16380,
 1588/45045, 1588/45045, 1588/45045, 932/45045}}
```

In[15]:=

```
Sh62=N[%[[2]],3]
```

Out[15]=

```
{0.117, 0.117, 0.117, 0.117, 0.0955, 0.0552, 0.0552, 0.0552,
 0.0552, 0.0454, 0.0454, 0.0353, 0.0353, 0.0353, 0.0207}
```

In[16]:=

```
Timing[ShPowerPlus[votesUE,65]]
```

Out[16]=

```
{4.229*Second, {21733/180180, 21733/180180, 21733/180180,
 21733/180180, 4216/45045, 2039/36036, 2039/36036, 2039/36036,
 2039/36036, 3587/90090, 3587/90090, 2987/90090,
 2987/90090, 2987/90090, 1667/90090}}
```

In[17]:=

```
Sh65=N[%[[2]],3]
```

Out[17]=

```
{0.121, 0.121, 0.121, 0.121, 0.0936, 0.0566, 0.0566, 0.0566,
 0.0566, 0.0398, 0.0398, 0.0332, 0.0332, 0.0332, 0.0185}
```

In[18]:=

```
TableForm[Transpose[{VUE,Ban62,Ban65,Sh62,Sh65}],
TableHeadings->{countries,{''VUE'',''Ban 62'',''Ban 65'',
''Sh 62'',''Sh 65''}}]
```

Out[18]=

Country	VUE	Ban 62	Ban 65	Sh 62	Sh 65
Germany	.115	.112	.11	.117	.121
U. Kingdom	.115	.112	.11	.117	.121
France	.115	.112	.11	.117	.121
Italy	.115	.112	.11	.117	.121
Spain	.092	.0924	.0927	.0955	.0936
Netherlands	.0575	.0587	.0602	.0552	.0566
Greece	.0575	.0587	.0602	.0552	.0566
Belgium	.0575	.0587	.0602	.0552	.0566
Portugal	.0575	.0587	.0602	.0552	.0566
Sweden	.046	.0479	.0455	.0454	.0398
Austria	.046	.0479	.0455	.0454	.0398
Denmark	.0345	.0359	.0369	.0353	.0332
Finland	.0345	.0359	.0369	.0353	.0332
Ireland	.0345	.0359	.0369	.0353	.0332
Luxembourg	.023	.0226	.0248	.0207	.0185

Table 11.1

The next table shows the *time in seconds* for the new functions and the classical algorithms BanzhafIndex (see page 283) and ShapleyValue3, based on the potential of Hart and Mas-Colell [123] which is implemented by Carter [34].

q	BanzhafIndex	BanzhafPowerPlus	ShapleyValue3	ShPowerPlus
62	6285.24	0.824	442.15	4.174
65	6050.04	0.824	464.07	4.229

Table 11.2

11.5 Power in 2-weighted voting games

To study the meet of the weighted voting games given by the votes in the EU Council and the population of the EU countries, we introduce the index PUE and the integer weights (obtained by roundoff) according to this population.

We use the following notebook of Mathematica to calculate power indices in *2-weighted voting games.*

In[1]:=

```
votesUE={10,10,10,10,8,5,5,5,5,4,4,3,3,3,2};
```

In[2]:=

```
population={80.61,57.96,57.53,56.93,39.11,15.24,10.35,
10.07,9.86,8.69,7.91,5.18,5.06,3.56,0.4};
```

In[3]:=

```
PUE=N[%/Plus @@ %,3];
```

Out[3]=

```
{0.219, 0.157, 0.156, 0.155, 0.106, 0.0414, 0.0281, 0.0273,
 0.0268, 0.0236, 0.0215, 0.0141, 0.0137, 0.00966, 0.00109}
```

The fuction Round[x] gives the integer closest to x. For numbers $x.5$ the round is x.

In[4]:=

```
popUE=Round[PUE*100]
```

Out[4]=

```
{22, 16, 16, 15, 11, 4, 3, 3, 3, 2, 2, 1, 1, 1, 0}
```

In[5]:=

```
Apply[Plus,%]
```

Out[5]=

```
100
```

First, the meet game $v_1 \wedge v_2$ in the EU Council is defined by

$$\begin{aligned} v_1 &= [62;\ 10,10,10,10,8,5,5,5,5,4,4,3,3,3,2]\,, \\ v_2 &= [p;\ 22,16,16,15,11,4,3,3,3,2,2,1,1,1,0]\,, \end{aligned}$$

where $p \in \{51, 75\}$.

Next, we define the generating function BanzhafTwoG for the meet of two weighted voting games. To obtain the normalized Banzhaf index of player i, we define the function BanzhafTwoIndex, and BanzhafTwoPower computes the vector of these indices for all players.

In[6]:=

```
BanzhafTwoG[weights_List,pop_List]:=Times @@ (1+x^weights y^pop)
```

The complexity bound C for the function BanzhafTwoG in the above two weighted voting game is given by

In[7]:=

```
Length[BanzhafTwoG[votesUE,popUE]//Expand]
```

Out[7]=

```
1644
```

In[8]:=

```
BanzhafTwoIndex[i_,weights_List,pop_List,q_Integer,p_Integer]:=
Module[{delwe,delpo,g,sw,sp,coefi,s1,s2},
delwe=Delete[weights,i];
delpo=Delete[pop,i];
g=BanzhafTwoG[delwe,delpo];
sw=Apply[Plus,delwe]+1;
sp=Apply[Plus,delpo]+1;
coefi=CoefficientList[g,{x,y}]/.{} -> Table[0,{sp}];
s1=Apply[Plus,Flatten[coefi[[
Range[Max[1,q-weights[[i]]+1],sw],
Range[Max[1,p-pop[[i]]+1],sp]]]]];
s2=If[((q+1)>sw) || ((p+1)>sp),0,Apply[Plus,
Flatten[coefi[[Range[q+1,sw],Range[p+1,sp]]]]]];
s1-s2]
```

In[9]:=

```
BanzhafTwoPower[weights_List,pop_List,q_,p_]:=# /(Plus @@ #)& @
Table[BanzhafTwoIndex[i,weights,pop,q,p],{i,Length[weights]}]
```

In[10]:=

```
Timing[BanzhafTwoPower[votesUE,popUE,62,51]]
```

Out[10]=

```
{14*Second, {1849/16565, 1849/16565, 1849/16565, 1849/16565,
 1531/16565, 973/16565, 973/16565, 973/16565, 973/16565,
 793/16565, 793/16565, 119/3313, 119/3313, 119/3313, 75/3313}}
```

In[11]:=

```
BanTwo51=N[%[[2]],3]
```

Out[11]=

```
{0.112, 0.112, 0.112, 0.112, 0.0924, 0.0587, 0.0587, 0.0587,
 0.0587, 0.0479, 0.0479, 0.0359, 0.0359, 0.0359, 0.0226}
```

In[12]:=

```
Timing[BanzhafTwoPower[votesUE,popUE,62,75]]
```

Out[12]=

```
{13.79*Second, {1013/6672, 775/6672, 775/6672, 193/1668,
 329/3336, 355/6672, 117/2224, 117/2224, 117/2224, 23/556,
 23/556, 33/1112, 33/1112, 33/1112, 125/6672}}
```

In[13]:=

```
BanTwo75=N[%[[2]],3]
```

Out[13]=

```
{0.152, 0.116, 0.116, 0.116, 0.0986, 0.0532, 0.0526, 0.0526,
 0.0526, 0.0414, 0.0414, 0.0297, 0.0297, 0.0297, 0.0187}
```

For computing the Shapley-Shubik index for 2-weighted voting games, the functions are denoted by ShTwoG and ShTwoPower. These functions are defined as follows.

In[14]:=

```
ShTwoG[weights_List,pop_List]:=Times @@ (1+x^weights y^pop z)
```

The complexity bound C for the function ShTwoG in the European Union two weighted voting game is

In[15]:=

```
Length[ShTwoG[votesUE,popUE]//Expand]
```

Out[15]=

2206

In[16]:=

```
ShTwoPower[weights_List,pop_List,q_Integer,p_Integer]:=
Module[{n=Length[weights],delwe,delpo,g,sw,sp,coefi,s1,s2,gg},
Table[delwe=Delete[weights,i];
delpo=Delete[pop,i];
g=ShTwoG[delwe,delpo];
sw=Apply[Plus,delwe]+1;
sp=Apply[Plus,delpo]+1;
coefi=CoefficientList[g,{x,y}]/.{} -> Table[0,{sp}];
s1=Apply[Plus,Flatten[coefi[
[Range[Max[1,q-weights[[i]]+1],sw],
Range[Max[1,p-pop[[i]]+1],sp]]]]];
s2=If[((q+1)>sw) || ((p+1)>sp),0,
Apply[Plus,Flatten[coefi[
[Range[q+1,sw],Range[p+1,sp]]]]]];
gg=s1-s2;
Sum[Coefficient[gg,z,j] j! (n-j-1)!,{j,0,n-1}]/n!,{i,n}]]
```

In[17]:=

```
Timing[ShTwoPower[votesUE,popUE,62,51]]
```

Out[17]=

```
{28.45*Second, {7/60, 7/60, 7/60, 7/60, 860/9009, 19883/360360,
 19883/360360, 19883/360360, 19883/360360, 743/16380, 743/16380,
 1588/45045, 1588/45045, 1588/45045, 932/45045}}
```

In[18]:=

```
ShTwo51=N[%[[2]],3]
```

Out[18]=

```
{0.117, 0.117, 0.117, 0.117, 0.0955, 0.0552, 0.0552, 0.0552,
 0.0552, 0.0454, 0.0454, 0.0353, 0.0353, 0.0353, 0.0207}
```

In[19]:=

```
Timing[ShTwoPower[votesUE,popUE,62,75]]
```

Out[19]=

```
{27.956*Second, {607/3003, 4835/36036, 4835/36036, 1202/9009,
 5561/45045, 1427/32760, 13207/360360, 13207/360360,
 13207/360360, 1597/60060, 1597/60060, 487/25740,
 487/25740, 487/25740, 829/90090}}
```

In[20]:=

```
ShTwo75=N[%[[2]],3]
```

Out[20]=

```
{0.202, 0.134, 0.134, 0.133, 0.123, 0.0436, 0.0366, 0.0366,
 0.0366, 0.0266, 0.0266, 0.0189, 0.0189, 0.0189, 0.0092}
```

In[21]:=

```
TableForm[Transpose[{PUE,BanTwo51,BanTwo75,ShTwo51,ShTwo75}],
TableHeadings->{countries,{''PUE'',''2Ban 51'',''2Ban 75'',
''2Sh 51'',''2Sh 75''}}]
```

Out[21]=

Country	PUE	2Ban 51	2Ban 75	2Sh 51	2Sh 75
Germany	.219	.112	.152	.117	.202
U. Kingdom	.157	.112	.116	.117	.134
France	.156	.112	.116	.117	.134
Italy	.155	.112	.116	.117	.133
Spain	.106	.0924	.0986	.0955	.123
Netherlands	.0414	.0587	.0532	.0552	.0436
Greece	.0281	.0587	.0526	.0552	.0366
Belgium	.0273	.0587	.0526	.0552	.0366
Portugal	.0268	.0587	.0526	.0552	.0366
Sweden	.0236	.0479	.0414	.0454	.0266
Austria	.0215	.0479	.0414	.0454	.0266
Denmark	.0141	.0359	.0297	.0353	.0189
Finland	.0137	.0359	.0297	.0353	.0189
Ireland	.00966	.0359	.0297	.0353	.0189
Luxembourg	.00109	.0226	.0187	.0207	.0092

Table 11.3

Chapter 12

Computing values with Mathematica

The package *DiscreteMath'Combinatorica'* extends the program Mathematica to combinatorics and graph theory. The best guide for this package is the book by Skiena [221]. The package *Cooperative* included in Carter [34] presents tools for solving cooperative games, including solution concepts as the core, the Shapley value and the nucleolus.

12.1 Computing values of restricted games

Let $(N, \mathcal{L})$ be a partition convex geometry and let (N, v) be a game. In this section we obtain algorithms to compute the Shapley and Banzhaf values of the players in the restricted game $v^{\mathcal{L}}$, in terms of the original game v. First, we need to obtain the following families of convex sets

$$\begin{aligned}
\mathcal{L}_i &= \{T \in \mathcal{L} : i \in T\}, \\
\mathcal{L}_i^+ &= \{T \in \mathcal{L} : i \in \mathrm{ex}(T),\ (T \setminus i)^+ = T^+\}, \\
\mathcal{L}_i^\star &= \{T \in \mathcal{L} : i \notin T,\ T \cup i \in \mathcal{L},\ T^+ \neq (T \cup i)^+\},
\end{aligned}$$

for all $i \in N$. Next, we apply the formulas proved in Theorem 5.5.3.

In the following notebook, we use the function ShapleyValue3 for computing the Shapley value for games (N, v) and we define the functions GeomShapley and GeomBanzhaf for obtaining the Shapley and Banzhaf values for restricted games $(N, v^{\mathcal{L}})$.

We introduce the following packages:

In[1]:=

```
<<DiscreteMath`Combinatorica`
```

In[2]:=

```
<<Cooperat`Cooperat`
```

We will now compute the voting power in the Spanish parliament (1996-2000). The players are the following political parties: PP(1), PSOE(2), IU(3), CiU(4), PNV(5), CC(6), BNG(7), HB(8), ERC(9), EA(10) and UV(11). The power game of the parties is defined by

$$[176;\ 156, 141, 21, 16, 5, 4, 2, 2, 1, 1, 1],$$

where the quota needed for a coalition to win is $q = 176$.

In[3]:=

```
SpanishParliament[q_Integer]:=(Clear[T,w,p,v];
T=Range[11];
w[1]:=156; w[2]:=141; w[3]:=21; w[4]:=16; w[5]:=5;
w[6]:=4; w[7]:=2; w[8]:=2;w[9]:=1; w[10]:=1; w[11]:=1;
p[S_List]:=Apply[Plus,w/@ S]; v[{}]:=0;
v[S_/; p[S]>=q]:=1; v[S_/; p[S]<q]:=0;)
```

In the next input, the function N[*expr*, *n*] gives the numerical value of *expr* with *n*-digit precision.

In[4]:=

```
N[ShapleyValue3[SpanishParliament[176]],3]
```

Out[4]=

```
{0.452,0.184,0.184,0.119,0.0172,0.0172,
 0.00804,0.00804,0.00368,0.00368,0.00368}
```

We consider the convex geometry $\mathcal{L} = Co(\mathbf{11})$, that is, the convex sets comprise the political parties that are adjacent to one another in the left-right ordering given by

HB—BNG—ERC—IU—PSOE—EA—PNV—CiU—PP—CC—UV.

In[5]:=

```
ShowLabeledGraph[Path[11],{''HB'',''BNG'',''ERC'',''IU'',
''PSOE'',''EA'',''PNV'',''CiU'',''PP'',''CC'',''UV''}];
```

Figure 12.1 Linear order in the Spanish parliament

In[6]:=

```
graph=FromUnorderedPairs[{{8,7},{7,9},{9,3},{3,2},{2,10},
  {10,5},{5,4},{4,1},{1,6},{6,11}}];
```

The family of convex sets $\mathcal{L}$ is denoted by F in the next Input.

In[7]:=

```
Clear[F];
F=Select[Rest[Subsets[Range[V[graph]]]],
 (ConnectedQ[InduceSubgraph[graph,#]])&];
```

In[8]:=

```
SupInt[S_List,F_List]:=Select[T,(MemberQ[F,Union[S,{#}]])&]
```

In[9]:=

```
Extreme[S_List,F_List]:=Select[S,(MemberQ[F,
       Complement[S,{#}]])&]
```

In[10]:=

```
InfInt[S_List,F_List]:=Complement[S,Extreme[S,F]]
```

In[11]:=

```
Geom1[i_,F_List]:=Geom1[i,F]=Select[F,(MemberQ[#,i])&]
```

In[12]:=

```
Geom2[i_,F_List]:=Geom2[i,F]=
Select[F,((Intersection[{i},Extreme[#,F]]=={i})
&& (SupInt[Complement[#,{i}],F] == SupInt[#,F]))&]
```

In[13]:=

```
Geom3[i_,F_List]:=Geom3[i,F]=
Select[F,((DeleteCases[#,i] == #)
&& (MemberQ[F,Union[#,{i}]])
&& (SupInt[#,F] != SupInt[Union[#,{i}],F]))&]
```

In[14]:=

```
co1[S_List]:=co1[S]=(Length[S]-1)!
*(Length[SupInt[S,F]]-Length[S])!/Length[SupInt[S,F]]!;
```

In[15]:=

```
co2[S_List]:=co2[S]=co1[S]*Length[S]
/(Length[SupInt[S,F]]-Length[S]);
```

In[16]:=

```
ba[S_List]:=ba[S]=2^(1-Length[SupInt[S,F]])
```

In[17]:=

```
GeomShapley[game_:Null]:=Module[{value},
value=Table[Apply[Plus,If[#=={},0,
co1[#]*(v[#]-v[DeleteCases[#,i]])]& /@ Geom2[i,F]]
+ Apply[Plus,If[#=={},0,co1[#]
* v[#]]&/@ Complement[Geom1[i,F],Geom2[i,F]]]
- Apply[Plus,If[#=={},0,co2[#]*v[#]]& /@ Geom3[i,F]],
{i,Length[T]}];Return[value]];
```

In[18]:=

```
GeomBanzhaf[game_:Null]:=Module[{value},
value=Table[Apply[Plus,If[#=={},0,
ba[#]*(v[#]-v[DeleteCases[#,i]])]& /@ Geom2[i,F]]
+ Apply[Plus,If[#=={},0,ba[#]
* v[#]]&/@ Complement[Geom1[i,F],Geom2[i,F]]]
- Apply[Plus,If[#=={},0,ba[#]*v[#]]& /@ Geom3[i,F]],
{i,Length[T]}];Return[value]];
```

In[19]:=

```
N[GeomShapley[SpanishParliament[176]],3]
```

Out[19]=

```
{0.25,0.0333,0.0333,0.45,0.117,0.0833,0,0,0,0.0333,0}
```

In[20]:=

```
N[GeomBanzhaf[SpanishParliament[176]],3]
```

Out[20]=

```
{0.344,0.0312,0.0312,0.406,0.156,0.125,0,0,0,0.0312,0}
```

A new voting game in the Spanish parliament is defined by

$$[176;\ 156(\text{PP}),\ 141(\text{PSOE}),\ 21(\text{IU}),\ 16(\text{CiU}),\ 6(\text{PNV-EA}),\ 4(\text{CC})].$$

In[21]:=

```
SpanishParties[q_Integer]:=(Clear[T,w,p,v];
T=Range[6];
w[1]:=156; w[2]:=141; w[3]:=21; w[4]:=16; w[5]:=6; w[6]:=4;
p[S_List]:=Apply[Plus, w/@S]; v[{}]:=0;
v[S_/; p[S]>=q]:=1; v[S_/; p[S]<q]:=0;)
```

In[22]:=

```
N[ShapleyValue3[SpanishParties[176]],3]
```

Out[22]=

```
{0.467,0.183,0.183,0.133,0.0167,0.0167}
```

The feasible coalitions are the connected subgraphs of the following graph.

In[23]:=

```
graph1=FromUnorderedPairs[
      {{1,4},{1,6},{4,5},{4,6},{5,3},{5,2},{2,3}}];
```

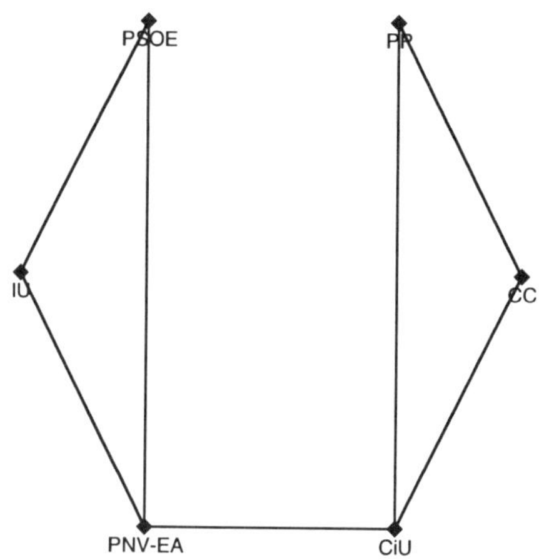

Figure 12.2

In[24]:=

```
Clear[F];
F=Select[Rest[Subsets[Range[V[graph1]]]],
  (ConnectedQ[InduceSubgraph[graph1,#]])&];
```

In[25]:=

```
N[GeomShapley[SpanishParties[176],3]
```

Out[25]=

```
{0.217,0.05,0.05,0.467,0.133,0.0833}
```

In[26]:=

```
N[GeomBanzhaf[SpanishParties[176],3]
```

Out[26]=

```
{0.312,0.0625,0.0625,0.437,0.187,0.125}
```

12.2 Computing convex Shapley values

We present a notebook of the program Mathematica for computing the Shapley value of games $v : \mathcal{L} \to \mathbb{R}$, where $\mathcal{L} \subseteq 2^N$ is a convex geometry (see Bilbao [11]).

First, we introduce the following packages:

In[1]:=

```
<<DiscreteMath`Combinatorica`
```

In[2]:=

```
<<Cooperat`Cooperat`
```

We use the function ShapleyValue3 for computing the ordinary Shapley value and we present the function ConvexShapleyValueF for the Shapley value on a convex geometry F.

Let $F = Co\left(2^2\right)$ the convex geometry of the convex sets of the boolean algebra $2^{\mathbf{2}}$.

In[3]:=

```
F={{},{1},{2},{3},{4},{1,2},{1,3},{2,3},
   {2,4},{3,4},{1,2,3},{2,3,4},{1,2,3,4}};
```

Now, we define the lattice of the convex sets with the function

In[4]:=

```
lattice[F_List]:=
MakeGraph[F,((Intersection[#2,#1]==#1) && (#1!=#2))&];
```

In[5]:=

```
labeling[F_]:=
InputForm[ShowLabeledGraph[HasseDiagram[lattice[F]],F]]
```

In[6]:=

```
labelF=labeling[F];
```

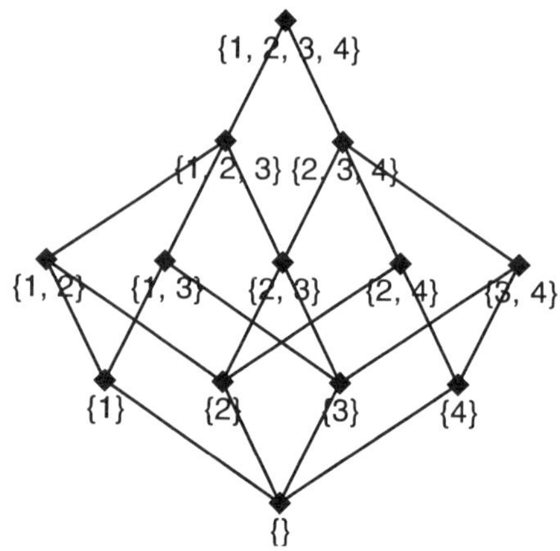

Figure 12.3 The convex geometry $Co\left(2^{\mathbf{2}}\right)$

We obtain the relationship between the convex sets and the vertices of the Hasse diagram with

In[7]:=

```
rulesF=Map[(Part[Table[labelF[[1,1,i,1]],
{i,Length[labelF[[1,1]]]-Length[F]+1,Length[labelF[[1,1]]]}],#]
-> Part[Table[i,{i,Length[F]}],#])&, Range[Length[F]]]
```

Out[7]=

```
{{}->1,{1}->2,{2}->3,{3}->4,{4}->5,{1,2}->6,{1,3}->7,{2,3}->8,
{2,4}->9,{3,4}->10,{1,2,3}->11,{2,3,4}->12,{1,2,3,4}->13}
```

We use the function GraphPower (Skiena [221, page 230]) to obtain the matrix of the maximal chains $c([T,S])$, where $c([S,S]) = 1$.

In[8]:=

```
chainMatrixF=Edges[GraphPower[HasseDiagram[lattice[F]],
Length[F]]] + IdentityMatrix[Length[F]];
```

In[9]:=

```
chainMatrixF[[1]]
```

Out[9]=

```
{0,1,1,1,1,2,2,2,2,2,6,6,12}
```

In[10]:=

```
chainNumberF[x_,y_]:=chainMatrixF[[x/.rulesF,y/.rulesF]]
```

In[11]:=

```
chainNumberF[{},Last[F]]
```

Out[11]=

```
12
```

In[12]:=

```
Extreme[S_List,F_List]:=Sort[Select[S,
(MemberQ[F,Complement[S,{#}]])&]]
```

In[13]:=

```
Extreme[Last[F],F]
```

Out[13]=

```
{1,4}
```

The family of all convex sets S such that $i \in \text{ex}(S)$ is given by

In[15]:=

```
Convex[i_,F_List]:=Sort[Select[F,
   (Intersection[{i},Extreme[#,F]]=={i})&]]
```

In[16]:=

```
Convex[2,F]
```

Out[16]=

```
{{2},{1,2},{2,3},{2,4},{1,2,3},{2,3,4}}
```

The convex Shapley value of the game $v : \mathcal{L} \to \mathbb{R}$ is given by, for all $i \in N$,

$$\Phi_i(v) := \sum_{\{S \in \mathcal{L}:\ i \in \text{ex}(S)\}} \frac{c(S \setminus i)\, c([S, N])}{c(N)} \left(v(S) - v(S \setminus i)\right).$$

In[17]:=

```
ConvexShapleyValueF[game_,i_]:=Apply[Plus,If[#=={},0,
(chainNumberF[#,T]*chainNumberF[{},DeleteCases[#,i]]
/chainNumberF[{},T])*(v[#]-v[DeleteCases[#,i]])]&/@ Convex[i,F]]
```

In[18]:=

```
Attributes[ConvexShapleyValueF]={Listable};
```

In[19]:=

```
ConvexShapleyValueF[game_]:=ConvexShapleyValueF[game,T]
```

We compute the ordinary Shapley value and the convex Shapley value in the following games.

In[20]:=

```
Game1:=(T=Range[4]; Clear[x,y,v];
x[1]:=18; x[2]:= 159; x[3]:=17; x[4]:=141;
y[S_List]:=Plus @@ x/ @ S; v[{}]:=0;
v[S_/;y[S]>=176]:=1; v[S_/;y[S]<176]:=0);
```

In[21]:=

```
ShapleyValue3[Game1]
```

Out[21]=

```
{1/6,1/2,1/6,1/6}
```

In[22]:=

```
ConvexShapleyValueF[Game1]
```

Out[22]=

```
{1/12,2/3,1/6,1/12}
```

In[23]:=

```
Game2:=(T=Range[4];Clear[v];v[{}]:=0;
v[S_]:=If[Length[S]>=3,1,0]);
```

In[24]:=

```
ShapleyValue3[Game2]
```

Out[24]=

```
{1/4,1/4,1/4,1/4}
```

In[25]:=

```
ConvexShapleyValueF[Game2]
```

Out[25]=

```
{1/6,1/3,1/3,1/6}
```

In[26]:=

```
ShapleyValue3[CorporationGame]
```

Out[26]=

```
{1/12,1/4,1/4,5/12}
```

In[27]:=

```
ConvexShapleyValueF[CorporationGame]
```

Out[27]=

```
{1/6,1/4,1/4,1/3}
```

We will also analyze Example 1 of Faigle and Kern [78].

In[28]:=

```
F:={{},{a},{b},{a,b},{b,d},{a,b,c},{a,b,d},{a,b,c,d}};
```

In[29]:=

```
labelF=labeling[F];
```

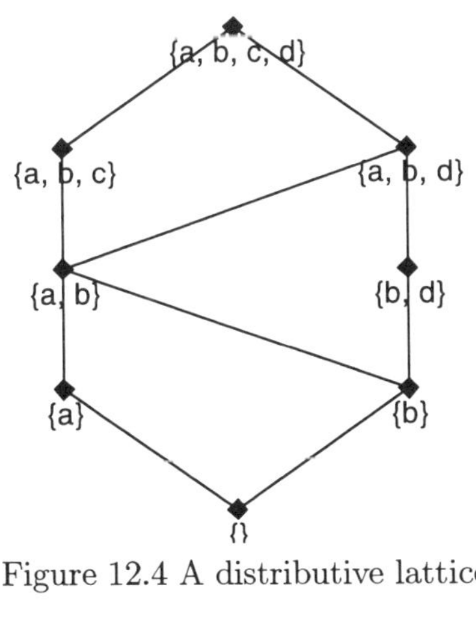

Figure 12.4 A distributive lattice

In[30]:=

```
rulesF
```

Out[30]=

```
{{}->1,{a}->2,{b}->3,{a,b}->4,{b,d}->5,
 {a,b,c}->6,{a,b,d}->7,{a,b,c,d}->8}
```

In[31]:=

```
chainMatrixF//MatrixForm
```

Out[31]=

$$\begin{matrix} 0 & 1 & 1 & 2 & 1 & 2 & 3 & 5 \\ 0 & 0 & 0 & 1 & 0 & 1 & 1 & 2 \\ 0 & 0 & 0 & 1 & 1 & 1 & 2 & 3 \\ 0 & 0 & 0 & 0 & 0 & 1 & 1 & 2 \\ 0 & 0 & 0 & 0 & 0 & 0 & 1 & 1 \\ 0 & 0 & 0 & 0 & 0 & 0 & 0 & 1 \\ 0 & 0 & 0 & 0 & 0 & 0 & 0 & 1 \\ 0 & 0 & 0 & 0 & 0 & 0 & 0 & 0 \end{matrix}$$

In[32]:=

```
Game3:=(T=Range{a,b,c,d}; Clear[v];v[{}]:=0;
        v[S_]:=If[Length[S]>=3,1,0]);
```

In[33]:=

```
ConvexShapleyValueF[Game3]
```

Out[33]=

```
{1/5,0,2/5,2/5}
```

12.3 Computing convex Banzhaf indices

First, we define the following functions for calculating the normalized Banzhaf indices for games on 2^N.

```
ba[S_List]:=1;
```

```
Banzhaf[game_, i_]:=Plus @@ (ba[#]& /@
Select[Coalitions,(v[#]==1 && v[DeleteCases[#,i]]==0)&])
```

```
Attributes [Banzhaf]={Listable};
```

```
Banzhaf[game_]:=Banzhaf[game,T]
```

```
BanzhafIndex[game_]:=(Banzhaf[game])/(Plus @@ Banzhaf[game])
```

We present a *notebook* which contains the functions ConvexBanzhafIndex and ConvexBanzhafProb for games on a convex geometry. The family of convex sets $\mathcal{L} \subseteq 2^N$ is denoted by F in *Input[3]*.

In[1]:=

```
<<DiscreteMath'Combinatorica'
```

In[2]:=

```
<<Cooperat'Cooperat'
```

In[3]:=

```
F={{},{1},{2},{3},{4},{5},{1,2},{2,3},{3,4},{4,5},{1,2,3},
   {2,3,4},{3,4,5},{1,2,3,4},{2,3,4,5},{1,2,3,4,5}};
```

In[4]:=

```
lattice[F_List]:=MakeGraph[F,
((Intersection[#2,#1]==#1) && (#1!=#2))&];
```

In[5]:=

```
ShowLabeledGraph[HasseDiagram[lattice[F]],F];
```

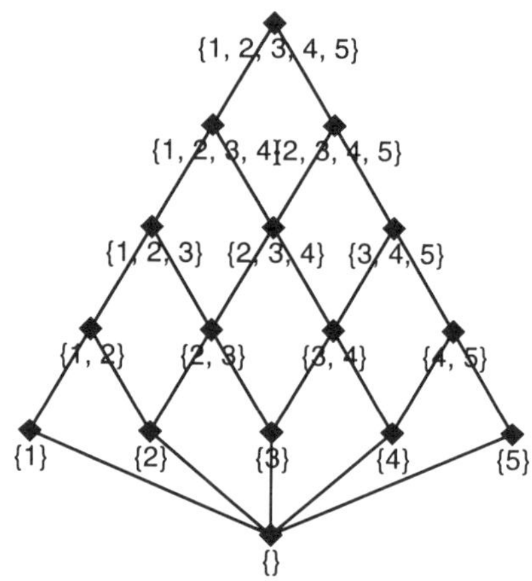

Figure 12.5 The convex geometry $Co(\mathbf{5})$

In[6]:=

```
Extreme[S_List,F_List]:=Sort[Select[S,
  (MemberQ[F,Complement[S,{#}]])&]]
```

In[7]:=

```
Extreme[{1,2,3},F]
```

Out[7]=

```
{1,3}
```

In[8]:=

```
Convex[i_,F_List]:=Sort[Select[F,
    (Intersection[{i},Extreme[#,F]]=={i})&]]
```

In[9]:=

```
Convex[1,F]
```

Out[9]=

```
{{1},{1,2},{1,2,3},{1,2,3,4},{1,2,3,4,5}}
```

In[10]:=

```
ConvexBanzhaf[game_,i_]:=Length[Select[Convex[i,F],
(v[#]==1 && v[DeleteCases[#,i]==0)&]]
```

In[11]:=

```
Attributes[ConvexBanzhaf]={Listable};
```

In[12]:=

```
ConvexBanzhaf[game_]:=ConvexBanzhaf[game,T]
```

In[13]:=

```
ConvexBanzhafIndex[game_]:=ConvexBanzhaf[game_]
/(Plus @@ ConvexBanzhaf[game])
```

In[14]:=

```
ConvexBanzhafProb[game_,i_]:=Length[Select[Convex[i,F],
(v[#]==1&&v[DeleteCases[#,i]]==0)&]]/Length[Convex[i,F]
```

In[15]:=

```
Attributes[ConvexBanzhafProb]={Listable};
ConvexBanzhafProb[game_]:=ConvexBanzhafProb[game,T]
```

We will analyze two voting games, the first is the simple majority rule in the Spanish parliament (176 out of 350 votes), the second is the simple majority rule in the Catalan parliament (68 out of 135 votes). In both cases, the convex sets comprise parties that are adjacent to one another in the left-right ordering. First, we obtain the convex Banzhaf indices for the 11 political parties with seats in the present Spanish parliament (1996–2000). The voting game is defined by

[176; 156(PP), 141(PSOE), 21(IU), 16(CiU), 5(PNV), 4(CC), 2(BNG), 2(HB), 1(ERC), 1(EA), 1(UV)],

and the ideological positions in the left-right ordering are

HB—BNG—ERC—EA—IU—PSOE—PNV—CiU—PP—CC—UV

In[16]:=

```
Game1:=(T=Range[11]; Clear[x,y,v]; x[1]:=156;
x[2]:=141; x[3]:=21; x[4]:=16; x[5]:=5; x[6]:=4;
x[7]:=2; x[8]:=2; x[9]:=1; x[10]:=1 ;x[11]:=1;
y[S_List]:=Plus @@ x/@S; v[{}]:=0;
v[S_/; y[S]>=176]:=1; v[S_/; y[S]<176]:=0);
```

In[17]:=

```
Banzhaf[Game1]
```

Out[17]=

```
{743,281,281,231,25,25,15,15,7,7,7}
```

In[18]:=

```
BanzhafIndex[Game1]
```

Out[18]=

```
{743/1637,281/1637,281/1637,231/1637,25/1637,
 25/1637,15/1637,15/1637,7/1637,7/1637,7/1637}
```

In[19]:=

```
ConvexBanzhaf[Game1]
```

Out[19]=

```
{2,0,1,7,1,1,0,0,0,0,0}
```

In[20]:=

```
ConvexBanzhafIndex[Game1]
```

Out[20]=

```
{1/6,0,1/12,7/12,1/12,1/12,0,0,0,0,0}
```

In[21]:=

```
ConvexBanzhafProb[Game1]
```

Out[21]=

```
{2/11,0,1/11,7/11,1/11,1/11,0,0,0,0,0}
```

The voting game in the Catalan parliament is defined by

$$[68;\ 60(\mathrm{CiU}),\ 34(\mathrm{PSC}),\ 17(\mathrm{PP}),\ 13(\mathrm{ERC}),\ 11(\mathrm{IC\text{-}LV})],$$

and the ideological positions in the left-right ordering are

$$\mathrm{ERC—IC/LV—PSC—CiU—PP}$$

In[22]:=

```
CatalanGame:=(T=Range[5];Clear[x,y,v];
x[1]:=60; x[2]:=34; x[3]:=17; x[4]:=13; x[5]:=11;
y[S_List]:=Plus @@ x/@S; v[{}]:=0;
v[S_/; y[S]>=68]:=1; v[S_/; y[S]<68]:=0);
```

In[23]:=

```
BanzhafIndex[CatalanGame]
```

Out[23]=

```
{7/11,1/11,1/11,1/11,1/11}
```

In[24]:=

```
ConvexBanzhafIndex[CatalanGame]
```

Out[24]=

```
{2/3,1/6,1/6,0,0}
```

In[25]:=

```
ConvexBanzhafProb[CatalanGame]
```

Out[25]=

```
{4/5,1/5,1/5,1/5,0,0}
```

12.4 The Banzhaf power in the European Union

We will now study the voting power in the Council of Ministers of the European Union. Herne and Nurmi [125], Widgrén [247] and Lane and Mæland [152] have studied the power indices (Shapley-Shubik and Banzhaf) in the EU Council. The *Journal of Theoretical Politics* 11 (1999) has recently published several articles about the modelling of decision making in the European Union. Voting power indices are defined to measure a priori voting power of players. Garrett and Tsebelis [103] [104] critized the classical voting power method in the context of the EU because the power indices do not take into account the preferences of players and the institutional structure of the EU. Lane and Berg [153] assert that

> "Cooperative theory solution concepts, on the other hand, assume that players have specified preferences and can make the types of binding commitments typically requiered to cement together a particular coalition in support of a particular outcome. Cooperative solutions, including power indices, therefore, are directly applicable to policy analysis only in conjunction with assumptions about preferences (spatial or otherwise) and in circumstances that suggest that binding agreements may be feasible."

Holler and Widgrén [128] propose some ideas for combining spatial voting games and power index models. Steunenberg, Schmidtchen and Koboldt [226], define the strategic power index, a new approach which is based on spatial and sequential models of decision making. In our model, we consider that preferences of players are modeled by the convex geometry of the connected subgraphs of the *communication situation* defined by the graph in Figure 12.6. There are two blocks containing the following countries:

$$\{\text{GE, BE, NE, LU, AU, SW, FI, DE}\} \quad \text{and} \quad \{\text{FR, IT, SP, GR, PO, IR}\},$$

and a bridge between them with {IR, UK, DE}.

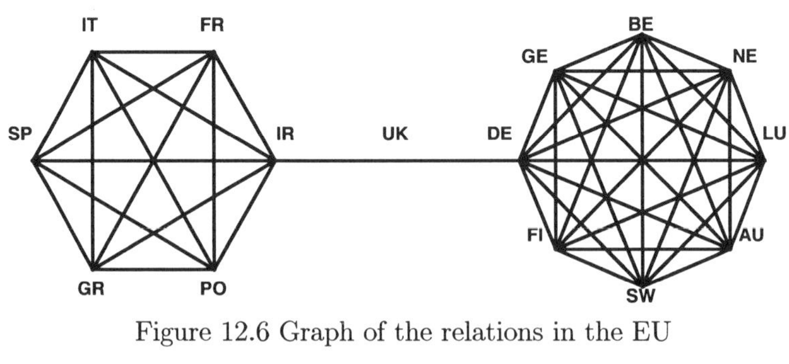

Figure 12.6 Graph of the relations in the EU

The following Table 12.1 contains the population, votes, normalized and convex Banzhaf power indices. The majority voting rule is assumed to be 62 and 65 out of 87 votes in the EU Council.

Country	Population	Votes	B 62	B 65	CoB 62	CoB 65
Germany	.219	.115	.112	.110	.140	.140
U. Kingdom	.157	.115	.112	.110	0	0
France	.156	.115	.112	.110	.140	.140
Italy	.155	.115	.112	.110	.140	.140
Spain	.106	.092	.092	.093	.113	.112
Netherlands	.041	.058	.059	.060	.069	.071
Greece	.028	.058	.059	.060	.069	.071
Belgium	.027	.058	.059	.060	.069	.071
Portugal	.027	.058	.059	.060	.069	.071
Sweden	.024	.046	.048	.045	.060	.057
Austria	.022	.046	.048	.045	.060	.057
Denmark	.014	.035	.036	.037	0	0
Finland	.014	.035	.036	.037	.043	.043
Ireland	.010	.035	.036	.037	0	0
Luxembourg	.001	.023	.023	.025	.025	.026

Table 12.1

12.5 Computing the number of subtrees

The following generating function for computing the subtrees of a tree was obtained by Jamison [132] [133]. We present a notebook of Mathematica for computing the number of subtrees t_k of a tree (V, T), where $1 \leq k \leq |V|$. First, we introduce the following combinatorial package:

In[1]:=

```
<<DiscreteMath`Combinatorica`
```

In[2]:=

```
edges={{1,2},{2,3},{2,4},{2,5},{2,6},{2,7},{1,8},
{8,9},{8,10},{8,11},{8,12},{8,13},{1,14},{14,15},
{14,16},{14,17},{14,18},{14,19}};
```

In[3]:=

```
tree=FromUnorderedPairs[edges];
ShowLabeledGraph[tree];
```

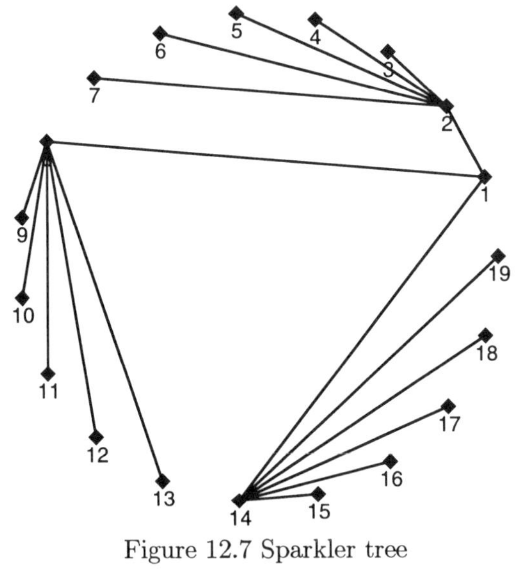

Figure 12.7 Sparkler tree

Next, we define the functions f[i] and SubtreesNumber.

In[4]:=

```
f[{i_}]:=f[{i}]=Module[{A,adjacents,factor,value},
A=Select[edges,(MemberQ[#,i])&];
adjacents=Complement[Flatten[A],{i}];
edges=Complement[edges,A];
factor=(1+f[#])& /@ Partition[adjacents,1];
value=x*Apply[Times,factor];
factor=. & /@ Partition[adjacents,1];
Return[value]];
```

In[5]:=

```
f[{1}]
```

Out[5]=

```
x (1 + x (1 + x)^5)^3
```

In[6]:=

```
RankPoly=Expand[Table[{i,f[{i}]},{i,V[tree]}]]//TableForm
```

Out[6]=

```
1     x + 3 x^2 + 18 x^3 + 61 x^4 + 180 x^5 + 480 x^6  +

      1088 x^7 + 2121 x^8 + 3633 x^9 + 5365 x^10  +

      6570 x^11 + 6465 x^12 + 5008 x^13 + 3003 x^14 +

      1365 x^15 + 455 x^16 + 105 x^17 + 15 x^18 + x^19


2     x + 5 x^2 + 10 x^3 + 10 x^4 + 5 x^5 + x^6


3     x

4     x

5     x
```

```
6    x

7    x

8    x + 5 x^2 + 10 x^3 + 10 x^4 + 5 x^5 + x^6

9    x

10   x

11   x

12   x

13   x

14   x + 5 x^2 + 10 x^3 + 10 x^4 + 5 x^5 + x^6

15   x

16   x

17   x

18   x

19   x
```

In[7]:=

```
SubtreesNumber=Rest[CoefficientList[Apply[Plus,
Table[RankPoly[[1,i,2]],{i,V[tree]}]],x]]
```

Out[7]=

```
{19, 18, 48, 91, 195, 483, 1088, 2121, 3633, 5365,
 6570, 6465, 5008, 3003, 1365, 455, 105, 15, 1}
```

We remove the generating funcion and we introduce a new tree, known as the (5,7) Baton tree.

In[8]:=

```
Clear[edges,f];
edges={{1,2},{2,3},{3,4},{4,5},{5,6},{6,7},
{7,8},{8,9},{9,10},{9,11},{9,12},{9,13},
{9,14},{1,15},{1,16},{1,17},{1,18},{1,19}};
```

In[9]:=

```
tree=FromUnorderedPairs[edges];
ShowLabeledGraph[tree];
```

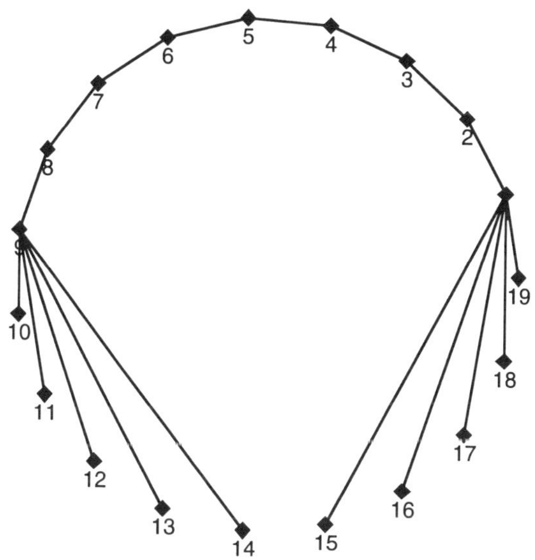

Figure 12.8 Baton tree

In[10]:= Go to *In[4]* and Return

In[11]:=

```
RankPoly=Expand[Table[{i,f[{i}]},{i,V[tree]}]]//TableForm
```

Out[11]=

```
1    x + 6 x^2 + 16 x^3 + 26 x^4 + 31 x^5 + 32 x^6 +
     32 x^7 + 32 x^8 + 32 x^9 + 36 x^10 + 61 x^11 +
     126 x^12 + 211 x^13 + 252 x^14 + 210 x^15 +
     120 x^16 + 45 x^17 + 10 x^18 + x^19

2    x + x^2 + x^3 + x^4 + x^5 + x^6 + x^7 + x^8 +
     5 x^9 + 10 x^10 + 10 x^11 + 5 x^12 + x^13

3    x + x^2 + x^3 + x^4 + x^5 + x^6 + x^7 + 5 x^8 +
     10 x^9 + 10 x^10 + 5 x^11 + x^12

4    x + x^2 + x^3 + x^4 + x^5 + x^6 + 5 x^7 +
     10 x^8 + 10 x^9 + 5 x^10 + x^11

5    x + x^2 + x^3 + x^4 + x^5 + 5 x^6 + 10 x^7 +
     10 x^8 + 5 x^9 + x^10

6    x + x^2 + x^3 + x^4 + 5 x^5 + 10 x^6 + 10 x^7 +
     5 x^8 + x^9

7    x + x^2 + x^3 + 5 x^4 + 10 x^5 + 10 x^6 +
     5 x^7 + x^8

8    x + x^2 + 5 x^3 + 10 x^4 + 10 x^5 + 5 x^6 + x^7
```

```
9     x + 5 x^2 + 10 x^3 + 10 x^4 + 5 x^5 + x^6

10    x

11    x

12    x

13    x

14    x

15    x

16    x

17    x

18    x

19    x
```

In[12]:=

```
SubtreesNumber=Rest[CoefficientList[Apply[Plus,
Table[RankPoly[[1,i,2]],{i,V[tree]}]],x]]
```

Out[12]=

```
{19, 18, 37, 56, 65, 66, 65, 64, 63, 62, 77, 132,
 212, 252, 210, 120, 45, 10, 1}
```

Remark 12.5.1 *In the next Section, we shall apply the above generating function to compute the Myerson value for voting games restricted by stars.*

12.6 The Myerson value in the European Union

We here present a notebook that will be used for computing the Myerson value of the voting power in the Council of Ministers of the European Union. We study the case where the power game of the countries in the EU Council is defined by

$$[q;\ 10, 10, 10, 10, 8, 5, 5, 5, 5, 4, 4, 3, 3, 3, 2]\,,$$

where $q \in \{62, 65, 70\}$.

This game is restricted by a communication structure given by the star $K_{1,14}$ such that the degree of Germany is 14 and the other countries are degree 1. Then the collection of connected coalitions is

$$\mathcal{F} = \{S \subseteq N : \text{Germany} \in S \ \ \texttt{or} \ \ |S| = 1\}\,.$$

In[1]:=

```
<<DiscreteMath'Combinatorica'
```

In[2]:=

```
<<Cooperat'Cooperat'
```

In[3]:=

```
<<Graphics'Graphics3D'
```

In[4]:=

```
graph=Star[15];
countries={''UK'',''FR'',''IT'',''SP'',''NE'',''GR'',''BE'',
     ''PO'',''SW'',''AU'',''DE'',''FI'',''IR'',''LU'',''GE''};
```

In[5]:=

```
ShowLabeledGraph[graph,countries];
```

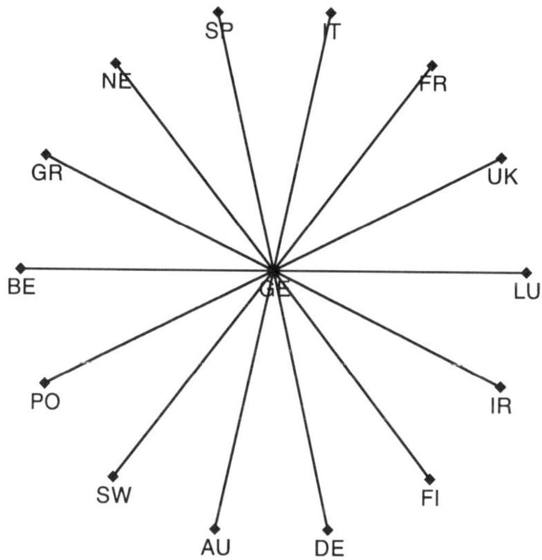

Figure 12.9 The European Union Star

In[6]:=

```
Clear[feasibles];
```

In[7]:=

```
feasibles=Sort[Select[Subsets[Range[15]],
  (MemberQ[#,15] || Length[#]==1)&]];
```

In[8]:=

```
T=Last[feasibles];
```

In[9]:=

```
weights={10,10,10,8,5,5,5,5,4,4,3,3,3,2,10};
```

We define the generating function f of the *weighted subtrees* of the star, by using an extension of the algorithm implemented by Jamison [132] [133].

In[10]:=

```
Clear[edges,f];
edges=Select[feasibles,(Length[#]==2)&];
```

In[11]:=

```
f[{i_},weights_]:=Module[{A,adjacents,factor,value},
A=Select[edges,(MemberQ[#,i])&];
adjacents=Complement[Flatten[A],{i}];
edges=Complement[edges,A];
factor=(1+f[#,weights])& /@ Partition[adjacents,1];
value=x*y^(weights[[i]])*Times @@ factor;
factor=. & /@ Partition[adjacents,1];
Return[value]];
```

From equation (5.4), page 124, and the generating function f, we obtain the following formulas of the Myerson value for voting games restricted by stars.

In[12]:=

```
MyersonStarPlayer[i_,weights_List,q_]:=
Module[{n=Length[weights],g,coefi,m,gg},
g=Expand[f[{i},weights]];
coefi=CoefficientList[g,y];
m=Exponent[g,y]+1;
gg=Apply[Plus,coefi[[Range[q+1,Min[m,q+weights[[i]]]]]]];
Sum[Coefficient[gg,x^j]*(j-1)!*(n-j)!,{j,2,n}]/n!]
```

In[13]:=

```
MyersonStarCenter[weights_List,q_]:=
Module[{n=Length[weights],g,coefi,m,gg},
edges=Select[feasibles,(Length[#]==2)&];
g=Expand[f[{Last[T]},weights]];
coefi=CoefficientList[g,y];
m=Exponent[g,y]+1;
gg=Apply[Plus,coefi[[Range[q+1,m]]]];
Sum[Coefficient[gg,x^j]*(j-1)!*(n-j)!,{j,2,n}]/n!]
```

In[14]:=

```
MyersonStarExtreme[weights_List,q_]:=Module[{value},
value=Table[edges=Select[feasibles,(Length[#]==2)&];
MyersonStarPlayer[i,weights,q],{i,Length[T]-1}];
Return[value]]
```

In[15]:=

```
MyersonStar[weights_,q_]:=
Prepend[MyersonStarExtreme[weights,q],
MyersonStarCenter[weights,q]]
```

In[16]:=

```
MyersonStar[weights,62]//N
```

Out[16]=

```
{0.345987,0.0882839,0.0882839,0.0882839,0.0708236,0.0406288,
 0.0406288,0.0406288,0.0406288,0.0328172,0.0328172,0.0251637,
 0.0251637,0.0251637,0.0146964}
```

In[17]:=

```
MyersonStar[weights,65]//N
```

Out[17]=

```
{0.308997,0.0964258,0.0964258,0.0964258,0.0739816,0.0439616,
 0.0439616,0.0439616,0.0439616,0.0306582,0.0306582,0.025197,
 0.025197,0.025197,0.0149906}
```

In[18]:=

```
MyersonStar[weights,70]//N
```

Out[18]=

```
{0.248601,0.105744,0.105744,0.105744,0.084016,0.0476024,
 0.0476024,0.0476024,0.0476024,0.0326174,0.0326174,0.0270729,
 0.0270729,0.0270729,0.0132867}
```

The following Table 12.2 and Figure 12.10 contain the population, votes, and Myerson values for the voting game in the Council of the European Union restricted by the German-Star. The majority voting rule is assumed to be 62, 65 and 70 out of 87 votes in the EU Council.

In[19]:=

```
population={80.61,57.96,57.53,56.93,39.11,15.24,10.35,
10.07,9.86,8.69,7.91,5.18,5.06,3.56,0.4};
PUE=N[%/Plus @@ %,4];
```

In[20]:=

```
votes={10,10,10,10,8,5,5,5,5,4,4,3,3,3,2};
VUE=N[%/Plus @@ %,4];
```

Country	Population	Votes	Myerson 62	Myerson 65	Myerson 70
Germany	.219	.115	.346	.309	.249
U. Kingdom	.157	.115	.088	.096	.106
France	.156	.115	.088	.096	.106
Italy	.155	.115	.088	.096	.106
Spain	.106	.092	.071	.074	.084
Netherlands	.041	.058	.041	.044	.048
Greece	.028	.058	.041	.044	.048
Belgium	.027	.058	.041	.044	.048
Portugal	.027	.058	.041	.044	.048
Sweden	.024	.046	.033	.031	.033
Austria	.022	.046	.033	.031	.033
Denmark	.014	.035	.025	.025	.027
Finland	.014	.035	.025	.025	.027
Ireland	.010	.035	.025	.025	.027
Luxembourg	.001	.023	.015	.015	.013

Table 12.2

In[21]:=

```
figure1=BarChart3D[{Reverse[PUE],Reverse[VUE],
Reverse[%16],Reverse[%17],Reverse[%18]},
Boxed -> True,  BoxRatios -> {0.5,1,0.5},
ViewPoint -> {0.9,-2,1},  Ticks ->
{{{1,''Population''},{2,''Votes''},{3,''Myerson 62''}
,{4,''Myerson 65''},{5,''Myerson 70''}},
{{1,''Luxembourg''},{2,''Ireland''},{3,''Finland''},
{4,''Denmark''},{5,''Austria''},{6,''Sweden''},
{7,''Portugal''},{8,''Belgium''},{9,''Greece''},
{10,''Netherlands''},{11,''Spain''},{12,''Italy''},
{13,''France''},{14,''U. Kingdom''},{15,''Germany''}},
Automatic},PlotLabel -> FontForm['''',{''Helvetica-Bold'',9}],
DefaultFont->{''Arial'',7}];
```

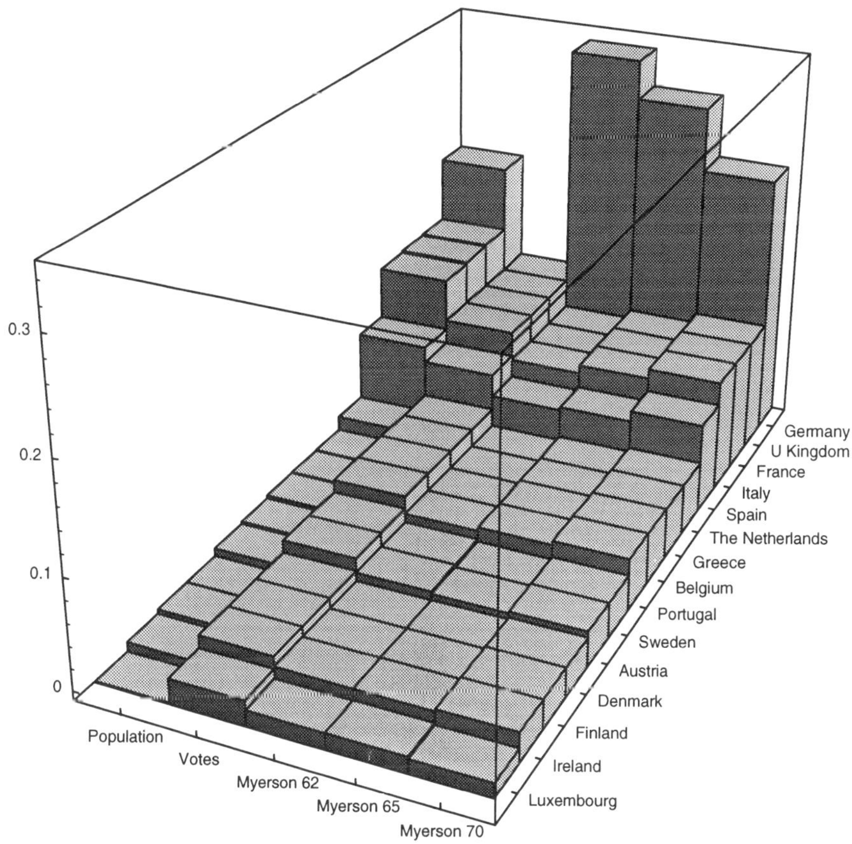

Figure 12.10 Power in the European Union Star game

Bibliography

[1] Algaba, E. (1998): *Extensión de juegos definidos en sistemas de conjuntos,* Ph. D. Thesis, University of Seville, Spain, available at http://www.esi.us.es/~mbilbao/dvifiles/tesisen.dvi

[2] Algaba, E., J. M. Bilbao, P. Borm and J. López (1998): The position value for union stable systems, *Research Memorandum FEW 768*, Tilburg University, The Netherlands.

[3] Algaba, E., J. M. Bilbao, P. Borm and J. López (1999): The Myerson value for union stable systems, *Research Memorandum FEW 773*, Tilburg University, The Netherlands.

[4] Amenta, N. (1994): Helly-type theorems and generalized linear programming, *Discrete Comput. Geom.* 12, 241–261, available at http://www.geom.umn.edu/~nina/papers/dcg.ps

[5] Ando, K. and S. Fujishige (1996): On structures of bisubmodular polyhedra, *Math. Programming* 74, 293–317.

[6] Aubin, J. P. (1993): *Optima and equilibria: an introduction to nonlinear analysis*, Springer-Verlag, Berlin.

[7] Aumann, R. and M. Maschler (1964): The bargaining set for cooperative games, in M. Dresher, L. S. Shapley and A. Tucker, eds., *Advances in Game Theory,* Princeton University Press, Princeton, New Jersey, 443–471.

[8] Banzhaf, J. F. III (1965): Weighted voting doesn't work: a mathematical analysis, *Rutgers Law Review* 19, 317–343.

[9] Berge, C. (1972): Balanced matrices, *Math. Programming* 2, 19–31.

[10] Bilbao, J. M. (1998): Values and potential of games with cooperation structure, *Int. J. Game Theory* 27, 131–145.

[11] Bilbao, J. M. (1998): Axioms for the Shapley value on convex geometries, *Europ. J. Oper. Res.* 110, 368–376.

[12] Bilbao, J. M., A. Jiménez Losada and J. J. López (1998): The Banzhaf power index on convex geometries, *Math. Social Sciences* 36, 157–173.

[13] Bilbao, J. M. and J. E. Martínez-Legaz (1998): The least increment function of a TU-game, *Third Spanish Meeting on Game Theory and Applications,* Barcelona.

[14] Bilbao, J. M., E. Lebrón and N. Jiménez Jiménez (1999): The core of games on convex geometries, *Europ. J. Oper. Res.* 119, 365–372.

[15] Bilbao, J. M., E. Lebrón, N. Jiménez Jiménez and H. Peters (2000): The selectope for games with partial cooperation. To appear in *Discrete Math.*

[16] Bilbao, J. M. and P. H. Edelman (2000): The Shapley value on convex geometries. To appear in *Discrete Appl. Math.*

[17] Bilbao, J. M., T. Driessen, E. Lebrón and A. Jiménez Losada (1999): The Shapley value for games on matroids: the static model, *Research Memorandum*, University of Seville, Spain, available at http://www.esi.us.es/~mbilbao/pdffiles/vmatro1.pdf

[18] Bird, C. G. (1976): On cost allocation for a spanning tree: A game theoretic approach, *Networks* 6, 335–350.

[19] Birkhoff, G. (1946): Tres observaciones sobre el álgebra lineal, *Rev. Fac. Ci. Exactas, Puras y Aplicadas Univ. Nac. Tucumán, Ser. A* 5, 147–151.

[20] Birkhoff, G. and M. K. Bennett (1985): The convexity lattice of a poset, *Order* 2, 223–242.

[21] Bixby, R. E., W. H. Cunningham and D. M. Topkis (1985): The partial order of a polymatroid extreme point, *Math. Oper. Res.* 10, 367–378.

[22] Bixby, R. E. and W. H. Cunningham (1995): Matroid optimization and algorithms, in R. Graham, M. Grötschel and L. Lovász, eds., *Handbook of Combinatorics, Vol. I,* Elsevier Science B.V., Amsterdam, 551–609.

[23] Blum, L., F. Cucker, M. Shub and S. Smale (1998): *Complexity and Real Computation*, Springer-Verlag, New York.

[24] Bondareva, O. N. (1963): Some applications of linear programming methods to the theory of cooperative games, *Problemy Kibernet.* 10, 119–139 (in Russian).

[25] Borel, É. (1921): La Théorie du Jeu et les Équations Intégrales à Noyau Simétrique Gauche, *C. R. Acad. Sci. Paris* 173, 1304–1308.

[26] Borm, P., G. Owen and S. H. Tijs (1992): On the position value for communication situations, *SIAM J. Discrete Math.* 5, 305–320.

[27] Borm, P., A. van den Nouweland and S. H. Tijs (1994): Cooperation and communication restrictions: a survey, in R. P. Gilles and P. H. M. Ruys, eds., *Imperfections and Behavior in Economic Organizations,* Kluwer Academic Publishers, Dordrecht.

[28] Bouchet, A. (1987): Greedy algorithm and symmetric matroids, *Math. Programming* 38, 147–159.

[29] Bouchet, A. and W. H. Cunningham (1995): Delta-matroids, jump systems and bisubmodular polyhedra, *SIAM J. Disc. Math.* 8, 17–32.

[30] Bovet, D. P. and P. Crescenzi (1994): *Introduction to the Theory of Complexity,* Prentice Hall, New York.

[31] Brams, S. F. and P. J. Affuso (1976): Power and size: a new paradox, *Theory and Decision* 7, 29–56.

[32] Burkard, R. E. (1997): Efficiently solvable special cases of hard combinatorial optimization problems, *Math. Programming* 79, 55–69.

[33] Carreras, F. (1991): Restriction of simple games, *Math. Social Sciences* 21, 245–260.

[34] Carter, M. (1993): Cooperative games, in H. R. Varian, ed., *Economic and Financial Modeling with Mathematica,* Springer-Verlag, Berlin, 167–191.

[35] Chandrasekaran, R. and S. N. Kabadi (1988): Pseudomatroids, *Discrete Math.* 71, 205–217.

[36] Christof, T. and G. Reinelt (1996): Combinatorial optimization and small polytopes, *Top* 4, 1–64.

[37] Christof, T. (1999): *PORTA (Polyhedron Representation Transformation Algorithm) 1.3.2:* sources, examples, man-pages (tar-file), available at http://www.iwr.uni-heidelberg.de/iwr/comopt/soft/PORTA/

[38] Clarkson, K. L. (1986): Linear programming in $O\left(n3^{d^2}\right)$ time, *Inform. Process. Lett.* 22, 21–24.

[39] Clarkson, K. L. (1995): Las Vegas algorithms for linear and integer programming when the dimension is small, *J. Assoc. Comput. Mach.* 42, 488–499, available at http://cm.bell-labs.com/who/clarkson/lp2.html

[40] Claus, A. and D. J. Kleitman (1973): Cost allocation for a spanning tree, *Networks* 3, 289–304.

[41] Cobham, A. (1965): The intrinsic computational difficulty of functions, in Y. Bar-Hillel, ed., *Proc. 1964 Internat. Congress for Logic, Methodology and Philosophy of of Science*, North-Holland, Amsterdam, 24–30.

[42] Coleman, J. S. (1973): Loss of power, *American Sociological Review* 38, 1–17.

[43] Cook, S. A. (1971): The complexity of theorem-proving procedures, *Proc. 3rd Annual ACM Symp. Theory of Computing*, Shaker Hts., Ohio, 151–158.

[44] Cook, W. J., J. Fonlupt and A. Schrijver (1986): An integer analogue of Carathéodory's theorem, *J. Combi. Theory (B)* 40, 63–70.

[45] Cook, W. J., W. H. Cunningham, W. R. Pulleyblank and A. Schrijver (1998): *Combinatorial Optimization*, Wiley-Interscience, New York.

[46] Cormen, T. H., C. E: Leiserson and R. L. Rivest (1990): *Introduction to Algorithms*, MIT Press, Cambridge, Massachusetts.

[47] Crescenzi, P. and V. Kann (2000): *A compendium of NP-optimization problems*, available at http://www.nada.kth.se/theory/problemlist.html

[48] Curiel, I. J. (1988): *Cooperative game theory and applications*, Ph. D. Thesis, University of Nijmegen, The Netherlands.

[49] Curiel, I. J., G. Pederzoli and S. H. Tijs (1989): Sequencing games, *Europ. J. Oper. Res.* 40, 344–351.

[50] Curiel I. J. and S. H. Tijs (1991): The minimarg and the maximarg operators. *J. Optimization Theory and Applications* 71, 277–287

[51] Curiel, I. J. (1997): *Cooperative Game Theory and Applications,* Kluwer Academic Publishers, Dordrecht.

[52] Damme, E. van (1998): On the State of the Art in Game Theory: An Interview with Robert Aumann, *Games Econ. Behav.* 24, 181–210.

[53] Deng, X. and C. H. Papadimitriou (1994): On the complexity of cooperative solution concepts, *Math. Oper. Res.* 19, 257–266.

[54] Derks, J. (1992): A short proof of the inclusion of the core in the Weber set, *Int. J. Game Theory* 21, 149–150.

[55] Derks, J., H. Haller and H. Peters (1998): The selectope for cooperative games, *Meteor RM 97-16,* University of Maastricht, The Netherlands.

[56] Dragan, I. (1996): New mathematical properties of the Banzhaf value, *Europ. J. Oper. Res.* 95, 451–463.

[57] Driessen, T. S. H. (1988): *Cooperative Games, Solutions and Applications*, Kluwer Academic Publishers, Dordrecht.

[58] Driessen, T. S. H. and C. Rafels (1996): k-Convexity for transferable utility games: several characterizations, *Working paper,* University of Barcelona.

[59] Dubey, P. and L. S. Shapley (1979): Mathematical properties of the Banzhaf power index, *Math. Oper. Res.* 4, 99–131.

[60] Duchet, P. (1995): Hypergraphs, in R. Graham, M. Grötschel and L. Lovász, eds., *Handbook of Combinatorics, Vol. I*, Elsevier Science B. V., Amsterdam, 381–432.

[61] Dudás, T., B. Klinz and G. J. Woeginger (1997): The computational complexity of the k-minimum spanning tree problem in graded matrices, *SFB-106, Optimierung und Kontrole,* Technische Universität Graz, Austria, anonymous ftp from ftp.tu-graz.ac.at, directory /pub/papers/math

[62] Dunstan, F. D. J. and D. J. A. Welsh (1973): A greedy algorithm for solving a certain class of linear programmes, *Math. Programming* 5, 338–353.

[63] Dyer, M. E. (1986): On a multidimensional search technique and its application to the euclidean one-center problem, *SIAM J. Comput.* 15, 725–738.

[64] Dyer, M. E. and A. M. Frieze (1989): On the complexity of computing the volume of a polyhedron, *SIAM J. Comput.* 18, 205–226.

[65] Edelman, P. H. and R. E. Jamison (1985): The theory of convex geometries, *Geom. Dedicata* 19, 247–270.

[66] Edelman P. H. and M. E. Sacks (1988): Combinatorial representation and convex dimension of convex geometries, *Order* 5, 23–32.

[67] Edelman P. H. (1997): A note on voting, *Math. Social Sciences* 34, 37–50.

[68] Edmonds, J. (1965): Maximum matching and a polyhedron with 0,1-vertices, *J. Res. Nat. Bur. Standards B* 69, 125–130.

[69] Edmonds, J. (1965): Paths, trees, and flowers, *Can. J. Math.* 17, 449–467.

[70] Edmonds, J. (1967): Systems of distinct representatives and linear algebra, *J. Res. Nat. Bur. Standards B* 71, 241–245.

[71] Edmonds, J. (1970): Submodular functions, matroids and certain polyhedra, in R. K. Guy, H. Hanani, N. Sauer and J. Schönheim, eds., *Combinatorial Structures and Their Applications,* Gordon and Beach, New York, 69–87.

[72] Edmonds, J. (1971): Matroids and the greedy algorithm, *Math. Programming* 1, 127–136.

[73] Edmonds, J. and R. Giles (1977): A min-max relation for submodular functions on graphs, in P. L. Hammer et al., eds., *Studies in Integer Programming, Annals of Discrete Mathematics* 1, 185–204.

[74] Einy, E. and D. Wettstein (1996): Equivalence between bargaining sets and the core in simple games, *Int. J. Game Theory* 25, 65–71.

[75] Einy, E., R. Holzman and D. Monderer (1999): On the least core and the Mas-Colell bargaining set, *Games Econ. Behav.* 28, 181–188.

[76] Fang, L., K. W. Hipel and M. Kilgour (1993): *Interactive decision making,* John Wiley & Sons, New York.

[77] Faigle, U. (1989): Cores of games with restricted cooperation, *ZOR–Meth. Mod. Oper. Res.* 33, 405–422.

[78] Faigle, U. and W. Kern (1992): The Shapley value for cooperative games under precedence constraints, *Int. J. Game Theory* 21, 249–266.

[79] Faigle, U. and W. Kern (1993): On some approximately balanced combinatorial cooperative games, *ZOR–Meth. Mod. Oper. Res.* 38, 141–152.

[80] Faigle, U. and W. Kern (1996): Submodular linear programs on forests, *Math. Programming* 72, 195–206.

[81] Faigle, U., W. Kern, S. P. Fekete and W. Hochstättler (1997): On the complexity of testing membership in the core of min-cost spanning tree games, *Int. J. Game Theory* 26, 361–366.

[82] Faigle, U. and W. Kern (1997): On the core of submodular cost games, University of Twente. To appear in *Math. Programming.*

[83] Faigle, U. and W. Kern (1998): An algebraic framework for the greedy algorithm with applications to the core and Weber set of cooperative games, *Working paper*, University of Twente, The Netherlands.

[84] Faigle, U., W. Kern and J. Kuipers (1998): An efficient algorithm for nucleolus and prekernel computation in some classes of TU-games, *Working paper*, University of Twente, The Netherlands.

[85] Faigle, U., W. Kern and J. Kuipers (1998): Computing the nucleolus of min-cost spanning tree games is NP-hard, *Int. J. Game Theory* 27, 443–450.

[86] Felsenthal, D. S. and M. Machover (1997): Ternary voting games, *Int. J. Game Theory* 26, 335–351.

[87] Felsenthal, D. S. and M. Machover (1998): *The Measurement of Voting Power: Theory and Practice, Problems and Paradoxes*, Edward Elgar, Cheltenham.

[88] Feltkamp, V., A. van den Nouweland, P. Borm, S. H. Tijs and A. Koster (1993): Linear production with transport of products, resources and technology, *ZOR–Meth. Mod. Oper. Res.* 38, 153–162.

[89] Fernández García, J. (2000): *Complejidad y algoritmos en juegos cooperativos,* Ph. D. Thesis, University of Seville, Spain, available at http://www.esi.us.es/~mbilbao/dvifiles/juliophd.dvi

[90] Frank, A. (1982): An algorithm for submodular functions on graphs, *Ann. Discrete Math.* 16, 97–120.

[91] Frank, A. (1999): Increasing the rooted-connectivity of a diagraph by one, *Math. Programming* 84, 565–576.

[92] Fredholm, I. (1903): Sur une classe d'equations fonctionelles, *Acta Math.* 27, 365–390.

[93] Fredman, M. L. and R. E. Tarjan (1987): Fibonacci heaps and their uses in improved network optimization algorithms, *J. Assoc. Comput. Mach.* 34, 596–615.

[94] Fujishige, S. (1984): Theory of submodular programs: A Fenchel-type min-max theorem and subgradients of submodular functions, *Math. Programming* 29, 142–155.

[95] Fujishige, S. (1984): On the subdifferential of a submodular function, *Math. Programming* 29, 348–360.

[96] Fujishige, S. (1991): *Submodular Functions and Optimization,* North-Holland, Amsterdam.

[97] Fujishige, S. (1997): A min-max theorem for bisubmodular polyhedra, *SIAM J. Disc. Math.* 10, 294–308.

[98] Fulkerson, D. R., A. J. Hoffman and R. Oppenheim (1974): On balanced matrices, *Math. Programming Study* 1, 120–132.

[99] Gabow, H. N., Z. Galil, T. Spencer and R. E. Tarjan (1986): Efficient algorithms for finding minimum spanning trees in undirected and directed graphs, *Combinatorica* 6, 109–122.

[100] Gabow, H. N., M. X. Goemans and D. P. Williamson (1998): An efficient approximation algorithm for the survivable network design problem, *Math. Programming* 82, 13–40.

[101] Gács, P. and L. Lovász (1999): *Complexity of Algorithms,* Lecture Notes, Yale University, available at http://www.cs.yale.edu/~lovasz/notes.html

[102] Garey, M. R. and D. S. Johnson (1979): *Computers and Intractability: A Guide to the Theory of NP-Completeness,* W. H. Freeman, San Francisco, California.

[103] Garrett, G. and G. Tsebelis (1999): Why resist the temptation to apply power indices to the European Union? *Journal of Theoretical Politics* 11, 291–308.

[104] Garrett, G. and G. Tsebelis (1999): More reasons to resist the temptation of power indices in the European Union, *Journal of Theoretical Politics* 11, 331–338.

[105] Gärtner, B. (1995): A subexponential algorithm for abstract optimization problems, *SIAM J. Comput.* 24, 1018–1035, available at http://www.inf.fu-berlin.de/inst/pubs/tr-b-93-05.abstract.html

[106] Gillies, D. B. (1953): *Some theorems on n-persons games*, Ph. D. Thesis, Princeton University, New Jersey.

[107] Girlich, E. and M. M. Kovalev (1996): Classification of polyhedral matroids, *Math. Meth. Oper. Res.* 43, 143–159.

[108] Gonzaga, C. L. (1989): An algorithm for solving linear programming programs in $O\left(n^3L\right)$ operations, in N. Megiddo [175], 1–28.

[109] Granot, D. and G. Huberman (1981): Minimum cost spanning tree games, *Math. Programming* 21, 1–18.

[110] Granot, D. and F. Granot (1992): Computational complexity of a cost allocation approach to a fixed cost spanning forest problem, *Math. Oper. Res.* 17, 765–780.

[111] Granot, D., M. Maschler, G. Owen and W. R. Zhu (1996): The kernel/nucleolus of a standard tree game, *Int. J. Game Theory* 25, 219–244.

[112] Granot, D., F. Granot and W. R. Zhu (1998): Characterization sets for the nucleolus, *Int. J. Game Theory* 27, 359–374.

[113] Gratzer, G. (1978): *General Lattice Theory*, Academic Press, New York.

[114] Greenberg, J. (1992): On the sensitivity of van Neumann and Morgenstern abstract stable sets. The stable set and the individual stable bargaining set, *Int. J. Game Theory* 21, 41–55.

[115] Grötschel, M., C. L. Monma and M. Stoer (1992): Facets for polyhedra arising in the design of communications networks with low connectivity constraints, *SIAM Journal on Optimization* 2, 474–504.

[116] Grötschel, M., L. Lovász and A. Schrijver (1993): *Geometric Algorithms and Combinatorial Optimization,* second edition, Springer-Verlag, Berlin.

[117] Hamers, H. (1995): *Sequencing and Delivery Situations: a Game Theoretic Approach*, Ph. D. Thesis, Tilburg University, The Netherlands.

[118] Hammer, P. L. and S. Rudeanu (1968): *Boolean methods in operations research and related areas,* Springer-Verlag, Berlin.

[119] Hammer, P. L., U. N. Peled and S. Sorensen (1977): Pseudo-Boolean functions and Game Theory I. Core elements and Shapley value, *Cahiers du CERO* 19, 159–176.

[120] Hamiache, G. (1999): A value with incomplete communication, *Games Econ. Behav.* 26, 59–78.

[121] Hamiache, G. (1999): A new axiomatization of the Owen value for games with coalition structures, *Math. Social Sciences* 37, 281–305.

[122] Harsanyi, J. C. (1963): A simplified bargaining model for the n-person cooperative game, *Int. Economic Review* 4, 194–220.

[123] Hart, S. and A. Mas-Colell (1988): The potential of the Shapley value, in A. E. Roth [208], 127–137.

[124] Hart, S. and A. Mas-Colell (1989): Potential, value and consistency, *Econometrica* 57, 589–614.

[125] Herne, K. and H. Nurmi (1993): The distribution of a priori voting power in the EC Council of Ministers and the European Parliament, *Scandinavian Political Studies* 16, 269–284.

[126] Hertel, E. (1994): Convexity in finite metric spaces, *Geom. Dedicata* 52, 215–220.

[127] Hoffman, A. J. and J. B. Kruskal (1956): Integral boundary points of convex polyhedra, in H. W. Kuhn and A. W. Tucker, eds., *Linear Inequalities and Related Systems*, Princeton University Press, Princeton, New Jersey, 223–246.

[128] Holler, M. and M. Widgrén (1999): Why power indices for assessing European Union decision-making? *Journal of Theoretical Politics* 11, 321–330.

[129] Ichiishi, T. (1981): Supermodularity: applications to convex games and to the greedy algorithm for LP, *J. Economic Theory* 25, 283–286.

[130] Isbell, J. R. (1958): A class of simple games, *Duke Math. J.* 25, 423–429.

[131] Jamison, R. E. (1982): A perspective on abstract convexity: classifying alignments by varieties, in D. C. Kay and M. Breen, eds., *Convexity and related combinatorial geometry*, Marcel Dekker, New York.

[132] Jamison, R. E. (1987): Alternating Whitney sums and matchings in trees, Part I, *Discrete Math.* 67, 177–189.

[133] Jamison, R. E. (1989/90): Alternating Whitney sums and matchings in trees, Part II, *Discrete Math.* 79, 177–189.

[134] Jiménez Jiménez, N. (1998): *Solution concepts for games on closure spaces,* Ph. D. Thesis, University of Seville, Spain, available at http://www.esi.us.es/~mbilbao/dvifiles/nievephd.dvi

[135] Jiménez Losada, A. (1998): *Valores para juegos sobre estructuras combinatorias,* Ph. D. Thesis, University of Seville, Spain, available at http://www.esi.us.es/~mbilbao/dvifiles/tesisan.dvi

[136] Kalai, E. and E. Zemel (1982): Totally balanced games and games of flow, *Math. Oper. Res.* 7, 476–478.

[137] Kalai, G. (1997): Linear programming, the simplex algorithm and simple polytopes, *Math. Programming* 79, 217–233.

[138] Kaneko, M. and M. H. Wooders (1982): Cores of partitioning games, *Math. Social Sciences* 3, 313–327.

[139] Kannai, Y. (1992): The core and balancedness, in R. J. Aumann and S. Hart, eds., *Handbook of Game Theory, vol. I*, North-Holland, Amsterdam, 355–395.

[140] Karp, R. M. (1972): Reducibility among combinatorial problems, in R. Miller and J. Thatcher, eds., *Complexity of Computer Computations*, Plenum Press, New York, 85–103.

[141] Khachiyan, L. G. (1979): A polynomial algorithm in linear programming, *Dokl. Akad. Nauk USSR* 244, 1093–1096.. English translation : *Soviet Math. Dokl.* 20, 191–194.

[142] Khmelnitskaya, A. B. (1999): Marginalist and efficient values for TU games, *Math. Social Sciences* 38, 45–54.

[143] Knuth, D. E. (1976): Big omicron and big omega and big theta, *ACM SIGACT News* 8, 18–24.

[144] Korte, B., L. Lovász and R. Schrader (1991): *Greedoids*, Springer-Verlag, Berlin.

[145] Kronsjö, L. (1985): *Computational Complexity of Sequential and Parallel Algorithms*, John Wiley & Sons, New York.

[146] Kruskal, J. B. (1956): On the shortest spanning subtree of a graph and the traveling salesman problem, *Proc. American Math. Society* 7, 48–50.

[147] Kuhn, H. W. (1956): The solvability and consistency for linear equations and inequalities, *Amer. Math. Monthly* 63, 217–232.

[148] Kuipers, J. (1993): On the core of information graph games, *Int. J. Game Theory* 21, 339–351.

[149] Kuipers, J. (1994): *Combinatorial Methods in Cooperative Game Theory*, Ph. D. Thesis, University of Maastricht, The Netherlands.

[150] Kuipers, J. (1996): A polynomial time algorithm for computing the nucleolus of convex games, *Report M 96-12*, University of Maastricht, The Netherlands.

[151] Kuipers, J., T. Solymosi and H. Aarts (1995): Computing the nucleolus of some combinatorially structured games, *Report M 95-08*, University of Maastricht, The Netherlands. To appear in *Math. Programming*.

[152] Lane, J. E. and R. Mæland (1995): Voting power under the EU Constitution. *Journal of Theoretical Politics* 7, 223–230.

[153] Lane, J. E. and S. Berg (1999): Relevance of voting power, *Journal of Theoretical Politics* 11, 309–320.

[154] Lawler, E. L., J. K. Lenstra, A. H. G. Rinnooy Kan and D. Shmoys, eds. (1985): *The Traveling Salesman Problem*, John Wiley & Sons, New York.

[155] Le Breton, M., G. Owen and S. Weber (1992): Strongly balanced cooperative games, *Int. J. Game Theory* 20, 419–427.

[156] Littlechild, S. C. and G. Owen (1973): A simple expression for the Shapley value in a special case, *Management Sci.* 20, 370–372.

[157] Littlechild, S. C. (1974): A simple expression for the nucleolus in a special case, *Int. J. Game Theory* 3, 21–29.

[158] López Vázquez, J. J. (1996): *Cooperación parcial en juegos de n personas,* Ph. D. Thesis, University of Seville, Spain, available at http://www.esi.us.es/~mbilbao/dvifiles/jorgephd.dvi

[159] Lovász, L. (1983): Submodular functions and convexity, in A. Bachem, M. Gröstschel and B. Korte, eds., *Mathematical programming: The state of the art,* Springer-Verlag, Berlin, 235–257.

[160] Lovász, L., D. B. Shmoys and É. Tardos (1995): Combinatorics in Computer Science, in R. Graham, M. Grötschel and L. Lovász, eds., *Handbook of Combinatorics, Vol. II,* Elsevier Science B.V., Amsterdam, 2003–2038.

[161] Lucas, W. F. (1975): Measuring power in weighted voting systems, in S. Brams, ed., *Game Theory and Politics*, Free Press, New York, 183–238.

[162] Mann, I. and L. S. Shapley (1962): Value of large games VI: evaluating the electoral college exactly, *RM-3158-PR, The Rand Corporation,* Santa Monica, California.

[163] Martin, R. K. (1999): *Large scale linear and integer optimization,* Kluwer Academic Publishers, Dordrecht.

[164] Martínez-Legaz, J. E. and I. Singer (1990): Dualities between complete lattices, *Optimization* 21, 481–508.

[165] Martínez-Legaz, J. E. (1996): Dual representation of cooperative games based on Fenchel-Moreau conjugation, *Optimization* 36, 291–319.

[166] Martínez-Legaz, J. E. (1997): A new characterization of totally balanced games, *Preprint*, Universitat Autònoma de Barcelona.

[167] Martínez-Legaz, J. E. (1998): A duality theory for totally balanced games, in *Proceedings of the IV Catalan Days of Applied Mathematics*, Universidad Rovira Virgili, Tarragona, 151–161.

[168] Maschler, M., B. Peleg and L. S. Shapley (1979): Geometric properties of the kernel, nucleolus and related solution concepts, *Math. Oper. Res.* 4, 303–338.

[169] Maschler, M. (1992): The bargaining set, kernel and nucleolus, in R. J. Aumann and S. Hart, eds., *Handbook of Game Theory, vol. I*, North-Holland, Amsterdam, 591–667.

[170] Mas-Colell, A. (1989): An equivalence theorem for a bargaining set, *J. Math. Econ.* 18, 129–139.

[171] Matousěk, J., M. Sharir and E. Welzl (1992): A subexponential bound for linear programming, *Technical Report B 92-17*, Freie Universität Berlin, http://www.inf.fu-berlin.de/inst/pubs/tr-b-92-17.abstract.html

[172] Meesen, R. (1988): *Communication games*, Master's Thesis (in Dutch), University of Nijmegen, The Netherlands.

[173] Megiddo, N. (1978): Computational complexity of the game theory approach to cost allocation for a tree, *Math. Oper. Res.* 3, 189–196.

[174] Megiddo, N. (1984): Linear programming in linear time when the dimension is fixed, *J. Assoc. Comput. Mach.* 31, 114–127.

[175] Megiddo, N., ed. (1989): *Progress in Mathematical Programming: Interior-Point and Related Methods*, Springer-Verlag, Berlin.

[176] Meyer, R. R. (1974): On the existence of optimal solutions to integer and mixed-integer programming problems, *Math. Programming* 7, 223–235.

[177] Motzkin, T. S. (1952): Contribution to the theory of linear inequalities, *The Rand Corporation Translation 22,* Santa Monica, California.

[178] Moulin, H. (1999): Incremental cost sharing: Characterization by strategy-proofness, *Social Choice and Welfare* 16, 279–320.

[179] Murota, K. (1996): Convexity and Steinitz's exchange property, *Advances in Mathematics* 124, 272–311.

[180] Murota, K. (1998): Fenchel-type duality for matroid valuations, *Math. Programming* 82, 357–375.

[181] Murota, K. (1998): Discrete convex analysis, *Math. Programming* 83, 313–371.

[182] Myerson, R. B. (1977): Graphs and cooperation in games, *Math. Oper. Res.* 2, 225–229.

[183] Myerson, R. B. (1980): Conference structures and fair allocation rules, *Int. J. Game Theory* 9, 169–182.

[184] Myerson, R. B. (1991): *Game Theory. Analysis of Conflict,* Harvard University Press, Cambridge.

[185] Nagamochi, H., D.-Z. Zeng, N. Kabutoya and T. Ibaraki (1997): Complexity of the minimum base game on matroids, *Math. Oper. Res.* 22, 146–164.

[186] Nemhauser, G. L. and L. A. Wosley (1988): *Integer and Combinatorial Optimization*, John Wiley & Sons, New York.

[187] Nouweland, A. van den and P. Borm (1991): On the convexity of communication games, *Int. J. Game Theory* 19, 421–430.

[188] Nouweland, A. van den, P. Borm and S. H. Tijs (1992): Allocation rules for hypergraph communication situations, *Int. J. Game Theory* 20, 255–268.

[189] Nouweland, A. van den (1993): *Games and graphs in economic situations*, Ph. D. Thesis, Tilburg University, The Netherlands.

[190] Nowak, A. S. and T. Radzik (1994): The Shapley value for n-persons games in generalized characteristic function form, *Games Econ. Behav.* 6, 150–161.

[191] Nowak, A. S. (1997): On an axiomatization of the Banzhaf value without the additivity axiom, *Int. J. Game Theory* 26, 137–141.

[192] Ordeshook, P. C. (1986): *Game theory and political theory,* Cambridge University Press, Cambridge.

[193] Ortmann, K. M. (1998): Conservation of energy in value theory, *Math. Meth. Oper. Res.* 47, 423–449.

[194] Owen, G. (1975): On the core of linear production games, *Math. Programming* 9, 358–370.

[195] Owen, G. (1986): Values of graph-restricted games, *SIAM J. Algebraic and Discrete Methods* 7, 210–220.

[196] Owen, G. (1988): Multilinear extension of games, in A. E. Roth [208], 139–151.

[197] Padberg, M. (1974): Perfect zero-one matrices, *Math. Programming* 6, 180–196.

[198] Papadimitriou, C. H. (1994): *Computational Complexity,* Addison-Wesley, Reading, Massachusetts.

[199] Peters, H. (1992): *Axiomatic Bargaining Game Theory,* Kluwer Academic Publishers, Dordrecht.

[200] Potters, J. A. M., I. J. Curiel and S. H. Tijs (1992): Traveling salesman games, *Math. Programming* 53, 199–211.

[201] Potters, J. A. M. and H. Reijnierse (1995): Γ-component additive games, *Int. J. Game Theory* 24, 49–56.

[202] Prim, R. C. (1957): Shortest connection networks and some generalizations, *Bell System Tech. J.* 36, 1389–1401.

[203] Queyranne, M., F. Spieksma and F. Tardella (1998): A general class of greedily solvable linear programs, *Math. Oper. Res.* 23, 892–908.

[204] Rasmusen, O. (1994): *Games and Information*, second edition, Blackwell Publishers, Cambridge.

[205] Ravi, R., R. Sundaram, M. V. Marathe, D. J. Rosenkrantz and S. S. Ravi (1996): Spanning trees short or small, *SIAM J. Discrete Math.* 9, 178–200.

[206] Reijnierse, H., M. Maschler, J. A. M. Potters and S. H. Tijs (1996): Simple flow games, *Games Econ. Behav.* 16, 238–260.

[207] Rockafellar, R. T. (1970): *Convex Analysis*, Princeton University Press, Princeton.

[208] Roth, A. E., ed. (1988): *The Shapley value: Essays in honor of Lloyd S. Shapley,* Cambridge University Press, Cambridge.

[209] Schmeidler, D. (1969): The nucleolus of a characteristic function game, *SIAM J. Appl. Math.* 17, 1163–1170.

[210] Schrijver, A. (1986): *Theory of Linear and Integer Programming,* John Wiley & Sons, New York.

[211] Schrijver, A. (1995): Polyhedral Combinatorics, in R. Graham, M. Grötschel and L. Lovász, eds., *Handbook of Combinatorics, Vol. II*, Elsevier Science B. V., Amsterdam, 1649–1704.

[212] Shapley, L. S. (1953): A value for n-person games, in H. W. Kuhn and A. W. Tucker, eds., *Contributions to the Theory of Games, Vol. II,* Princeton University Press, Princeton, New Jersey, 307–317.

[213] Shapley, L. S. and M. Shubik (1954): A method for evaluating the distribution of power in a committee system, *American Political Science Review* 48, 787–792.

[214] Shapley, L. S. (1962): Simple games: an outline of the descriptive theory, *Behavioral Sci.* 7, 59–66.

[215] Shapley, L. S. (1967): On balanced sets and cores, *Naval Res. Logist. Quart.* 14, 453–460.

[216] Shapley, L. S. and M. Shubik (1969): On market games, *J. Economic Theory* 1, 9–25.

[217] Shapley, L. S. (1971): Cores of convex games, *Int. J. Game Theory* 1, 11–26.

[218] Shapley, L. S. and M. Shubik (1972): The assignment game I: the core, *Int. J. Game Theory* 1, 111–130.

[219] Sharir, M. and E. Welzl (1992): A combinatorial bound for linear programming and related problems, in *Proc. 9th Annual Symposium on Theoretical Aspects of Computer Science (STACS),* Lecture Notes in Computer Science 577, Springer-Verlag, Berlin, 569–579.

[220] Shmoys, D. B. and É. Tardos (1995): Computational Complexity, in R. Graham, M. Grötschel and L. Lovász, eds., *Handbook of Combinatorics, Vol. II,* Elsevier Science B.V., Amsterdam, 1599–1645.

[221] Skiena, S. (1990): *Implementing Discrete Mathematics: Combinatorics and Graph Theory with Mathematica,* Addison-Wesley, Reading, Massachusetts.

[222] Solymosi, T. and T. E. S. Raghavan (1994): An algorithm for finding the nucleolus of assignment games, *Int. J. Game Theory* 23, 119–143.

[223] Solymosi, T., H. Aarts and T. Driessen (1998): On computing the nucleolus of a balanced connected game, *Math. Oper. Res.* 23, 983–1009.

[224] Stanley, R. P. (1986): *Enumerative Combinatorics, Vol. I,* Wadsworth, Monterey, California.

[225] Stanley, R. P. (1999): *Enumerative Combinatorics, Vol. II,* Cambridge University Press, Cambridge.

[226] Steunenberg, B., D. Schmidtchen and C. Koboldt (1999): Strategic power in the European Union: evaluating the distribution of power in policy games, *Journal of Theoretical Politics* 11, 339–366.

[227] Stoer, J. and C. Witzgall (1970): *Convexity and Optimization in Finite Dimensions I*, Springer-Verlag, Berlin.

[228] Swamy, M. N. S. and K. Thulasiraman (1992): *Graphs: Theory and Algorithms,* Wiley-Interscience, New York.

[229] Tannenbaum, P. (1997): Power in weighted voting systems, *The Mathematica Journal* 7, 58–63.

[230] Tutte, W. T. (1965): Lectures on matroids, *J. Res. Nat. Bur. Standards B* 69, 1–47.

[231] Tijs, S. H. (1981): Bounds for the core and the τ-value, in O. Moeschlin and D. Pallaschke, eds., *Game Theory and Mathematical Economics*, North-Holland, Amsterdam, 123–132.

[232] Tijs, S. H., T. Parthasarathy, J. A. M. Potters and V. Rajendra Prasad (1984): Permutation games: another class of totally balanced games, *OR Spektrum* 6, 119–123.

[233] Tijs, S. H. and T. S. H. Driessen (1986): Extension of solution concepts by means of multiplicative ε-tax games, *Math. Social Sciences* 12, 9–20.

[234] Tijs, S. H. (1993): *Game Theory*, Mimeo, Tilburg University, The Netherlands.

[235] Topkis, D. M. (1984): Adjacency on polymatroids, *Math. Programming* 30, 229–237.

[236] Traub, J. F. and H. Wozniakowski (1982): Complexity of linear programming, *Oper. Res. Lett.* 1, 59–62.

[237] Trefethen, L. N. (1997): *Maxims about numerical mathematics, science, computers and life on earth,* available at
http://www.comlab.ox.ac.uk/oucl/users/nick.trefethen/maxims.html

[238] Valiant, L. G. (1979): The complexity of computing the permanent, *Theoretical Comp. Sci.* 8, 189–202.

[239] Valiant, L. G. (1979): The complexity of enumeration and reliability problems, *SIAM J. Comput.* 8, 410–421.

[240] Van Deemen, A. (1997): *Coalition Formation and Social Choice,* Kluwer Academic Publishers, Dordrecht.

[241] Von Neumann, J. (1928): Zur Theorie der Gesellschaftsspiele, *Math. Ann.* 100, 295–320. English translation: On the theory of games of strategy, in A. W. Tucker and R. D. Luce, eds., *Contributions to the Theory of Games, Vol. IV,* Princeton University Press, Princeton, New Jersey, 13–42.

[242] Von Neumann, J. and O. Morgenstern (1944): *Theory of Games and Economic Behavior,* Princeton University Press, Princeton, New Jersey.

[243] Von Neumann, J. (1953): A certain zero-sum two-person game equivalent to the optimal assignment problem, in H. W. Kuhn and A. W. Tucker, eds., *Contributions to the Theory of Games, Vol. II,* Princeton University Press, Princeton, New Jersey, 5–12.

[244] Weber, R. J. (1988): Probabilistic values for games, in A. E. Roth [208], 101–119.

[245] Welsh, D. J. A. (1995): Matroids: fundamental concepts, in R. Graham, M. Grötschel and L. Lovász, eds., *Handbook of Combinatorics, Vol. I,* Elsevier Science B.V., Amsterdam, 481–550.

[246] Whitney, H. (1935): On the abstract properties of linear dependence, *Amer. J. Math.* 57, 509–533.

[247] Widgrén, M. (1994): Voting power in the EC decision making and the consequences of two different enlargements, *European Economic Review* 38, 1153–1170.

[248] Wolfram, S. (1991): *Mathematica. A System for Doing Mathematics by Computer,* Addison-Wesley, Reading, Massachusetts.

[249] Young, H. P. (1985): Monotonic solutions of cooperative games, *Int. J. Game Theory* 14, 65–72.

[250] Ziegler, G. M. (1995): *Lectures on Polytopes,* Springer-Verlag, Berlin.

Index

airport game, 93
algorithm, 89, 92
algorithm complexity, 89
all-negative circuit, 20
allocation rule, 136
  additive, 140
  basis monotonic, 138
  fair, 137
  influence property, 146
  point anonymity, 139
  superfluous player property, 139
  superfluous support property, 144
anti-exchange property, 16
antimatroid, 21
approximate algorithm, 100
arborescence, 9, 93
assignment games, 10
atomic family, 118, 205

Banzhaf index, 173, 254, 266
Banzhaf value, 5, 123
bargaining set
  Aumann-Maschler, 246
  Greenberg, 248
  Mas-Colell, 248
barrier method, 93
base polyhedron, 57
bases, 20
basic coalitions, 186
basis of a set system, 131
bin packing games, 12
bisubmodular system, 24
bit cost, 90
block, 8
bridge, 8

carrier axiom, 170
chain axiom, 171
characteristic function, 3
chord, 8
circuit, 20, 109
class
  #P, 100
  coNP, 95
  coNP-complete, 98
  NP, 95
  NP-complete, 98
  P, 92
clique, 8
closed set, 15
closure space, 15, 131
coalitions, 2
communication structure, 14
compatible-order value, 166
complementary slackness, 36
component, 50, 112, 133
component-dummy, 136
component-efficient, 136
computational complexity, 89, 258
computational problem, 89
concave conjugate function, 65
conference structure, 135, 158
convex Banzhaf index, 175
convex Banzhaf value, 179
convex conjugate function, 65
convex geometry, 16, 132, 158

convex set, 16, 158
convex swing, 174
convex swing probability, 174
core, 3, 57, 77, 207, 231, 240
cost allocation tree game, 93
counterobjection, 246
counting problem, 100
cube, 29
cut, 8
cutvertex, 8
cycle, 8

decision problem, 92
degree, 7
delivery games, 13
dependent sets, 18
digraph, 9
dimension, 27
direct market, 86
dividends, 6, 82, 85, 120, 208
dominance, 244
dual greedy algorithm, 55
duality theorem, 35
dummy axiom, 162, 176, 186
dummy player, 161, 175, 186

efficiency axiom, 165, 186
essential coalition, 47
excess, 4
extreme point, 16, 125
extreme power, 174
extreme power axiom, 177

face, 28
facet, 28
Farkas' lemmas, 33
feasible continuations, 155
Fenchel duality, 67
fixed cost spanning forest game, 94
forest, 8
forest polytope, 56
Fourier-Motzkin elimination, 32, 92, 223

game
- almost positive, 212
- arc, 142
- balanced, 3, 46, 77
- bicooperative, 23
- bimodular, 24
- bisubmodular, 24, 59
- bisupermodular, 24
- concave, 83, 106
- concave on a convex geometry, 228
- conference, 142
- convex, 83, 106, 150
- convex on a closure space, 238
- convex on a convex geometry, 228
- cooperative, 2
- cost, 3, 65
- dual, 4, 68, 106
- graph-component additive, 113
- graph-restricted, 14
- identity, 160
- locally concave, 106
- locally convex, 106
- marginal, 233
- market, 79, 86
- monotone, 79, 164, 232
- oligarchic, 109
- on a convex geometry, 158
- on a matroid, 185
- point, 142
- profit, 3, 65
- quasi-balanced, 6
- quasi-concave, 229
- quasi-convex, 229, 238
- quasi-subadditive, 234
- quasi-superadditive, 234
- restricted, 113, 122, 135
- simple, 109, 237, 253
- solution continuation, 155
- subadditive, 47
- superadditive, 47

symmetric, 192, 235
ternary voting, 23
totally balanced, 3, 49, 75
transferable utility, 2
unanimity, 5, 160, 251
weak, 238
zero-monotone, 107
zero-normalized, 138, 142
Gaussian elimination, 92
generalized coalition, 23
generalized linear program, 105
generating function, 255
graph, 7
2-connected, 8
bipartite, 7
block, 16, 118
chordal, 8
complete, 7
component, 8
connected, 8
cycle-complete, 16
Hamiltonian, 8
simple, 7
star, 296
underlying, 9
greedoid, 21, 105
greedy algorithm, 52, 103

Hart and Mas-Colell potential, 119
hierarchical situations, 158
hierarchical strength, 170, 190
hierarchical strength axiom, 170
Hilbert basis, 43

imputation set, 3, 240
independence system, 18, 52
independent sets, 18
indirect function, 75
influence of a player, 146
instance, 89
integer linear programming, 38
integral polyhedron, 38
interior point methods, 93
intersecting family, 118, 214
istmus player, 184

Kachiyan ellipsoide algorithm, 93

least core, 4, 248
least increment function, 80
lexicographic order, 4
linear production games, 11
linear programming, 32
linearity axiom, 160, 186
local submodularity, 105
loop, 21
losing coalition, 237
Lovász extension, 82, 87
lower value, 6
LP-relaxation, 39

marginalistic value, 180
market, 86
matching, 8, 43
matching games, 10, 50
matching polytope, 44
Mathematica
function
BanzhafIndex, 283
BanzhafPowerPlus, 261
BanzhafTwoPower, 267
ConvexBanzhafIndex, 285
ConvexBanzhafProb, 285
ConvexShapleyValueF, 280
GeomBanzhaf, 275
GeomShapley, 274
MyersonStar, 298
ShPowerPlus, 262
ShTwoPower, 268
notebook, 260, 265, 271, 277, 283, 290, 296
packages, 271
system, 260
matrix
adjacency, 100
balanced, 41

incidence, 43, 47
integral, 38
perfect, 41
totally balanced, 41
totally unimodular, 40
matroid, 18, 108, 181
basis, 19
contraction, 183
forest, 18
free, 18
graphic, 18
opposition, 184
polytope, 54
rank, 20, 182
maximal chain, 158, 217
maximal marginal operator, 233
meet game, 257
min-cost spanning tree games, 10, 94, 102
minimal marginal operator, 233
minimum base game on matroids, 20, 101
minimum forest game, 20, 101
model of computation, 90
monotonicity axiom, 164
multilinear extension, 127
multiplicative ε-core, 4
Myerson value, 14, 135, 298

nondeterministic polynomial time, 95
normalized convex Banzhaf index, 175
nucleolus, 5

objection, 246
optimization problem, 99
oracle, 102
oracle-polynomial time, 102
order ideal, 17
orientation, 9
orthant, 24

P = NP question, 97
partition convex geometry, 118
partition system, 112
path, 8
perfect matching, 8
perfect matching polytope, 45
permutation games, 12
players, 2
policy order, 173
polyhedral concave function, 66
polyhedral convex function, 66
polyhedron
bisubmodular, 24
bisupermodular, 24
convex, 27
of dimension n, 27
pointed, 30, 217
representations, 223
polymatroid, 57
polymatroid function, 62
polymatroid greedy algorithm, 59
polynomial complexity, 89
polynomial-time algorithm, 92
polynomial-time reduction, 98
polytope, 29, 207
PORTA, 222
position value, 142
pre-imputation set, 3
pre-nucleolus, 5
probabilistic convex Banzhaf index, 175
probabilistic Shapley value, 200, 203
probabilistic value, 160, 189, 194
problem
CLIQUE, 99
CORE, 103
EXACT COVER BY 3-SETS, 103
GRAPH RELIABILITY, 104
HAMILTONIAN GRAPH, 96
KNAPSACK, 104
MAX-CUT, 103
MEMBER CORE, 103

MINIMUM SPANNING TREE, 94
NODE COVER, 99
NUCLEOLUS, 104
PERMANENT, 100
PRIME NUMBER, 97
SATISFIABILITY, 98
SHAPLEY VALUE, 104
STABLE SET, 99
SUBADDITIVITY, 103
SUBMODULARITY, 103
TIJS VALUE, 104
TRAVELING SALESMAN, 95
WEIGHTED MATCHING, 94

quasi-probabilistic value, 194, 197

rational polyhedron, 38
real number model, 90
relative extreme power, 174
relative extreme power axiom, 177
restricted games, 14
restricted potential, 119
root, 9

search problem, 92
selectope, 209
selector, 208
consistent, 219
sequencing games, 14
set packing problem, 41
set system, 15
Shapley value, 5, 104, 123, 171, 187
Shapley-Shubik index, 254, 262
signed coalition, 23
simple flow games, 12
simplex, 28
simplex method, 37, 93
size of a game, 102
spanning tree, 8
stable set, 244
strong ε-core, 4
subdifferential, 70
subgame, 3, 119
subgradient, 70
submodular function, 57, 83
submodular polyhedron, 57
supermodular function, 57, 83
supermodular polyhedron, 57
support anonymous, 146
supporting hyperplane, 28, 46
supports of a set system, 131
survivable network, 10
swing, 174, 254
symmetry axiom, 186
system
integral, 38
rational, 38
totally dual integral, 39, 55
totally primal integral, 39

tax games, 11
tight set, 58
Tijs value, 6, 104
total swings axiom, 177
total swings probabilities axiom, 177
trail, 8
transfer axiom, 177
traveling salesman games, 13
tree, 8
Turing machine model, 90

union stable structure, 135, 142
union stable system, 130
unit cost, 90
unit marginal worth, 108
upper value, 6
utility function, 79, 86

vector
characteristic, 38, 82
depth, 82
efficient, 3
incidence, 38, 66, 182
integral, 38
marginal worth, 83, 218
vector rank function, 68

vertex
  isolated, 7
  pendant, 7
  simplicial, 8
veto player, 238

walk, 8
Weber set, 83, 218, 231, 242, 251
weighted chain, 82
weighted graph games, 101
weighted voting game, 101, 254
winning coalition, 237, 253